NOVEL APPROACHES AND THEIR APPLICATIONS IN RISK ASSESSMENT

Edited by **Yuzhou Luo**

Novel Approaches and Their Applications in Risk Assessment
Edited by Yuzhou Luo

Published by InTech

Janeza Trdine 9, 51000 Rijeka, Croatia

Publishing Process Manager Vana Persen

Technical Editor Teodora Smiljanic

Cover Designer InTech Design Team

Additional hard copies can be obtained from orders@intechopen.com

Novel Approaches and Their Applications in Risk Assessment, Edited by Yuzhou Luo
p. cm.
ISBN 978-953-51-0519-0

For EU product safety concerns:
IN TECH d.o.o., Prolaz Marije Krucifikse Kozulić 3, 51000 Rijeka, Croatia,
info@intechopen.com or visit our website at intechopen.com.

Meet the editor

Dr Yuzhou Luo is a Research Fellow at the Department of Land, Air and Water Resources at the University of California (Davis). He graduated from Thinghua University in Beijing, China, where he obtained his BEng in Environmental Engineering in 2000. He then proceeded to acquire his MS in Natural Resources from the University in Connecticut, USA, which was followed by a PhD in Environmental Engineering at the same University. His research interests include numerical modeling for unsteady-state biogeochemical processes of chemical transport in complex environmental systems, GIS technologies in hydrologic simulation, environmental assessment, and watershed management, integrated risk assessment for human health and ecosystem at watershed scales, impact of water resources and global climate change on hydrology, environmental quality, and crop production.

Contents

Preface

This book compiles chapters written by researchers who focus on the development and application of risk assessment approaches in various subjects of hydrology, water quality, human health, and industrial designing.

Risk assessment is a critical component in the evaluation and protection of natural or anthropogenic systems. Conventionally, risk assessment is involved with some essential steps such as the identification of problem, risk evaluation, and assessment review. Other novel approaches are also discussed in the book chapters. This book is compiled to communicate the latest information on risk assessment approaches and their effectiveness. Presented materials cover subjects from environmental quality to human health protection.

It's the editor's expectation that the chapters enclosed in this book will provide helpful information for risk assessment in both academic and industrial fields, and encourage the development and implementation of relevant applications for the protection of environment and human health.

Yuzhou Luo, Ph.D.
Department of Land, Air and Water Resources
University of California (Davis)
USA

Section 1

Risk Assessment in Environmental and Ecosystem Quality

1

A Practical Example of Risk Assessment – Risk Assessment to Phycotoxins in a Recreational Shellfish Harvester's Subpopulation

Cyndie Picot and Alain-Claude Roudot
Laboratoire de Toxicologie Alimentaire et Cellulaire (EA 3880), Université Européenne de Bretagne, Université de Bretagne Occidentale (UEB-UBO)
France

1. Introduction

The past few decades have seen an increase in the frequency, concentrations, and geographic distribution of marine algal toxins (phycotoxins), secondary metabolites produced by marine microalgae (phytoplankton). Among the 3400-4000 known phytoplankton species, only about 2 % are potentially harmful (Frémy & Lassus, 2001). Bivalve molluscs filter-feed on these micro-algae, accumulate toxins, and may be consumed by humans (Shumway et al,. 1995; Van Dolah, 2000). In order to determine whether phycotoxins are a matter of concern for human health, a risk assessment must be undertaken. It comprises four steps: hazard identification, hazard characterization, exposure assessment and risk characterization (WHO, 1985).

2. Risk assessment

Hazard identification is defined as follows: "the identification of biological, chemical and physical agents capable of causing adverse health effects and that may be present in a particular food or group of foods" (CAC, 2006). The outcome is a scientific judgment as to know whether the chemical being evaluated could, under given exposure conditions, cause an adverse effect in humans. In view of reported intoxications and deaths, phycotoxins are identified as a matter of concern for human health. Because of their high and increasing occurrence, their worldwide distribution and their different profile of toxicology and contamination, this chapter focused on two families of phycotoxins: okadaic acid (OA) and analogs, and spirolide (SPX) and analogs (Ade *et al.*, 2003; Hallegraeff, 2003; EFSA, 2008a ; EFSA, 2010).

Hazard characterization (also known as dose-response assessment) is defined as: "the qualitative and/or quantitative evaluation of the nature of the adverse health effects associated with biological, chemical, and physical agents that may be present in food" (CAC, 2006). It describes the relationship between the ingested quantity of the substance and the incidence of an adverse health effect (CAC, 2006). The aim is to allocate two toxicological reference values: the Acute Reference Dose (ARfD) and the Tolerable Daily Intake (TDI) for acute and chronic risks, respectively. The ARfD and the TDI are the amount of a substance, in mg/kg bw, which can be ingested without adverse health effects, for one meal or on a daily basis during a life time, respectively. Concerning acute hazard characterization, based on human and animal data, provisional ARfD have been allocated to OA, but a lack of data does not allow to allocate an ARfD to SPX. Moreover, concerning chronic hazards, the lack of (sub)chronic data for animals prevented international expert committees from allocating TDI to these toxins (Toyofuku, 2006; EFSA, 2008a; EFSA, 2010).

Exposure assessment is the evaluation of the likely intake of chemicals *via* food. It combines the level of the chemical in the diet and the consumption rates of the foods containing the chemical (Kroes *et al.*, 2002; EFSA, 2008b). Exposure assessment must firstly concern at risk subpopulation. Recreational shellfish harvesters appear to be an at risk subpopulation because a priori they consumed a larger quantity of seafood than the general population, because their practice is both recreational and a free source of food (Burger *et al.*, 1998, Gagnon *et al.*, 2004; Leblanc, 2006). Unfortunately the shellfish harvester subpopulation is generally not taken into account (USEPA, 1998) and no exposure assessment to phycotoxins by recreative shellfish harvesters is available. Indeed, achieving a meaningful exposure assessment on phycotoxins by recreational shellfish harvesters is extremely difficult because of: i) the consumption data often do not distinguish between fish and shellfish, between purchased shellfish and recreationally harvested shellfish between the general population and specific subpopulations, and between shellfish species; ii) the contamination data reveal great variations between shellfish species, toxin profiles, inter- and intra-country levels; iii) consumption and contamination data are derived from different and unrelated studies.

Therefore it is critical to assess the phycotoxin dietary intakes from shellfish consumption in this at risk subpopulation. These considerations led us to monthly monitor these phycotoxins in harvested shellfish and to conduct a one-year survey of shellfish consumption by recreational shellfish harvesters. Then these data were combined with a probabilistic method to assess the exposure.

In the **risk characterization** phase, acute and chronic exposure intakes are compared with ARfD and TDI, respectively, to assess whether or not the presence of contaminant is a matter of concern (WHO, 1985).

3. Practical example: Risk assessment to phycotoxins in a recreational shellfish harvester's subpopulation

3.1 Hazard identification

Tibbetts (1998) assesses than phycotoxins could be responsible for some 60,000 incidents per year the world over with an overall mortality rate of 1.5%. Incidence and severity are different according to types of toxins. Thus, in view of reported intoxications and deaths, phycotoxins are identified as a matter of concern for human health.

3.1.1 Okadaic acid and analogs

Okadaic acid (OA) and its congeners (dinophysistoxins) are produced by two species of dinoflagellates Dynophisis *spp* and Prorocentrum *spp* (Hallegraeff, 2003). Historically, OA and analogs are classified in Diarrheic Shellfish Poisoning (DSP) because of the symptoms they cause (gastrointestinal distress, diarrhea, nausea, vomiting and abdominal pain). Prior to the 1980's DSP incidents affected mainly Europe (first outbreak in 1961 in the Netherlands) and Japan (the second incident is reported in Japan in the late 1970's), whereas currently diarrheic shellfish toxin outbreaks are documented all over the world. Tens of thousands cases of intoxication have been reported (Picot *et al.*, 2011a), all over the world, but because of nonspecific clinical symptoms, DSP cases are probably underdiagnosed and underreported (Economou *et al.*, 2007). No death has been attributed to OAs.

3.1.2 Spirolides

Spirolides (SPXs) are cyclic imines produced by the dinoflagellate Alexandrium ostenfeldii. SPXs are included in the "emerging toxin" group because they have been recently isolated and characterised: in early 1990s, (in scallops and mussels harvested in Nova Scotia, Canada) (Hu *et al.*, 1995). In Europe SPXs have only recently been found in their producer dinoflagellate and/or shellfish in Scotland, Italy, Denmark, Ireland, Norway, Spain and France (in the 2000s). Nowadays, no report of human illness due to SPXs, have been identified. Episodes of toxicity, involving non-specific symptoms such as gastric distress and tachycardia were recorded in individuals in Nova Scotia, (Canada) consuming shellfish during times when SPXs were known to be present, but these could not be definitively ascribed to SPXs and are not consistent with the signs of toxicity in mice (Richard *et al.*, 2001). In mice, acute toxicity of SPXs is characterised by the rapid onset of systemic neurotoxicity following i.p. (intra-peritoneal) injection and death within minutes. Thus SPXs are therefore often denoted "fast acting toxins".

3.2 Hazard characterization

As mentioned before, the aim is to allocate two toxicological reference values: the ARfD and the TDI for acute and chronic risks, respectively. The two international organizations which have evaluated toxicological studies and proposed these toxicological reference values are: the JECFA (Joint FAO/WHO (Food and Agriculture Organization / World Health Organization) Expert Committee on Food Additives) *ad hoc* Expert Consultation on Biotoxins in Bivalve Molluscs and the European Food Safety Authority (EFSA).

3.2.1 Okadaic acid and analogs

Human data from Japan (eight people from three families, ages 10–68) indicate a LOAEL (Lowest Observed Adverse Effect Level) of 1.2–1.6 µg/kg bw. In a second study from Norway, 38 of 70 adults were affected at levels ranging from 1.0 to 1.5 µg/kg bw (Aune and Yndestad, 1993). Based on the LOAEL of 1.0 µg OA/kg bw and a chosen safety factor of 3 because of documented human cases involving more than 40 people and because DSP symptoms are readily reversible, the JEFCA established a provisional ARfD of 0.33 µg OA equ/kg bw. In 2008, the EFSA proposed an ARfD of 0.30 µg OA equ/kg bw based on a LOAEL equals to 0.9 µg OA/kg bw and a chosen safety factor of 3 to the use of a LOAEL instead of a NOAEL (No Observed Adverse Effect Level). The JECFA and the EFSA determined that no TDI could be established because of insufficient data on the chronic effects of OA (EFSA, 2008a; Toyofuku, 2006).

3.2.2 Spirolides

The toxins of the group of SPXs are characterised by binding to and blocking of AChR receptors in the central- and peripheral nervous system including neuromuscular junctions. The acute toxic signs have a rapid onset, in particular following i.p. (intra-peritoneal) administration. With regard to oral toxicity the reported toxicity varies greatly depending on whether the toxin is administered by gavage or in feed and whether the animal is fasted. In general, gavage administration shows lower LD50 (Lethal Dose) values for the various toxins. In humans, no quantitative data on toxicity exist. In view of the acute toxicity of SPXs the EFSA considered that an ARfD should be established for the SPXs, but due to the lack of adequate quantitative data on acute oral toxicity (i.e. no-observed-adverse-effect levels (NOAELs)) this was not possible. However, the toxicology working group of the European Union Community Reference Laboratory for marine biotoxins (CRLMB) had proposed a guidance level of 400 µg sum of SPXs/kg shellfish meat (CRLMB, 2005; Pigozzi *et al.*, 2008). But currently there are no regulations on SPXs in shellfish in Europe or in other regions of the world.

There are no long term studies on the group of SPXs in experimental animals. Thus no TDI had been allocated to SPXs.

3.3 Exposure assessment

3.3.1 Input data

To determine recreational shellfish harvester exposure, shellfish contamination data have to be combined with their shellfish consumption rates. These two kinds of data are lacking. Consumption data often do not distinguish fish and shellfish consumption and do not take into consideration harvested shellfish consumption. Moreover, getting base levels of these phycotoxins i in the most concerned shellfish i.e. bivalve molluscs, is a prerequisite to any phycotoxin-exposure assessment. Such data are missing because shellfish contamination is only analysed in case of phytoplankton bloom. These considerations led us to monthly monitor these phycotoxins in such species and to conduct a one-year survey of shellfish consumption by recreational shellfish harvesters from the same area (i.e. contamination- and consumption-data collected in the same area and relative to the same subpopulation).

3.3.1.1 Shellfish consumption data

The population of interest was a group of recreational shellfish harvesters set along the coasts of Finistère (Western Brittany, France, see figure 1).

Their shellfish consumption was investigated from February 2008 to February 2009 through two complementary methods: a Food Frequency Questionnaire (FFQ) and a food diary. The FFQ was conducted through face-to-face interviews at the harvesting sites. As this tool provides long-term consumption data, but relies upon memory, this drawback was counteracted by using the records versus time of each shellfish meal (with quantities) kept in the food diary. Moreover, this diary gave additional information such as the origin of consumed shellfish (harvest, shop, restaurant...), consumption by different household members and the way shellfish had been prepared. Data were validated for bivalve and gastropod groups (for more details, see Picot *et al.*, 2011b). The consumption data about five of the most consumed bivalve species in the area of interest are expressed as follows in Table 1: portion sizes and daily shellfish consumption rates (both with the mean and 95th percentile (P95)) as well as the raw consumption in percentage.

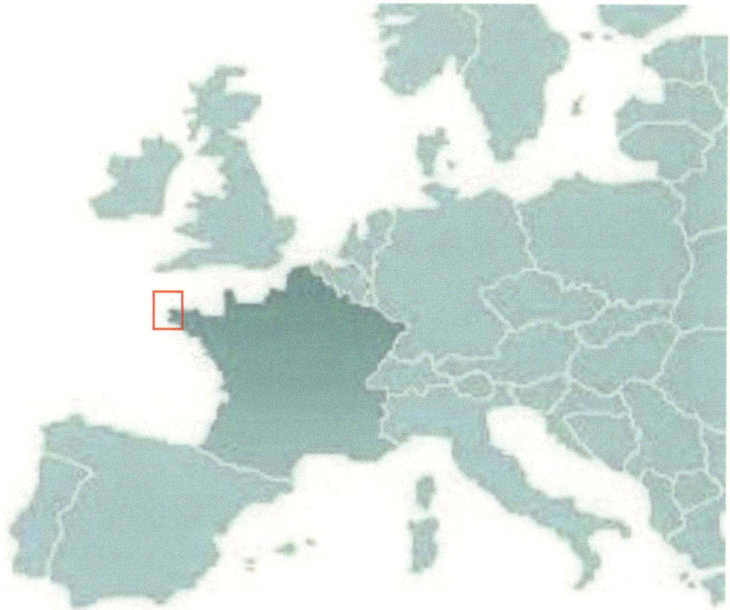

Fig. 1. Description of the area of interest: Finistère, Western Brittany, France.

Bivalve species	Consumption derived only from harvested bivalve					Consumption derived only from purchased bivalve					% of raw consumption
	% of consumers	Portion size (g/portion)		Daily consumption rate (g/day)		% of consumers	Portion size (g/portion)		Daily consumption rate (g/day)		
		Mean	P95	Mean	P95		Mean	P95	Mean	P95	
Oyster	20.3	36.6	172.8	1.68	7.58	27.2	34.4	102.4	2.02	8.12	97.8
Mussel	27.3	69.4	396.0	1.66	10.1	33.6	80.0	264.0	4.04	11.6	0.00
Cockle	63.7	109	271.5	3.15	13.8	2.6	2.7	nc	0.09	0.26	2.38
Carpet shell clam	74.6	73.7	259.5	2.43	10.7	4.7	3.2	nc	0.14	0.47	31.7
Razor clam	23.4	27.6	167.8	0.57	3.31	0.0	0.00	0.00	0.00	0.00	0.00
King scallop	0.39	0.13	nc	0.0012	nc	20.3	15.2	85.5	0.45	2.50	3.69

nc : not calculable because of an insufficient number of consumers.

Table 1. Shellfish consumption data according to shellfish species and origin (harvest or purchase). Consumption rates derived from the total population, including non consumers.

3.3.1.2 Shellfish contamination data

To counteract the lack of databases about phycotoxin base levels in shellfish, samples were harvested monthly, from June 2009 to June 2010, on beaches of Finistère selected from three criteria: i) the presence of several bivalve species, ii) regular shellfish harvesters and iii) regular phycotoxin events. The analyses were made on only two among the five bivalve species, which had been previously identified as being either the most consumed species or the most contaminated from a consumption survey and a test about inter-species variability (Picot *et al.*, submitted). OAs and SPXs were analysed after methanolic extraction from samples, purification by solid phase extraction and quantification by High Performance Liquid Chromatography - tandem mass spectrometry (HPLC-MS/MS) (Picot *et al.*, submitted). The results of contamination are presented in the figures 2 and 3.

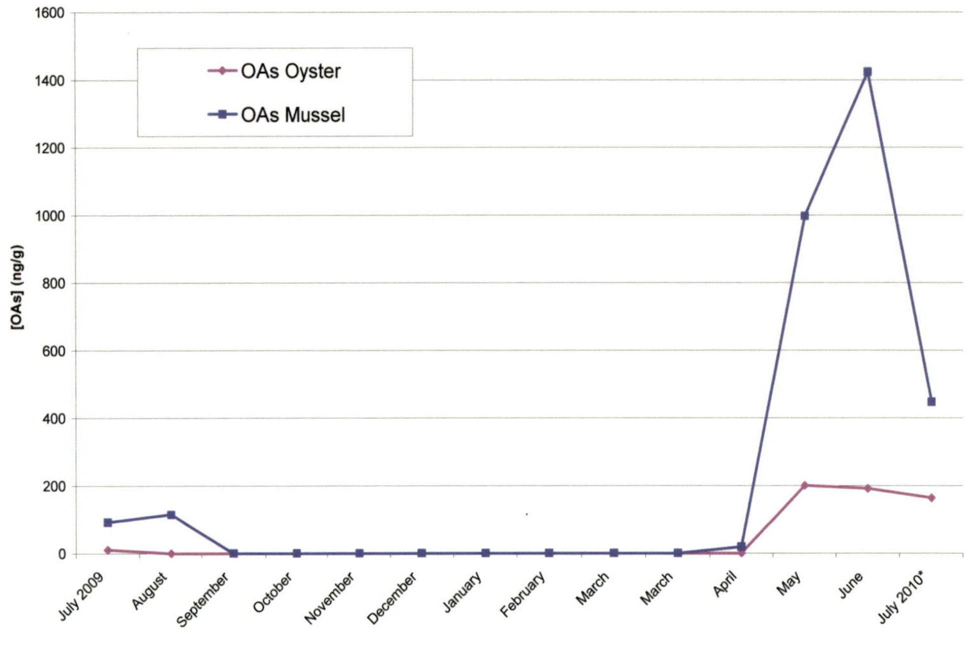

Fig. 2. Okadaic Acid profile of contamination.

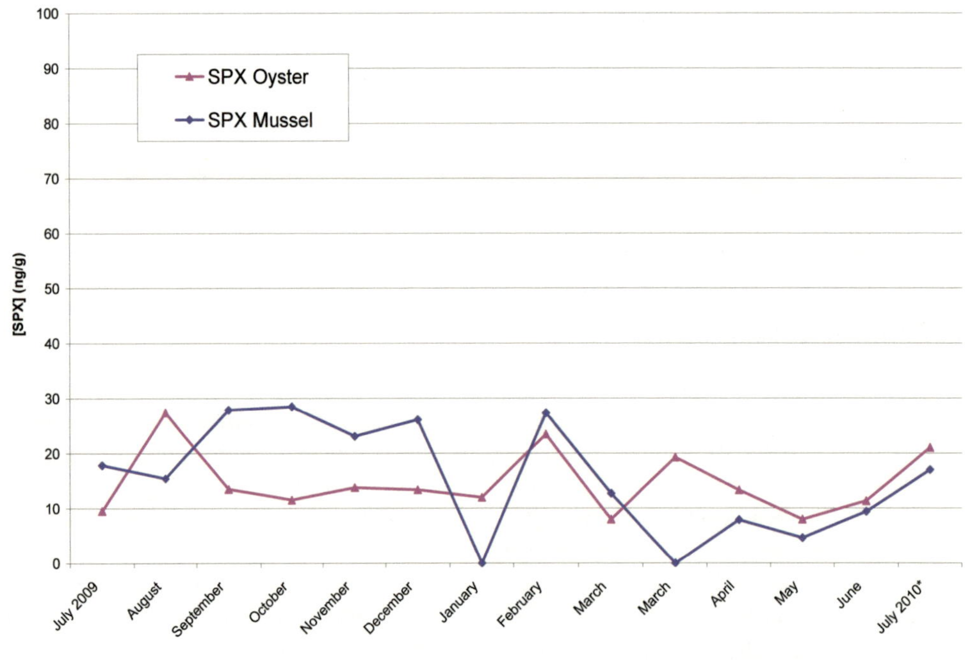

Fig. 3. Spirolides profile of contamination

Contamination data are often left-censored because of the limits of detection (LOD) and quantification (LOQ) of analytical methods. The GEMS/Food-Euro framework proposed different treatments according to the prevalence of censored-data (WHO, 1995):

- the number of censored data is less than or equal to 60 %, then, the censored data are replaced by the corresponding LOD or LOQ divided by 2 (T1);
- the number of censored data is greater than 60 % and then:
  - either the censored data are replaced by zero (T2a)
  - or they are replaced by the corresponding LOD or LOQ (T2b).

The contamination data are described in Table 2. As, for OAs, the censored data accounted for more 60 % of values, zero and LOD (or LOQ) values were used in two separate estimations of the distributions and calculations (mean, median, percentiles...). On the other hand, as the censored values about SPXs were less than 60 %, they were replaced by the half of the corresponding LOD or LOQ.

Phycotoxin	Bivalve species	Number of analyses (n=)	< LOD (%)	< LOQ (%)	Mean			Median			Max		
					T1	T2a	T2b	T1	T2a	T2b	T1	T2a	T2b
OAs	Oysters	13	76.9	0	nc	31.0	38.7	nc	0.0	10.0	nc	200.6	200.6
OAs	Mussels	13	61.5	0	nc	203.6	209.7	nc	0.0	10.0	nc	1423	1423
SPX	Oysters	13	0	0	14.2	nc	nc		nc	nc		nc	nc
SPX	Mussels	13	15.4	0	15.5	nc	nc		nc	nc		nc	nc

nc : non concerned; LOD : Limit Of Detection; LOQ : Limit Of Quantification; T1 : parameters are estimated after replacing values below LOD with LOD/2 and valued lying between LOD and LOQ with (LOQ - LOD)/2; T2a : parameters are estimated after replacing values below LOD by zero and values below LOQ by LOD; T2b : parameters are estimated after replacing the censored data by the corresponding LOD or LOQ.

Table 2. Okadaic acid and spirolide contamination data set by shellfish species (ng/g) according to the censored data treatment.

3.3.2 Exposure modelling

3.3.2.1 General mode exposure

An acute exposure corresponds to a short exposure to a harmful compound at high dose. Let us consider a phycotoxin denoted by m. The acute phycotoxin exposure is the amount of m ingested in a single meal. It is obtained by multiplying the edible portion size of one shellfish species by the concentration of m in this portion. For each phycotoxin, acute intakes were calculated individually for each shellfish species.

Chronic exposure is a repeated exposure to low, or very low, doses for a long time. The chronic phycotoxin exposure is the amount of m ingested daily from the daily consumption of all shellfish species. The general exposure model used, here, to assess individual phycotoxin intake from shellfish consumption can be expressed as follows:

$$E_m = \frac{\sum (C_{mj} * CR_j * P_j)}{BW}$$

where Em is the individual exposure (mg/kg bw/day) to the phycotoxin, m, from the ingested shellfish species, j, Cmj is the concentration (mg/kg) of the same phycotoxin in the edible portion of the same species, CRj is the daily consumption rate (kg/day) of this species, Pj is the proportion of a given shellfish species in a consumer diet (unitless), and BW is the consumer body weight (kg) assumed, in this study, to be 60 kg (USEPA, 2000).

The difference in acute- and chronic-exposure assessments stands in the consumption parameter to be used: the former takes into account the portion size of a given shellfish species, whereas the latter considers the daily consumption rate of all shellfish species.

In this study, focus was on five of the bivalve species the most consumed in the geographical area under study: oysters, mussels, cockles, carpet shell clams and razor clams.

Moreover, the approach in use for exposure calculation usually depends on the nature of the available data. This study was based on a probabilistic approach described in Kroes *et al.* (2002) and briefly recalled hereafter.

3.3.2.2 Probabilistic approach

Given that a shellfish consumer will not eat, at each time, the same portion size and that the toxin level in the eaten portion will not be alike, the probabilistic calculation considers all of the combinations of phycotoxin occurrence and consumption data. Distributions for both the food consumption data and the contamination data were used in the model to simulate dietary intakes by repeatedly drawing random values for each input distribution. The description of input variables in terms of distributions allows one to characterise their variability and/or uncertainty. Monte Carlo simulation techniques are used by the model to generate output distributions of dietary intakes liable to be ultimately considered in risk characterization. Output distributions i) give several exposure data (mean, median, minimum, maximum and all percentiles) and ii) include a comprehensive analysis of the sensitivities of the resulting exposure with respect to uncertainties in parameters (Counil *et al.*, 2005; Kroes *et al.*, 2002; Tressou *et al.*, 2004).

- Model assumptions

The subjective assumptions in use in our simulation model were about factors liable to affect contamination rates such as regulatory limits, inter-species variability and cooking process. Because of their possible impact on the results obtained in exposure assessments, they have to be taken into account to generate the model outputs.

Given that recreational shellfish harvesters can transgress bans, in this study, only the contamination distributions in purchased shellfish were right truncated at the regulatory limits when they exist (0.16 µg/g for OAs), whereas distributions of harvested shellfish contamination were not truncated (thus contamination higher than regulatory limits can be used).

- Each phycotoxin level is species-specific since the contamination rates are bivalve species-dependent, but the analyses were not made on all species. For each toxin, contamination levels were monthly determined in only two species. For the other ones, a 3-month preliminary study was conducted to gain insight into the variability of inter-species contamination. It allowed us to evaluate, for each toxin, the relationship between the levels of contamination in the most contaminated species and in the other ones, (for more details, see Picot *et al.*, submitted). According to these results, a normal distribution was applied to the contamination levels in the species under study to describe the distributions of the levels in the non-analysed bivalve species.

- As the phycotoxin levels are affected by the cooking process in use, this parameter has to be considered. The analyses were made on raw bivalves. To take into account the cooking process impact, for each toxin, the ratio between the phycotoxin rates in raw samples and in cooked samples was determined in a preliminary study (for more details, see Picot *et al.*, submitted). Then, a Normal distribution reflecting the difference between the raw and the cooked contamination rates was assigned to the contamination levels of raw bivalves in order to obtain the cooked contamination based on the raw contamination rates.

- Model simulation

The @Risk package, version 4.5 (Palisade, USA) with the Microsoft Excel spreadsheet under XP (Microsoft, USA) was used to perform risk analysis from Monte Carlo simulations and probability distributions so as to develop the exposure model on taking into account uncertainty and variability. Each simulation was run for 10 000 iterations to mimic the inherent uncertainty in shellfish-contamination and -consumption as well as the uncertainty in the mathematical process. The probability of existence of a phycotoxin in shellfish, its level in the shellfish and the probability of human exposure were all outputs of the mathematical model. To help in the identification of critical points in the process, the model sensitivity was analysed.

3.3.3 Results (output data)

Acute- and chronic-exposures to each of the phycotoxins under study were assessed through probabilistic approach. With this approach, the exposure assessment model produced, for each phycotoxin, a probability density distribution of dietary intakes from all the bivalves under study.

3.3.3.1 Acute exposure

Acute- exposure corresponds to the phycotoxin intake by an individual over a meal composed of a single portion of bivalves. For each bivalve species, the exposure is, thus, the quantity obtained by multiplying the portion size by the contamination data. Table 3 presents the main results about acute exposure issued from the probabilistic assessment.

	OA			SPX		
	Exposure* (ng/kg.bw/portion)			Exposure* (ng/kg.bw/portion)		
	Mean	Median	P95	Mean	Median	P95
Harvested oysters	19.4	7.62	78.1	8.60	5.46	27.4
Harvested mussels	456	151	1912	26.0	13.3	94.8
Harvested cockles	702	243	2808	78.4	49.6	250
Harvested carpet shell clams	378	149	1466	19.8	12.3	60.00
Harvested razor clams	133	44.9	569	8.58	5.50	27.2
Purchased oysters	17.7	7.1	72	8.10	5.21	25.5
Purchased mussels	183	79	706	29.9	15.6	105
Purchased cockles	4.00	1.7	18	1.90	1.20	6.00
Purchased carpet shell clams	3.80	1.5	15	0.7	0.45	2.20
Purchased razor clams	0	0	0	0	0	0
ARfD	333[a] / 300[b]			No ARfD allocated		

ARfD: Acude Reference Dose. * Assuming a body weight equals to 60 kg. [a]According to the JECFA (Joint FAO/WHO Expert Committee on Food Additives). [b]According to the EFSA (European Food Safety Authority).

Table 3. Acute dietary intakes of okadaic acid and spirolide obtained by a probabilistic approach and comparison with toxicological reference values for each bivalve species.

- Okadaic acid and analogs

Concerning OAs, the exposure distribution led to a maximal (for harvested cockles) mean value, a median value and a 95th percentile value equal to 0.70, 0.24 and 2.81 µg/kg bw, respectively. Figure 4 shows the acute exposure for each shellfish species.

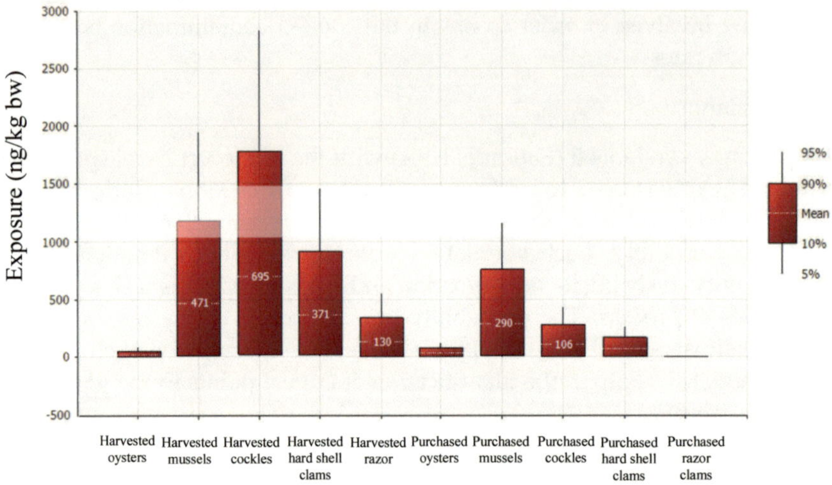

Fig. 4. Acute exposure to Okadaic Acid (ng/kg bw) for each shellfish species.

- Spirolides

For SPXs exposure distribution, the highest (for harvested cockles) mean value, median and 95th percentile were equal to 78.5, 49.6 and 250 ng/kg.bw, respectively. Figure 5 shows the acute exposure for each shellfish species.

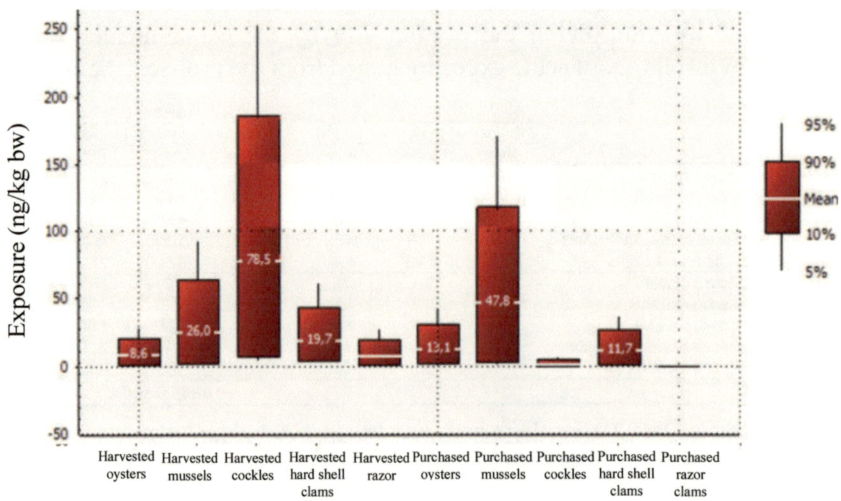

Fig. 5. Acute exposure to Spirolide (ng/kg bw) for each species.

3.3.3.2 Chronic exposure

The chronic exposure assessment corresponds to the level of exposure after a daily consumption of shellfish, thus the useful consumption data are the daily consumption rates. Table 4 illustrates the chronic-exposure levels issued from the probabilistic exposure approach for harvested-, purchased-bivalves and "all bivalves".

Censored values treatment	Bivalve sources	OA			SPX		
		Exposure* (ng/kg.bw/day)			Exposure* (ng/kg.bw/day)		
		Mean	Median	P95	Mean	Median	P95
T1 (SPX) or T2a (DA and OA)	Harvested bivalves	44.4	29.3	134	3.5	2.8	8.3
	Purchased bivalves	9.70	5.10	34.5	1.9	1.3	5.70
	All bivalves	54.1	39.0	149	5.4	4.60	11.9
T2b	Harvested bivalves	45.9	30.9	137			
	Purchased bivalves	10.3	5.40	36.6			
	All bivalves	56.2	41.1	155			
Tolerable Daily Intake		No Tolerable Daily Intake allocated					
ARfD		333[a] / 300[b]			No ARfD allocated		

ARfD: Acude Reference Dose. * Assuming a body weight equals to 60 kg. [a]According to the JECFA (Joint FAO/WHO Expert Committee on Food Additives). [b]According to the EFSA (European Food Safety Authority).

Table 4. Chronic dietary intakes of okadaic acid and spirolide obtained by a probabilistic approach for harvested, purchased and all bivalves; and comparison with toxicological reference values.

- Okadaic acid and analogs

Concerning OAs, the "all bivalves"-related exposure distribution presents maximal means of 54.1 and 56.2 ng/kg bw/day for T2a and T2b scenarios, respectively, as well as median values of 39.0 and 41.1 ng/kg bw/day and 95th percentiles of 149 and 155 ng/kg bw/day. One should note that the censored value scenario (T2a or T2b) has a very limited effect upon the chronic dietary exposure to phycotoxins. About the comparison of the contribution by harvested bivalves against the one by purchased bivalves, table 4 shows clearly that, for OAs, the intakes derived from harvest are about 5-fold those derived from purchase, mainly because the contamination distribution of harvested bivalves took into account levels above the regulatory limit. The figure 6 shows the exposure distribution of OAs for all species.

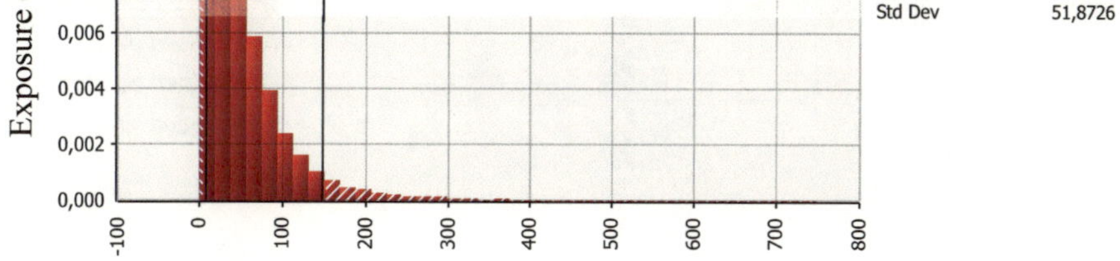

Fig. 6. Chronic exposure distribution to Okadaic Acid (ng/kg bw/day) for all species.

- Spirolides

Concerning SPX, the chronic distribution of exposure (for "all bivalves") leads to a mean value of 5.4 ng/kg bw/day, a median of 4.6 ng/kg bw/day, and a 95th percentile of 11.9 ng/kg bw/day.. For SPX, the intakes derived from harvest are about 2-fold those derived from purchase. The figure 7 shows the exposure distribution of OAs for all species.

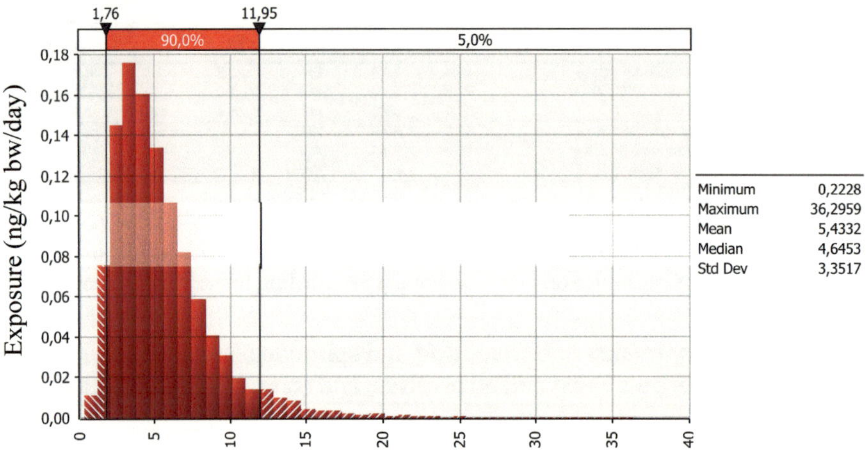

Fig. 7. Chronic exposure distribution to Spirolide (ng/kg bw/day) for all species.

- Species contribution

Figure 8 illustrates the contribution of each bivalve shellfish to the daily intakes of each of the phycotoxin under study. It evidences that the three greatest contributor species are the cockles and the carpet shell clams from harvest as well as the mussels from both harvest and purchase. Their high contributions come from the high consumption and contamination rates for harvested-species and to the high consumption rate for purchased mussels.

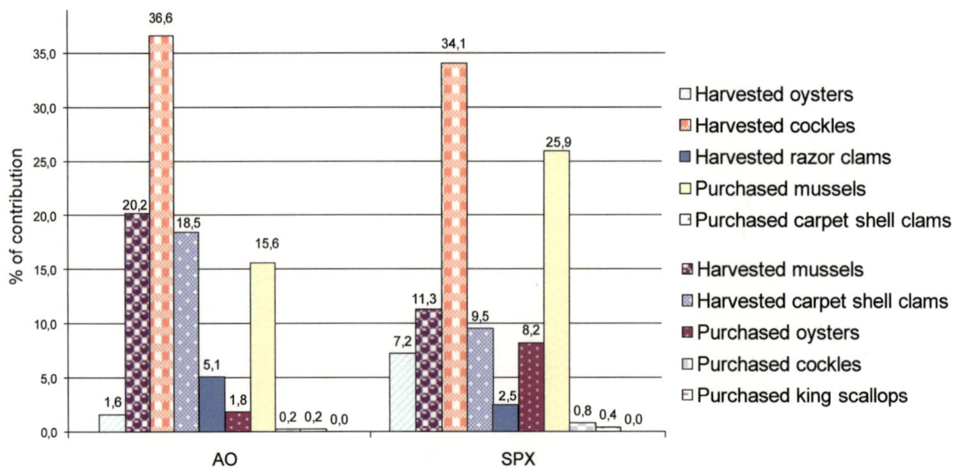

Fig. 8. Contribution of each bivalve species to the daily intakes of okadaic acid and spirolide determined through the probabilistic approach (T2a and T1 treatments of censored values).

3.4 Risk characterization

3.4.1 Acute risk characterization

For acute-risk characterization, probabilistic estimates of dietary exposure to phycotoxins have to be compared to the ARfD (Acute Reference Dose). For OAs, the provisional ARfDs established are 0.33 and 0.30 µg/kg bw by the JECFA and the EFSA, respectively, whereas no ARfD has been allocated for SPX (Toyofuku, 2006).

3.4.1.1 Okadaic acid and analogs

Concerning OAs, the highest probabilistic assessment (for harvested cockles) led to a mean exposure and a 95th percentile of, respectively, about 2.5-fold and 9-fold the OA ARfD, but to a median exposure almost 1.25-fold less than the OA ARfD. One should note that, for purchased bivalves, all exposures (means, medians and 95th percentiles) were below than the OA ARfD, excepted for high consumers of mussels, they are above the acute reference value.

3.4.1.2 Spirolides

Characterizing the SPX acute risk by comparison of the acute exposure with the ARfD cannot be realized because of the unavailability of SPX ARfD value explained by the lack of quantitative data on acute oral toxicity by this phycotoxin. However, only in order to have an idea of the order of magnitude of the margin of exposure, the acute exposure could be compared with the LD_0 (Lethal Dose)[1]. Munday (2008) studied acute oral exposure to SPX by mice. In the case of oral administration of SPX to mice, it results in two LD_0 equal to 53 and 400 µg/kg bw, for gavage and feeding with cheese cream containing SPX, respectively.

In the case of the most representative way of administration (mice feeding with cheese cream containing SPX), the highest probabilistic assessment (for harvested cockles) led to a mean exposure and a 95th percentile of, respectively, about 5000-fold and 1600-fold less than the SPX LD_0. Comparing with the most protective LD_0 (= 53 µg/kg.bw), the highest probabilistic assessment (for harvested cockles) led to a mean exposure and a 95th percentile of, respectively, about 675-fold and 210-fold less than the SPX LD_0.

Since the margin of exposure is higher than 100, it seems appear that acute SPX is not a matter of concern for human health. But this conclusion has to be confirmed with other relevant toxicological studies, allowing to establish an ARfD for SPX.

3.4.2 Chronic risk characterization

For chronic risk characterization, probabilistic estimates of dietary exposure to phycotoxins have to be compared to the TDI. But, as no TDI has been allocated to phycotoxins by international committees, we used two other methods, not satisfactory but the only ones possible: comparison with the corresponding ARfD and with the Threshold of Toxicological Concern (TTC). The TTC is a principle, which refers to the establishment of a human exposure threshold value for all chemicals, below which there would be no appreciable risk

[1] LD_0 (Lethal Dose 0): the amount of a chemical that if administered to an animal will kill 0 % of the sample population.

to human health. This threshold value is equal to 2.5 ng/kg bw/day for genotoxic substances and 25 ng/kg bw/day for all other substances (Kroes *et al.*, 2004; EFSA, 2011).

3.4.2.1 Okadaic acid and analogs

- Comparison with TTC

Concerning OAs, the levels of exposure are equal to 54; 39 and 149 ng/kg bw/day, for the mean, median and 95th percentile, respectively. These levels of exposure are not below the TTC, thus it cannot be excluded that there would be a risk to human health.

- Comparison with ARfD

For OAs, the values of the mean and 95th percentile intake issued from the probabilistic approach (0.054 and 0.15 µg/kg bw/day, respectively) are only about 5- and 2-fold less than the most protective OA ARfD (0.30 µg/ kg bw). Thus chronic OA intakes were close to ARfD. TDI is, by definition, less than ARfD. The former is, indeed, derived from a NOAEL value or a LOAEL one determined from long-term toxicological studies, whereas the latter is determined from acute toxicological studies. Moreover, in addition to the traditional security factors employed for ARfD, the establishment of TDI requires the use of a few other ones such as, for example, an uncertainty factor of 10 to extrapolate subchronic to chronic exposure (Lewis, 1995), leading to TDIs much lower than ARfD. The finding, in this study, of a chronic exposure to OA via shellfish consumption (only 2 to 5-fold below the ARfD) suggests that OA should be considered as a possible cause for concern about human health.

3.4.2.2 Spirolides

- Comparison with TTC

For SPX, the values of the mean, the median and 95th percentile intake issued from the probabilistic approach (5.4; 4.6 and 11.9 ng/kg bw/day, respectively) are higher than the TTC (2.5 ng/kg bw/day). We made the comparison with the most protective TTC because no (sub)chronic and genotoxic data are available for SPX. Thus, the only exposure data do not allow to reject a chronic risk due to SPX.

- Comparison with ARfD

As neither ARfD nor TDI have been allocated to SPX by international committees, no comparison can be made. Though there is no toxicological reference value, the calculations made in this study highlighted the regular exposure of humans to low SPX doses. Thus, in the case where toxicological data indicate chronic impact by SPX on health, it would be worth taking into account exposure to SPX.

4. Conclusion

Further to the increasing number of reports about phycotoxin-induced intoxications and deaths, these compounds have become a matter of concern for human health. But, phycotoxin exposure assessments are almost non-existent because related data about consumption and contamination are missing. This led us to study, in the same geographical area, shellfish consumption by humans and shellfish contamination by phycotoxins to assess exposure of humans to these compounds. The acute- and chronic-exposure assessments made a probabilistic approach showed that: i) in terms of acute risk, OAs appear to be a cause for concern about high consumers in cases of high contamination levels that may exceed the OA ARfD. For instance, a high and ban-transgressing consumer could be exposed to an OA acute intake up to 9-fold the ARfD; ii) about chronic risk, the finding of daily OA intakes close to the ARfD, known to be, by definition, much greater than the TDI, suggests that, among the phycotoxins under study, OA is the one to be considered. Moreover, it should be noted that bivalves contain regularly SPX at low concentrations. Chronic and subchronic data on SPX are missing, but in case of (sub)chronic toxicity, SPX exposure should be taken into consideration.

These phycotoxin-exposure assessments were aimed at making a first realistic evaluation of human exposure to phycotoxins. Their interest stands in the facts that: i) they were based on consumption- and contamination-data in the same subpopulation and area, ii) the recreational shellfish harvesters under study constitute an at-risk subpopulation iii) inter-species variability in contamination and consumption data was taken into account, iv) the impact of cooking process on phycotoxin levels was also considered.

To gain more comprehensive insight into this health issue, in the future, it would be worth: i) increasing the number of shellfish species to be investigated, ii) considering the contamination data relative to recorded cases of intoxication further to ingestion of fish and crustaceans, iii) extending the contamination database to several years and iv) studying co-exposure to several phycotoxins.

5. References

Ade P., Funari E., Poletti R. (2003). Risk to human health associated with marine toxic algae. Ann Ist Super Sanita, 39, 53–68.

Aune, T., Yndestad, M. (1993). Diarrhetic shellfish poisoning. Dans : Falconer IR (éditions), Algal Toxins in Seafood and Drinking Water, Academic Press, San Diego, California, USA., 87-104 p.

Burger J., Sanchez J., Gochfeld M. (1998). Fishing, consumption, and risk perception in fisherfolk along an east coast estuary. Environmental Research, 77, 25-35.

CAC (Codex Alimentarius Commission). (2006). Codex Alimentarius Commission Procedural Manual. 16th edition. Joint FAO/WHO Food Standards Programme. Available at: ftp://ftp.fao.org/codex/Publications/ProcManuals/Manual 16e.pdf (accessed April 4, 2010).

Counil, E., Verger, P., Volatier, J.L. (2005). Handling of contamination variability in exposure assessment : A case study with ochratoxin A. Food and Chemical Toxicology, 43, 1541-1555.

CRLMB (Community Reference Laboratory for Marine Biotoxins). (2005). Report on toxicology working group meeting, Cesenatico, Italy, 24-25 October, 2005. Available from http://www.aesan.msps.es/en/CRLMB/web/home.shtml.

Economou, V., Papadopoulou, C., Brett, M., Kansouzidou, A., Charalabopoulos, K., Filioussis, G., *et al.* (2007). Diarrheic shellfish poisoning due to toxic mussel consumption: the first recorded outbreak in Greece. Food Additives and Contaminants, 24, 297-305.

EFSA (European Food Safety Agency). (2008a). Opinion of the Scientific Panel on Contaminants in the Food chain on a request from the European Commission on marine biotoxins in shellfish - okadaic acid and analogues. The EFSA Journal, 589, 1-62.

EFSA (European Food Safety Authority). (2008b.) Guidance Document for the Use of the Concise European Food Consumption Database in Exposure Assessment (adopted on 17 March 2008). EFSA/DATEX/2008/01, Parma, Italy.

EFSA (European Food Safety Agency). (2010). EFSA Panel on Contaminants in the Food Chain (CONTAM); Scientific Opinion on marine biotoxins in shellfish - Cyclic imines (spirolides, gymnodimines, pinnatoxins and pteriatoxins). The EFSA Journal, 1628, 1-39.

EFSA (European Food Safety Agency). (2011). EFSA Scientific Committee; Draft Scientific Opinion on Exploring options for providing preliminary advice about possible human health risks based on the concept of Threshold of Toxicological Concern (TTC). Available online: www.efsa.europa.eu

Frémy, D.J., Lassus, P. (2001). Toxines d'algues dans l'alimentation. Editions Quae, France, 558 p.

Gagnon F., Tremblay T., Rouette J., Cartier J. (2004). Chemical risks associated with consumption of shellfish harvested on the north shore of the St. Lawrence River's lower estuary. Environ Health Perspectives, 112, 883-888.

Hallegraeff, G.M. (2003). Harmful algal blooms: A global overview. In : Hallegraeff, G.M., Anderson, D.M., Cembella, A.D., (éditions), Manual on Harmful Marine Microalgae. Intergovernmental Oceanographic Commission of UNESCO, Paris, France.

Hu, T., Curtis, J.M., Oshima, Y., Quilliam, M.A., Walter, J.A., Watson-Wright, W.M., Wright, J.L.C. (1995). Spirolides B and D, two novel macrocycles isolated from the digestive glands of shellfish. Journal of the chemical society, Chemical communications, 20, 2159-2161.

Kroes R., Müller D., Lambe J., Löwik M.R.H., Van Klaveren J., Kleiner J., *et al.* (2002). Assessment of intake from the diet. Food Chemical Toxicology, 40, 327-385.

Kroes R., Renwick A.G., Cheeseman M., Kleiner J., Mangelsdorf I., Piersma A., *et al.* (2004). Structure-based thresholds of toxicological concern (TTC): guidance for application to substances present at low levels in the diet. Food and chemical toxicology, 65-83.

Leblanc JC. (2006). CALIPSO, Fish and seafood consumption study and biomakers of exposure to trace elements, pollutants and Omega-3. AFSSA (French food safety agency), 2006. Available at http://www.afssa.fr/Documents/PASER-Ra-Calipso.pdf (accessed June 19, 2010).

Lewis S.C. (1995). Subchronic to chronic exposure extrapolation: Toxicologic evidence for a reduced uncertainty factor. Human and Ecological Risk Assessement, 1, 516-526.

Munday R., 2008. Toxicology of cyclic imines: gymnodimine, spirolides, pinnatoxins, pteriatoxins, prorocentrolide, spiro-prorocentrimine, and symbioimines. In: Seafood and Freshwater toxins: Pharmacology, Physiology and Detection. 2nd edition. Ed LM Botana. CRC Press (Taylor and Francys Group), Boca Raton, FL, 581-594.

Picot C., Nguyen T.A., Roudot A.C., Parent-Massin D. (2011a). A Preliminary Risk Assessment of Human Exposure to Phycotoxins in Shellfish : A Review. Human and Ecological Risk Assessement: an international journal, 17, 328-366.

Picot C., Nguyen T.A., Carpentier F.G., Roudot A.C., and Parent-Massin D. (2011b). Relevant shellfish consumption data for dietary exposure assessment among high shellfish consumers, Western Brittany, France. International Journal of Environmental Health Research, 21, 86-105.

Picot C., Limon G., Durand G., Wesolek N., Parent-Massin D., and Roudot A.C. Domoic acid, okadaic acid and spirolides: inter-species variability in contamination and cooking effects. Environ Res, Submitted.

Pigozzi S., Bianchi L., Boschetti L., Cangini M., Ceredi A., Magnani F., *et al.* (2008). First evidence of spirolide accumulation in northwestern Adriatic shellfish. Copenhagen, 319-322.

Richard, D., Arsenault, E., Cembella, A.D., Quilliam, M.A. (2001). Investigations into the toxicology and pharmacology of spirolides, a novel group of shellfish toxins. Dans : Harmful Algal Blooms 2000, Hallegraef, G.M., Blackburn, S.I., Bolch, C.J., Lewis, R.J (editions), Intergovernmental of Oceanographic Commision of UNESCO, 383-386 p.

Shumway, S.E., Van Egmond, H.P., Hurst, J.W., *et al.* (1995). Management of shellfish resources. In : Hallegraeff, G.M., Anderson, D.M., et Cembella, A.D. (editions), Manual on Harmful Marine Microalgae. IOC Manuals and Guides no. 33, 433-474 p. Intergovernmental Oceanographic Commission of UNESCO, Paris, France.

Tibbetts J. (1998). Toxic tides. Environmental health perspectives, 106, A326–331.

Toyofuku, H. (2006). Joint FAO/WHO/IOC activities to provide scientific advice on marine biotoxins (research report). Marine pollution bulletin, 52, 1735-1745.

Tressou, J., Crépet, A., Bertail, P., Feinberg, M., Leblanc, J.C. (2004). Probabilistic exposure assessment to food chemicals based on extreme value theory. Application to heavy metals from fish and sea products. Food and Chemical Toxicology, 42, 1349-1358.

USEPA (United States Environmental Protection Agency). (1998). Guidance for Conducting Fish and Wildlife Consumption Surveys. United States Environmental Protection Agency, EPA-823-B-98-007, Office of Water, Washington, 157 p.

USEPA (United States Environmental Protection Agency). (2000). Guidance for Assessing Chemical Contaminant Data for Use. Fish Advisories - Fish Sampling and Analysis, Volume 1, Third Edition. United States Environmental Protection Agency, 823-B-00-007, Office of Water, Washington, 383 p.

Van Dolah, F.M. (2000). Marine Algal Toxins: Origins, Health Effects, and Their Increased Occurrence. Environmental Health Perspectives, 108, S133-141.

WHO (World Health Organization). (1985). Guidelines for the study of dietary intakes of chemical contaminants, WHO Offset Publication, n° 87, Geneva, Switzerland, 104 p.

WHO (World Health Organization). (1995). GEMS/Food-EURO Second Workshop on Reliable Evaluation of Low-Level Contamination of Food, Kulmbach, Germany, EUR/ICP/EHAZ.94.12/WS04-FSR/KULREP95.

2

Spatial Cadmium Distribution in the Charente Watershed and Potential Risk Assessment for the Marennes Oleron Bay (Southwest France)

Coynel Alexandra, Khoury Alaa, Dutruch Lionel,
Blanc Gérard, Bossy Cécile, Derriennic Hervé and Schäfer Jörg
University of Bordeaux, EPOC, UMR 5805, Talence
France

1. Introduction

In recent years, high levels of pollutants in coastal ocean waters and in the marine food chain have been attributed to inputs either via the atmosphere (Nriagu, 1989) or by direct inputs from rivers and estuaries (e.g. Millward et al., 1996; Baeyens et al., 1997; Chiffoleau et al., 1999). In fact, rivers are considered as a major pathway for the organic and inorganic contaminant transfers from the continent to the ocean, especially during flood events (e.g. Coynel et al., 2007).

The Marennes Oleron Bay (MOB) is the first oyster-producing area in France providing nearly half of the oysters commercialised (Goulletquer & Héral, 1997; Soletchnik et al., 1999, 2002; Bry & Hoflack, 2004). The MOB and its biological compartments are subject to chronic pollution by some metals, especially Cd, representing a potential risk for shellfish cultivation (Pigeot et al., 2006; Strady et al., 2011). The latest estimates suggest that legal restrictions of oyster production in the MOB would result in a shortfall of between 50% and 70% of oysters on the market in the next years. Given that oyster is one of the top ten seafoods consumed in France, the economic consequences would be catastrophic for this region.

The Gironde Estuary, located ~30 km southward of the MOB, is affected by historic metal (e.g. Cd, Zn, Hg, Ag; Blanc et al., 1999; Schäfer et al., 2006; Castelle et al., 2009; Larrose et al., 2010; Lanceleur et al., 2011) contamination due to former Zn ore treatment in the industrial Decazeville basin, that was stopped after a major pollution accident in 1986. In this watershed drained by the Riou Mort River, anthropogenic contributions to particulate element fluxes were estimated to ~90-95% for Cd, Zn and Hg (Coynel et al., 2009). Despite decreasing Cd emissions in the Decazeville area due to ongoing remediation efforts since the early 1990s, the Lot-Garonne River system still contributes up to 80% to the annual Cd gross fluxes into the Gironde Estuary (Schäfer et al., 2002; Audry et al., 2004). Additionally, intensive agriculture and ongoing urbanization also considerably contribute to metal (e.g. Zn, Cu, Ag) gross fluxes into the Gironde Estuary (e.g. Masson et al., 2006; Lanceleur et al., 2011). In contrast to the well-studied Garonne-Gironde system, the Charente fluvial-estuarine system remains poorly studied in spite of its great hydrological influence on the MOB. The Charente River is the major river discharging directly into the MOB. During summer, 90% of the freshwater inputs into the Bay come from the Charente River (Ravail-Legrand, 1993), which drains an area of 10,549 km² dominated by farming (Bry & Hoflack, 2004). Only few studies have previously assessed the importance of the Charente Estuary to the overall metal contamination of the MOB (Gonzalez et al., 1991; Boutier et al., 2000; Dabrin, 2009). Recently, Dabrin (2009) showed that dissolved and particulate Cd concentrations the outlet of the Charente watershed, i.e. at the entry of the Charente Estuary were similar to those in the Garonne River and contributed up to 60% to total Cd inputs into the MOB, highlighting the need to precisely identify the origin(s).

In this context, the objective of this study is to (i) characterize the Cd content in water (<0.2 µm; dissolved phase) and in particles (SPM and stream sediments) exported by the Charente sub-watersheds; (ii) assess their level of contamination by comparison with world references and ecotoxicological indices and (iii) identify point sources and/or diffuse sources in the Charente watershed. This first assessment of the sub-watershed contributions to the fluvial Cd export is essential to the control and reduction of Cd contamination in oysters, i.e. to successful environmental management in this vulnerable region.

2. Presentation of the study area

The Charente watershed (surface area = 10,549 km²; ~500,000 inhabitants) is surrounded by the Massif Central to the East, the Paris sedimentary basin to the North, the Aquitaine sedimentary basin to the South and by the Armorican Massif to the Northwest (Figure 1). It is essentially composed of limestone formations dating from the Secondary: (i) in the North, the Jurassic formations are composed of large limestone beds in various facies and, (ii) in the South, the Cretaceous formations are formed by clay, sand, chalk and decalcification clays. The Primary formations crop out in the most upstream catchment areas of the Charente watershed, represented by plutonic and metamorphic rocks (BRGM, 2003).

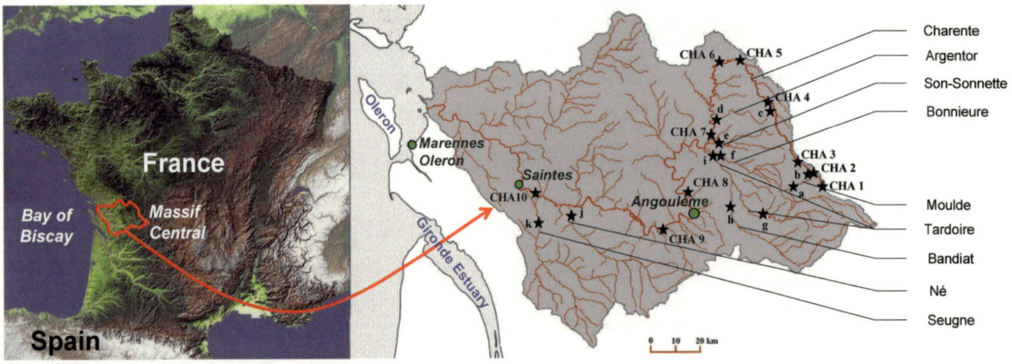

Fig. 1. Location of sampling sites

The Charente watershed can be divided into three main domains (Figure 1):

- The upstream Charente watershed features metamorphic rocks to the East. Two large reservoir dams, the Lavaud reservoir (400 ha; 10 Mm³) on the Charente River and the

Mas-Chaban reservoir (176 ha; 14.2 Mm^3) on the Moulde River, were built in 1989-1990. In this section, the Tardoire River is considered the main tributary of the Charente System.

- The downstream Charente River system is located between Angouleme and Saintes, and represented by Quaternary formations. The main tributaries of the Charente River in this section are the Ne and the Seugne Rivers.
- The Charente Estuary, downstream of Saintes, is located in areas where sedimentary rocks and limestone prevail. Two dams have been set up within the estuarine reaches submitted to tidal influence: one at Saint Savinien on the Charente River and the second on the Trezence River, the main tributary of the Boutonne River.

The Charente watershed is considered predominantly as a rural sedimentary basin. Agriculture activities cover about 60% of this area and about 11% of the cultivated area is irrigated. However, 52% of the water bodies in this river system are at risk of failing the European Water Framework Directive objectives to achieve good ecological and chemical status, due to diffuse pollution (nitrates, turbidity, and pesticides) attributed to agricultural activities and practices (EPTB-Charente, 2007; Vernier et al., 2010).

The Charente watershed meteorology is characterized by the Atlantic Ocean disturbances. The mean annual water discharge of the Charente River is ~74 m^3/s, corresponding to a specific discharge (average discharge/watershed area) of ~0.007 $m^3/s/km^2$, i.e. 3 times less than that of the adjacent Dordogne watershed (Regional Environment Agency- DIREN; BanqueHydro®; Schäfer et al., 2002). Two major factors contribute to this low specific discharge: on the one hand, relatively low supply from its major tributaries (the Tardoire and Bandiat Rivers) flowing over a karst formation and, on the other hand, intense agricultural irrigation throughout the basin. During our sampling campaign, the estimated Charente River water discharge at the watershed was ~100 m^3/s (Charente River at Chaniers [CHA10; Figure 1] + Boutonne River; data from DIREN), implying that the studied situation is representative of a moderate to a high hydrological situation.

3. Material and methods

3.1 Sampling campaign

A sampling campaign was conducted from April 6 to April 8 2010. Strategic sites were selected by Geographical Information System (GIS, ArcView ®) for testing different environmental characteristics (e.g. geology, land-use) and evaluating their impacts on Cd concentrations. In total, 20 strategic sites were selected on the Charente River (n=10 sites; notified by CHA; Figure 1; Table 1) and its tributaries (n=10 sites; notified by a letter; Figure 1; Table 1) characterized by contrasting geology, industrial and agricultural activities. An additional site was chosen on a small drain near Riou Mort River (former mining area; "c", Figure 1) for collecting a stream sediment deposit. Note that this Riou Mort River in the Charente watershed is not identical with the well-studied Riou Mort River draining the polluted Decazeville basin, responsible for important historical polymetallic (mainly Cd, Zn, Cu, Pb, Hg, Ag) pollution in the Lot-Garonne river continuum (e.g. Blanc et al., 1999; Schäfer et al., 2002; Audry et al., 2004; Coynel et al., 2009).

Item		River	Area km²	Q m³/s		Cond. µS/cm	pH	eH mV	O_2 %	SPM mg/l	Nitrate µmol/l	Dissolved Cd µg/l	Part. Cd SPM mg/kg	Part. Cd (<63µm) mg/kg
CHA1	*Saint Gervais*	Charente	40	0.14	X	86.4	7.3	101	98.4	15	112	0.030	3.78	1.69
CHA2	*Sansac*	Charente	44	1.1	X	96.6	7.4	95	92.5	4	170	0.010	2.11	0.73
CHA3	*Pont de Suris*	Charente	110	2.1	Diren	110	7.6	120	91.6	12	163	0.010	4.36	0.70
CHA4	*Chez Paire*	Charente	230	4.5	X	115	7.6	165	90.8	10	162	0.031	8.16	3.06
CHA5	*Charroux*	Charente	346	6.4	Diren	207	7.5	166	86.2	9	223	0.020	4.86	2.63
CHA6	*Saint-Saviol*	Charente	492	8.2	Diren	226	7.6	171	92.6	13	259	0.022	3.51	2.29
CHA7	*Aunac*	Charente	1090	22	X	373	7.8	153	93.6	13	415	0.011	3.07	1.39
CHA8	*Vindelle*	Charente	3750	47	Diren	374	8.0	227	89.7	11	425	0.009	2.33	0.90
CHA9	*Sireuil*	Charente	4070	82	X	378	8.0	205	91.0	12	381	0.008	2.96	2.18
CHA10	*Chaniers*	Charente	7412	91	Diren	461	8.0	197	90.3	9	410	0.010	4.97	2.11
a	*Massignac*	Upstream Moulde	52	0.46	X	98.3	7.4	85	101	6	114	0.009	1.20	0.82
b	*Chez Boige*	Downstream Moulde	54	0.96	X	103	7.5	135	99.8	2	134	0.008	2.72	0.76
c	*Riou Mort*	-	-	-	-	-	-	-	-	n.d	n.d	n.d	n.d	37.7
d	*Poursac*	Argentor	108	2.9	Diren	426	8.0	175	102	7	335	0.010	3.57	1.20
e	*Mouton*	Son-Sonnette	226	2.7	X	425	8.3	162	115	8	319	0.009	3.95	1.14
f	*Saint Ciers/Bonnieure*	Bonnieure	203	2.7	Diren	262	8.1	224	103	11	218	0.011	2.30	0.89
g	*Montbron*	Upstream Tardoire	389	9.4	Diren	87.5	7.6	94	89.4	9	93	0.013	4.59	1.78
h	*Feuillade*	Bandiat	333	5.2	Diren	176	7.8	189	90.0	9	100	0.009	6.17	1.70
i	*Coulgens*	Downstream Tardoire	1200	8.2	Diren	128	7.2	199	98.1	23	108	0.011	3.59	1.31
j	*Les Perceptiers*	Ne	602	7.4	Diren	651	8.1	202	110	11	477	0.009	1.80	0.72
k	*La Lijardière*	Seugne	902	9.9	Diren	654	7.9	212	97.0	12	470	0.009	2.13	0.72
Charente - CHA10 (2006-2007; Dabrin, 2009)				mean								0.024	4.89	
				min								0.011	1.21	
				max								0.048	9.86	

Table 1. Description of sampling sites, water discharge obtained by the Regional Environment Agency-DIREN or measured in this study (X), physical and chemical parameter values (conductivity, Eh, pH, dissolved oxygen saturation) and SPM, nitrate, dissolved and particulate Cd (in SPM and stream sediments < 63µm) concentrations

3.2 Discharge measurements

This sampling strategy, aiming at estimating instantaneous fluxes, required reliable discharge data for each of the selected observation sites. However, only 12 sites are equipped with permanent gauging stations maintained by the Regional Environment Agency-DIREN (BanqueHydro®). Therefore, for this study, 8 additional river gauging measurements were performed at the other sampling sites (Table 1). Standard instantaneous discharge measurements were made by measuring flow velocities at different depths along vertical profiles, each of them representing a segment of the river cross-section. The cross-sectional area of each segment was then multiplied by the corresponding integrated measured velocities to estimate water discharge in the segment. The sum of river discharges in all segments represents the estimated instantaneous water discharge of the river section. The uncertainty on the measurements was estimated between 5 and 10% (Regional Environment Agency- DIREN).

3.3 Water sampling

The general physical and chemical parameters (pH, conductivity, Eh and O_2) were measured in-situ at each site. Temperature and conductivity were measured using a TetraCon 96® probe (PROFILINE, WTW). Oxygen saturation was determined by an ISY 52® probe. Determinations of pH and Eh were performed using a Sentix® 41 probe (PROFILINE, WTW). At each site, water was sampled manually for SPM, nitrate and Cd concentrations using clean techniques: all materials in contact with the water samples were made of polypropylene (PP), carefully decontaminated as previously detailed in Canton et al. (2012) for nitrate and in Audry et al. (2004) for Cd.

Back in the laboratory, river water samples were homogenized and precise volumes (~500 ml) were filtered through pre-weighed 0.7 µm filters (Durieu®). Then the filters were dried to constant weight (45°C; 12 h) and re-weighed in order to obtain SPM concentrations. For dissolved nitrate and Cd, all river water samples were immediately filtered on-site through 0.2 µm Sartorius® polycarbonate filters. For nitrate, filtrates were collected in 14 ml polypropylene tubes and stored at -80°C until analysis; for dissolved Cd, filtrates were collected in pre-cleaned 30 ml polypropylene bottles, acidified (1/1000; HNO_3 suprapur grade) and stored at 4°C awaiting analysis.

Suspended particulate matter for Cd analyses was retrieved by pumping up to 80 L of river water (~50 cm from the bank at 10-20 cm depth) using a peristaltic pump with PP-tubing followed by centrifugation (Westfalia, Germany; 12,000 g). This technique is considered a practicable and reliable method for SPM sampling in all hydrological situations (e.g. Schäfer & Blanc, 2002).

3.4 Stream sediment sampling

Stream sediments are commonly used for geochemical prospecting. Collected just after a strong hydrological event, the geochemical composition of these samples corresponds to the maximum particulate transfer to the estuary and coastal zone. Unlike SPM, whose composition can rapidly fluctuate, stream sediments integrate metal contamination (Coynel et al., 2009).

The coordinates of the sampling locations were recorded with a differential GPS. At each site, representative samples, consisting of the uppermost 1 cm of sediment from several recent depositional pockets were collected with a plastic spatula within a distance of 5-10 meters to enhance representativeness. The preferential accumulation of metals, either of natural or anthropogenic origin, in the fine-grained sediment fractions may induce grain size effects and reduced sample representativeness (Förstner & Wittmann, 1981; Horowitz, 1991; Benoit & Rozan, 1998). Therefore, stream sediment samples were sieved (<63 µm; nylon sieves) to remove coarse material which was obviously not representative of typical grain size of suspended sediment (Coynel et al., 2009).

3.5 Nutrient analysis

The dissolved inorganic compounds were colorimetrically analyzed according to standardized techniques. Dissolved nitrates ($\Sigma NO_3^- = NO_3^- + NO_2^-$) were analyzed by Flow Injection Analysis (FIA) according to Canton et al. (2012). Precision was ±10% for ΣNO_3^-.

3.6 Particulate Cd extraction in SPM and stream sediment

Representative subsamples (~30 mg of dry, powdered and homogenized material) of SPM or sediment were digested in acid-cleaned closed reactors using 1.5 ml HCl s.p. (12 M), 2 ml HF s.p. (22 M) and 0.75 µl HNO_3 s.p. (14 M) at 110°C for 2 h using a temperature-controlled digestion system (DigiPREP MS®, SCP SCIENCE). After evaporation to dryness (10 h at 110°C), the residues were completely re-dissolved in 0.25 ml HNO_3 s.p. (14 M) and 5 ml Milli-Q® water on a heating plate (15 min at 60°C) and after cooling brought to 10 ml in volumetric flasks using Milli-Q® water.

3.7 Dissolved and particulate Cd analysis

Dissolved and particulate Cd concentrations were measured using ICP-MS (X7, THERMO) with external calibration under standard conditions. The applied analytical methods were continuously quality checked by analysis of international certified reference sediments (PACS-1, BCR 320, SL-1, SLRS-4). Accuracy was within 5% and 7% of the certified values in the dissolved and particulate fractions, respectively. The analytical error (relative standard deviation) was better than 5% (rsd) for both phases. The detection limit estimated as 3 sigma of method blanks was 2 ng/l for the dissolved phase and 0.04 mg/kg for the particulate phase.

4. Results and discussion

4.1 Physical and chemical parameters

Conductivity, redox potentials (Eh) and pH (to a lesser extent) in the Charente River and its tributaries tended to increase from upstream to downstream, probably reflecting local geological characteristics (metamorphic rocks in the upstream sections; limestone formations in the downstream watershed) and land use (vineyards). The dissolved oxygen saturation values were rather similar whatever the sites (Table 1).

4.2 Discharge monitoring

The spatial variability of the Charente River water discharge is linearly related to its drainage basin area for the section between CHA1 and CHA9, except for CHA8 (Figure 2).

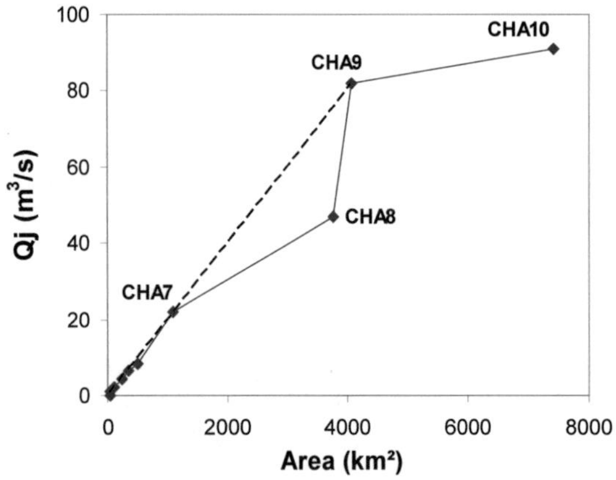

Fig. 2. Relationship between daily water discharges measured at different sites of the Charente River and the corresponding surface areas

The permanence of shoal sedimentation during the Middle and Upper Jurassic is a local feature which explains the development of the La Rochefoucauld karstic system, close to Angouleme city (Figure 1) and the relatively low water discharge measured at CHA8. The Bandiat and the Tardoire Rivers flow over the La Rochefoucauld karst accounting for more than 50% of the outlet discharges (Kurtulus & Razack, 2007). The La Touvre spring, located near of Angouleme city, constitutes an important discharge system of the aquifer varying between 2 m^3/s and 40 m^3/s. During our water monitoring campaign, the La Touvre water discharge was ~19.3 m^3/s. The apparent water discharge deficit at CHA10 compared to the general relation between discharge and area drained may reflect the intense agricultural irrigation (Figure 2).

4.3 SPM measurements

The SPM concentrations ranged from 2 mg/l in the Moulde River, just after the Mas-Chaban Reservoir to 23 mg/l in the Tardoire River (Table 1). The lowest value can be explained because of hard to erode rock sills combined with the settling of SPM due to the presence of the Mas-Chaban Reservoir. Based on the world river classification of SPM concentration proposed by Meybeck et al. (2003), these values can be considered as either "very low" (<20 mg/l), generally observed for watersheds located downstream of major or numerous lakes (e.g. Alpine Rhone River), or in very flat and humid regions with wetland predominating (e.g. the Central Amazon watershed) or "low" (20-100 mg/l) characteristic of plain watersheds. However, even if our sampling campaign is representative of a moderate to high discharge hydrological situation (Q=~100 m^3/s), SPM concentrations probably are much higher during flood events as previously observed by Dabrin (2009), during 2006-2007, at the outlet of the Charente watershed (Chaniers site; SPM = 200 mg/l during a flood event with Q=350 m^3/s)

4.4 Nitrate concentrations

The nitrate concentrations were 93-477 μM/l with an average ~246 μM/l (Table 1). Nitrate concentrations measured in the main hydrological section of the Charente River increased from upstream (~100 μM/l) to downstream (up to 425 μM/l at CHA8) and were positively correlated to water discharge at the watershed scale (Figure 3). This clear nitrate increase

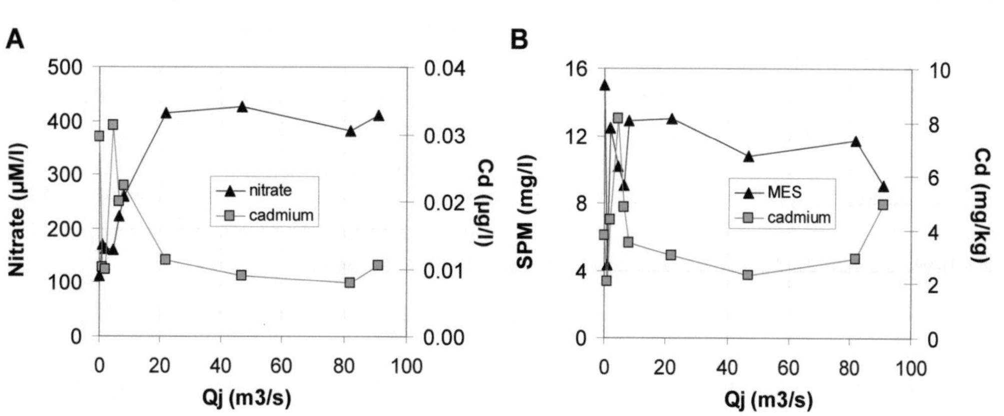

Fig. 3. (A) Nitrate and dissolved Cd concentrations versus daily water discharge in the Charente River; (B) Suspended Particulate Matter (SPM) and particulate Cd concentrations versus daily water discharge in the Charente River.

starts at the CHA5 site, i.e. where the Charente River drains maize areas. Based on this observation, we have determined typical nitrate concentrations for each land use in the Charente System: the mean nitrate concentrations are ~140 μM/l in the areas mainly occupied by pasture, ~350 μM/l for corn/maize production and ~430 μM/l in vineyard areas.

The Charente River faces high nitrate levels (Bry & Hoflack, 2004; EPTB-Charente, 2007). These high levels probably reflect the intensification of agriculture in the central and lower systems and increase the risk of eutrophication. The intensification of crops, particularly corn production, was accompanied by a high water demand for irrigation over the last thirty years. Vineyards are also located in the basin and can generate diffuse water pollution by nitrates. To assess whether the concentrations obtained on the Charente River are comparable to other systems, we compared our results with those recently published for rivers draining the Arcachon basin (SW France), which is affected by eutrophication due to nitrogen transfer from agricultural areas to the river system (Canton et al., 2012). These authors demonstrated that low concentrations (20 to 45 μM/l) occur in watersheds dominated by forest, whereas nitrate concentrations were considered as high in watersheds draining agricultural areas with an average of 140 μM/l (Canton et al., 2012). For these agricultural watersheds, nitrate concentrations increased from 200 to 500 μM/l in winter during high water discharges. Accordingly, our results obtained for the Charente River are similar to those in the rivers studied by Canton et al. (2012) supporting that the high nitrate concentrations may be attributed to agriculture as suggested previously (Bry & Hoflack, 2004; Vernier et al., 2010) and probably result in significant nitrate export to the Marennes-Oleron Bay and the adjacent coastal area.

4.5 Dissolved Cd concentrations

The dissolved Cd concentrations (Cd_D) in the Charente watershed ranged from 8 to 31 ng/l, with an average of ~17 ng/l (Table 1). Unlike the spatial nitrate evolution, the dissolved Cd concentrations in the mainstream decreased from upstream to downstream until CHA9; then a slight Cd_D increase occurred at CHA10 (Figure 3). Based on the Cd_D levels, a Cd_D distribution map aims at visualizing the spatial variation in the Charente watershed (Figure 4). The color classes were determined using the detection limit (2 ng/l), the Cd_D level (10 ng/l) representing "good status" of water quality as proposed by the Ministry of Environment and Sustainable Development and the Water Agencies quality guideline (SEQ-eau; MEDD and Agences de l'Eau, 2003) and the average Cd_D in world rivers proposed by Martin & Meybeck (1979) marked MA in Figure 4 (MA=50 ng/l).

In the downstream Charente watershed, the Seugne, Né, Sonnette and Bandiat Rivers showed Cd_D typical of good water quality (<10 ng/l), as well as the upstream sections of the Moulde, the Argentor and the Son-Sonnette Rivers. In contrast, the Cd_D concentrations in the upstream section of the Charente River (CHA1; CHA4-CHA6) suggested lower water quality (20-30 ng/l; Table 1). However, even the highest Cd_D measured in this study were lower than the Cd_D level defined for the world's major rivers (MA=50 ng/l; Figure 4), with low human influences (e.g. Amazon and Congo Rivers; Martin & Meybeck, 1979).

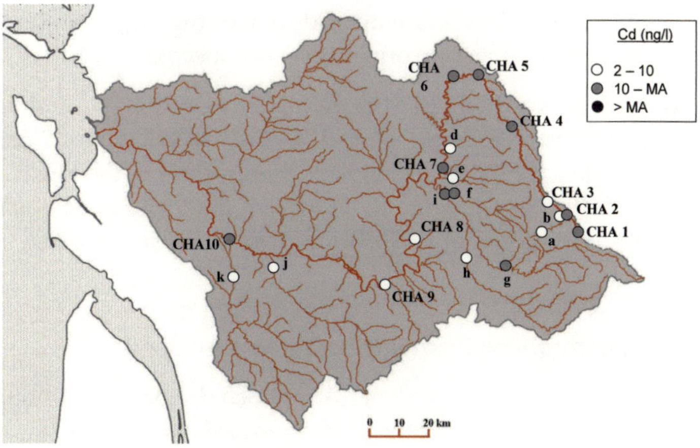

Fig. 4. Spatial distribution of dissolved Cd concentrations

4.6 Particulate Cd concentrations in SPM

The particulate Cd concentrations (Cd_P) were variable from one site to another with values comprised between 1.20 mg/kg and 8.16 mg/kg (Table 1). Most of these values were clearly higher than the average concentration measured in the SPM defined for Word Rivers (MA=1.55 mg/kg; Viers et al., 2009), except for the Upstream Moulde River (Table 1). As observed for Cd_D in the main hydrological section of the Charente watershed, the Cd_P concentrations in SPM decreased from upstream to downstream until CHA9; then a significant Cd increase occurred at CHA10 (Figure 3). The spatial evolution of the particulate Cd concentration is not correlated with that of the Charente SPM concentrations (Figure 3). We compared particulate Cd concentrations with consensus-based sediment quality guidelines proposed for freshwater sediments (MacDonald et al., 2000), i.e. concentration classes supporting the assessment of ecotoxicological risk potential (Figure 5).

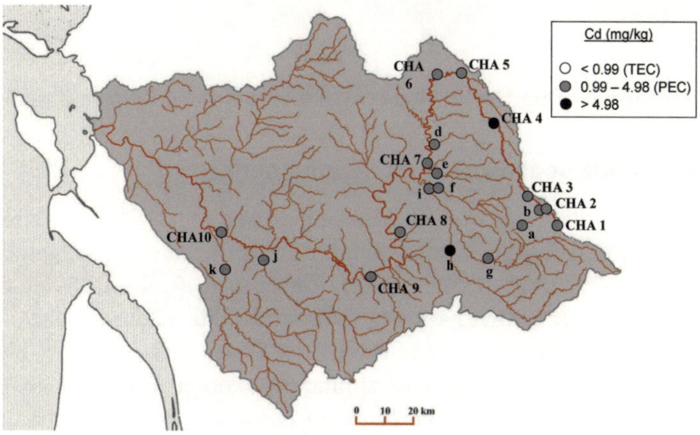

Fig. 5. Spatial distribution of particulate Cd concentrations in SPM

- The TEC (Threshold Effect Concentration; Cd_P=0.99 mg/kg) is defined as the Cd_P concentration below which no effect on organisms is expected;

- The PEC (Probable Effect Concentration; Cd_P =4.98 mg/kg) is defined as the Cd_P concentration above which effects on organisms are expected.

Most of the Cd_P in the Charente River SPM were higher than the TEC and at two sites the measured values exceeded the PEC level ("b" and "CHA4" with Cd_P = 6.17 and 8.16 mg/kg, respectively); two other sites (CHA10; CHA5) have values close to the PEC level (Figure 5). These results suggest that in the Charente River system toxic effects on aquatic organisms due to the presence of Cd (i) cannot be excluded at most of the sites studied and (ii) should be expected locally.

4.7 Particulate Cd concentrations in stream sediments (<63µm)

Excluding the Riou Mort site "c" (Cd = 37.7 mg/kg) which drains a former mineral resource deposit, Cd_P in sieved stream sediments ranged from 0.70 to 3.06 mg/kg (Table 1). The comparison with ecotoxicological indices showed that a majority of sites have Cd_P concentrations above the TEC, yet without exceeding the PEC level (Figure 6). This suggests that potential toxicity effects on water organisms in the Charente River due to the presence of Cd cannot be excluded.

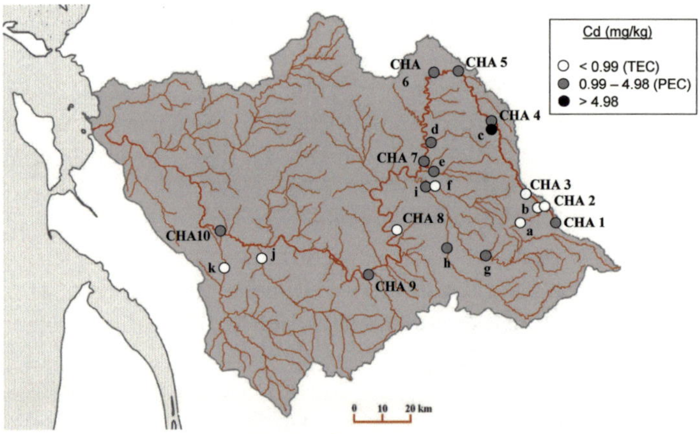

Fig. 6. Spatial distribution of particulate Cd concentrations in <63 µm sediments

4.8 Comparison with the geochemical monitoring performed during 2006-2007

The dissolved and particulate Cd concentrations measured at the Chaniers site (CHA10 site) in this study were compared with those obtained during the 2 year-geochemical monitoring achieved in the Inter-Regional project "Défi Cadmium" (Cadmium Challenge, Water Agency Adour-Garonne) performed during 2006 and 2007 (Dabrin, 2009; Table 1). The Chaniers site is considered as the outlet of the fluvial Charente watershed. The Cd_P in SPM from CHA10 measured in this study was similar to average Cd_P established during the "Défi Cadmium" project. In contrast, the Cd_D concentration was lower than the minimum Cd_D value measured by Dabrin (2009). It has been classically observed in river systems that the lowest metal concentrations in water and in SPM occurred during high flow rates. This phenomenon can be related to dilution by (i) rainwater for the dissolved phase and (ii) coarse particles (coarse silt to sand) with low metal adsorption capacity for the particulate phase (e.g. Schleichert, 1975; Dawson & Macklin, 1998; Cobelo-Garcia et al., 2004). In the Charente River system, Dabrin (2009) showed that the dissolved concentrations of some metals such as Cd decreased with water discharges, highlighting the close link between hydrology and geochemistry. As our campaign is representative of "medium to high water" conditions in the Charente River, our evaluation of the SPM quality defined throughout the watershed does not represent the worst geochemical situation. We can therefore assume that Cd_P may frequently exceed critical thresholds (e.g. PEC) at many sites during low waters and we recommand specific water monitoring during this period in order to evaluate ecotoxicological risk at the watershed scale.

4.9 Dissolved and particulate Cd fluxes

The instantaneous (daily) dissolved and particulate Cd fluxes were calculated for each sampling site (Table 2). The Cd_D levels significantly increased between the CHA3 and the CHA4 sites, inducing a clear increase in Cd_D flux that was not proportional to water discharge, suggesting an additional point source within this section (Figure 7A). The additional Cd point source (Δ1 ~8g/d; Figure 7A) may be linked to the presence of mineral deposits in the metamorphic formation areas and specifically, near the fault of Beaumont Confolent (BRGM, 2003). This is supported by high Cd_P measured in the sediment collected in the former mining area (site "c"; Figure 1). A second abnormal increase in Cd_D fluxes (Δ2 ~25g/d; Figure 7A) can be observed between the CHA9 and the CHA10 sites and cannot be explained by the Cd_D supply via the Ne and Seugne Rivers alone ("j" and "k"; Table 2). Additional campaigns will be needed to identify the source in this section by performing a longitudinal profile along the Charente River. Note that in this geographical area, we have located a manufactury which produces batteries for all types of civil and military aircrafts.

Item		River	SPM Flux t/j	Part Cd flux g/j	Diss Cd flux g/j	Part Cd flux contribution
CHA1	*Saint Gervais*	Charente	0.2	0.7	0.4	66%
CHA2	*Sansac*	Charente	0.4	0.8	0.9	47%
CHA3	*Pont de Suris*	Charente	2.2	10	1.8	85%
CHA4	*Chez Paire*	Charente	4.0	32	12	73%
CHA5	*Charroux*	Charente	5.0	24	11	69%
CHA6	*Saint-Saviol*	Charente	9.1	32	16	67%
CHA7	*Aunac*	Charente	25	76	21	78%
CHA8	*Vindelle*	Charente	44	102	36	74%
CHA9	*Sireuil*	Charente	83	245	56	81%
CHA10	*Chaniers*	Charente	71	353	82	81%
a	*Massignac*	Upstream Moulde	0.2	0.3	0.3	45%
b	*Chez Boige*	Downstream Moulde	0.2	0.5	0.7	40%
c	*Riou Mort*	-	-	-	-	-
d	*Poursac*	Argentor	1.8	6.3	2.4	72%
e	*Mouton*	Son-Sonnette	1.8	7.1	2.2	77%
f	*Saint Ciers/Bonnieure*	Bonnieure	2.5	5.9	2.6	69%
g	*Montbron*	Upstream Tardoire	7.2	33	11	76%
h	*Feuillade*	Bandiat	3.9	24	4.2	85%
i	*Coulgens*	Downstream Tardoire	16	59	8.1	88%
j	*Les Perceptiers*	Ne	7.3	13	5.7	70%
k	*La Lijardière*	Seugne	11	22	8.0	74%

Table 2. Daily SPM flux and dissolved and particulate Cd flux and particulate Cd flux contribution at strategic sites

The evolution of the daily Cd_P flux in the section of the Charente River was compared to that of the corresponding SPM flux. This comparison revealed a good statistical relationship between both parameters (Figure 7B). Based on this relationship, we can define a Cd_P average of ~2.90 mg/kg, corresponding to the regression slope and, probably, to the Cd_P level in agricultural lands which are mechanical eroded during runoff. Based on this, we have detected a small increase in Cd_P that was not due to the contribution of SPM at the CHA4 site and a strong Cd_P anomaly at the CHA10 site. Like for Cd_D, the minor Cd_P anomaly in CHA4 probably results from the presence of formerly exploited mineral deposits (Ba-Pb-Zn mineralization). Between the CHA9 and CHA10 sites, we have observed a SPM decrease related to sedimentation inducing decreasing SPM fluxes ('loss') in this section. Based on the observed relationship between SPM flux and Cd_P flux in the Charente River, the expected Cd_P flux at the Chaniers site for an instantaneous SPM flux of 71 t/d would be ~200 g/d (Figure 7B). The observed Cd_P flux of 353 g/d being ~1.8 times higher (Table 2) clearly suggests a significant Cd enrichment in SPM due to anthropogenic sources and reflects observations made for Cd_D.

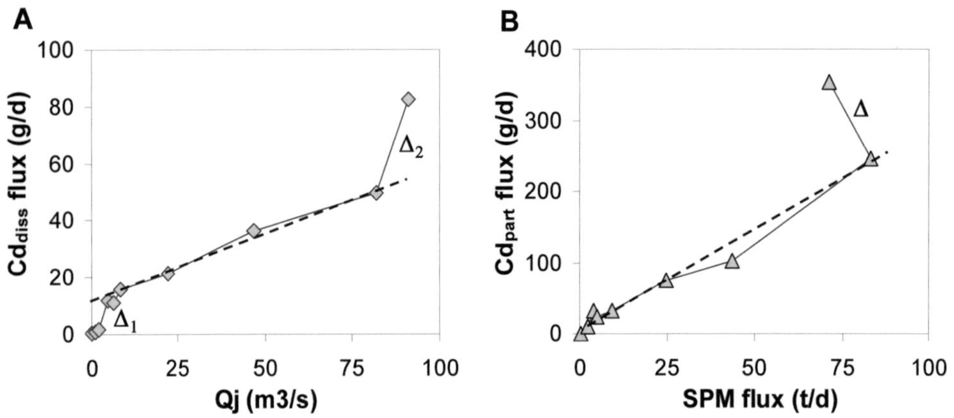

Fig. 7. Relationship between daily dissolved Cd flux and water discharge (A) and between daily particulate Cd flux and SPM flux (B) at strategic sites in the main channel of the Charente River

The dissolved and particulate flux estimates underline the Cd partitioning (i.e chemical form) in the Charente River system. In the main section of the Charente River, the instantaneous Cd flux was mainly due to SPM transport (reaching 80% at CHA10), except at CHA2 (downstream from the Lavaud Reservoir) and in the Moulde River (upstream and downstream part), where the particulate transport contributed 47%, 45% and 40%, respectively, to total (dissolved + particulate) Cd (Cd_T) fluxes. These results suggest that the particulate phase is the predominant Cd vector in the Charente River system and demonstrate that efficient reduction of Cd transport into the Marennes-Oleron Bay would imply limitation of mechanical erosion, mainly in the agricultural areas.

By extrapolating daily fluxes to the annual scale, we have obtained a first approximation of annual SPM and Cd exports to the Charente Estuary and in fine to the MOB. These estimates probably represent the orders of magnitude, although errors on flux estimates related to inadequate sampling frequency may be very high and are likely to result in severe under-estimation of fluxes (Coynel et al., 2004). Based on this extrapolation, annual SPM flux and Cd_T export at CHA10 were evaluated to ~26,000 t/yr and 159 kg/yr, respectively, corresponding to a specific sediment yield of ~2.45 t/km²/yr and a specific Cd_T flux ~ 15 g/km²/yr. Accordingly, the specific (potentially under-estimated) Cd flux of the Charente watershed would be 2 times lower than that of the Cd-polluted Garonne watershed in 2006/2007 (39 g/km²/yr; Dabrin, 2009). However, as explained above, this Cd estimation for the Charente River system is probably highly underestimated. Indeed, based on a permanent observation sampling in 2006-2007 with high temporal resolution, the annual specific CdT was evaluated to 54 g/km²/yr for the Charente River, i.e. higher of that reported for the Garonne River (Dabrin, 2009). Furthermore the totality of Cd exported from the Charente watershed via the Charente Estuary directly enters the MOB and may reach the oyster farms therein, whereas only part of the Cd exported from the Gironde system may reach the vulnerable zones of the MOB. Our work intends to demonstrate that the Charente River system should be further studied because of its significant impact on the MOB.

5. Conclusion

This study represents a major advance in the geochemical characterization of water and water-borne particles in the Charente River and contributes to a better understanding of the Cd transfer into the Marennes-Oleron Bay. The main results obtained have allowed to:

- develop a first spatially resolved geochemical database of dissolved and particulate Cd concentrations at the watershed scale, i.e. in three different compartments (water, sediment and SPM) at 21 strategic sites along the Charente River and its tributaries, taking into account the geology and land use characteristics;
- assess the quality of these three geochemical compartments using ecotoxicological criteria or world river references and locate sub-catchment with potential chemical risk;
- quantify the daily dissolved and particulate Cd fluxes representative of a moderate hydrological situation from the water discharge measurements provided by the Regional Environment Agency-DIREN and complemented by gauging measurements during this campaign;
- estimate the Cd contribution from subcatchments and identify major zones with point and/or diffuse Cd sources.

In terms of perspectives, further work should be carried out during low water because of the strong relationship between water quality and hydrology. It would also be necessary to determine more precisely the origins of the observed Cd anomaly, by taking samples from the former mining deposit and agricultural soils. This approach should be completed by a longitudinal sampling profile between CHA9 and CHA10, where significant additional dissolved and particulate Cd fluxes occurred. Our study includes the Charente River upstream from the Chaniers site, which is considered as the outlet of the fluvial system. Furthermore, a more complete mass balance should include fluxes via two other major tributaries, the Boutonne and Arnoult Rivers, entering the Charente Estuary in the tidal reaches.

6. References

Audry, S., Schäfer, J., Blanc, G. & Jouanneau, J.M. (2004). 50-year sedimentary record of heavy metal pollution (Cd, Zn, Cu, Pb) in the Lot River reservoirs (France), *Environmental Pollution*, 132, pp. 413-426.

Baeyens, W., Elskens, M., Van Ryssen, R. & Leermakers, M. (1997). The impact of the Scheldt input on the trace metal distribution in the Belgian coastal area (results of 1981-1983 and 1995-1996), *Hydrobiologia*, 366, pp. 91-108.

Benoit, G. & Rozan, T.F. (1999). The influence of size distribution on the particle concentration effect and trace metal partitioning in rivers, *Geochimica et Cosmochimica Acta*, 63, pp. 113-127.

Blanc, G., Lapaquellerie, Y., Maillet, N. & Anschutz, P. (1999). A cadmium budget for the Lot-Garonne fluvial system (France), *Hydrobiologia*, 410, pp. 331-341.

Boutier, B., Chiffoleau, J.-F., Gonzalez, J.-L., Lazure, P., Auger, D. & Truquet, I. (2000). Influence of the Gironde estuary outputs on cadmium concentrations in the coastal waters: consequences on the Marennes-Oléron bay (France), *Oceanologica Acta*, 23, pp. 745-757.

BRGM-Bureau de Recherches Géologiques et Minières (2003). Carte géologique de la France 1 : 1 000 000.

Bry, C. & Hoflack, P. (2004). Le bassin versant de la Charente : une illustration des problèmes posés par la gestion quantitative de l'eau, *Courrier de l'environnement de l'INRA*, 52. pp. 81-84.

Canton, M., Anschutz, P., Coynel, A., Polsenaere, P., Auby, I. & Poirier, D. (2012). Nutrient export to an Eastern Atlantic coastal zone: first modeling and nitrogen mass balance, *Biogeochemistry*, 107, pp. 361-377

Castelle, S., Schäfer, J., Blanc, G., Audry, S., Bossy, C. & Lissalde, J.P. (2007). 50-year record and solid state speciation of mercury in natural and contaminated reservoir sediment, *Applied Geochemistry*, 22, pp. 1359-1370.

Chiffoleau, J.M., Auger, D. & Chartier, E. (1999). Fluxes of selected trace metals from the Seine Estuary to the Eastern English Channel during the period August 1994 to July 1995, *Continental Shelf Research*, 19(15-16), pp. 2063-2082.

Cobelo-Garcia, A., Prego, R. & Labandeira, A. (2004). Land inputs of trace metals, major elements, particulate organic carbon and suspended solids to an industrial coastal bay of the NE Atlantic, *Water Research* 38, pp. 1753-1764.

Coynel, A., Schafer, J., Hurtrez, J.E., Dumas, J., Etcheber, H. & Blanc, G. (2004). Sampling frequency and accuracy of SPM flux estimates in two contrasted drainage basins, *Science of the Total Environment*, 330, 233-247.

Coynel, A., Schafer, J., Blanc, G. & Bossy, C. (2007). Scenario of particulate trace metal and metalloid transport during a major flood event inferred from transient geochemical signals, *Applied Geochemistry*, 22, pp. 821-836.

Coynel, A., Blanc, G., Marache, A., Schäfer, J., Dabrin, A., Maneux, E., Bossy, C., Masson, M. & Lavaux, G. (2009). Assessment of metal contamination in a small mining-affected watershed - high resolution monitoring coupled with spatial analysis by GIS, *Journal of Environmental Monitoring*, 11, pp. 962-976.

Dabrin, A. (2009). Mécanismes de transfert des éléments traces métalliques (ETM) et réactivité estuarienne - Cas des systèmes Gironde, Charente, Seudre et Baie de Marennes Oléron. Thèse de doctorat, Univ. Bordeaux 1, 375 p.

Dawson, E.J. & Macklin, M.G. (1998). Speciation of heavy metals on suspended sediment under high flow conditions in the River Aire, West Yorkshire, UK, *Hydrological Processes* 12, pp. 1483-1494.

EPTB-Charente (2007). Association française des Etablissements Publics Territoriaux de bassin. http://www.eptb.asso.fr/les-eptb/eptb-charente.

Förstner, U. & Wittmann, G.T.W. (1981). Metal Pollution in the Aquatic Environment. New York, Springer-Verlag, 486 pp.

Gonzalez, J.L., Boutier, B., Chiffoleau, J.F., Auger, D., Noel, J. & Truquet, I. (1991). Distribution of Cd and Hg in the Bay of Marennes-Oleron, *Oceanologica Acta*, 14, pp. 559-568.

Goulletquer, P. & Heral, M. (1997). Marine molluscan production trends in France: from fisheries to aquaculture in: C.L. Mackenzie, Jr. V.G. Burnell, Jr., A. Rosenfield, & W.L. Hobart (Eds). The history, present condition and future of the molluscan fisheries of North and Central America and Europe Volume 3, Europe. NOAA Technical Report NMFS, 129, pp. 137-164.

Horowitz, A.J. (1991). A primer on sediment-trace element chemistry. Chelsea, Lewis Publishing Co, 2nd edition, 136 p.

Kurtulus, B. & Razack, M. (2007). Evaluation of the ability of an artificial neural network model to simulate the input-output responses of a large karstic aquifer. The La Rochefoucauld aquifer (Charente, France), *Hydrogeology Journal*, 15, pp. 241-254.

Lanceleur, L., Schäfer, J., Bossy, C., Coynel, A., Larrose, A., Masson, M. & Blanc, G. (2011). Silver fluxes to the Gironde Estuary - Eleven years (1999–2009) of monitoring at the watershed scale, *Applied Geochemistry*, 26, pp. 797-808.

Larrose, A., Coynel, A., Schafer, J., Blanc, G., Masse, L. & Maneux, E. (2010). Assessing the current state of the Gironde Estuary by mapping priority contaminant distribution and risk potential in surface sediment, *Applied Geochemistry*, 25, pp. 1912-1923.

MacDonald, D.D., Ingersoll, C.G. & Berger, T.A. (2000). Development and Evaluation of consensus-Based Sediment Quality Guidelines for Freshwater Ecosystems, *Archives of Environmental Contamination and Toxicology*, 39, pp. 20–31.

Martin J.M., Meybeck M. (1979). Elemental mass-balance of material carried by major world rivers, *Marine Chemistry*, 7, pp. 173-206.

Masson, M., Blanc, G. & Schäfer, J. (2006). Geochemical signals and source contributions to heavy metal (Cd, Zn, Pb, Cu) fluxes into the Gironde Estuary via its major tributaries, *Science of the Total Environment*, 370, pp. 133-146.

MEDD & Agences de l'Eau (2003). Système d'évaluation de la qualité de l'eau des cours d'eau (SEQ-EAU; version 2), http://rhin-meuse.eaufrance.fr/IMG/pdf/grilles-seq-eau-v2.pdf; 40p.

Meybeck, M., Laroche, L., Dürr, H.H., & Syvitski, J.P.M. (2003). Global variability of daily Total Suspended Solids and their fluxes in rivers. *Global and Planetary Change*, 39, pp. 65-93.

Millward, G. E., Allen, J.I., Morris, A.W., & Turner, A. (1996). Distributions and fluxes of non-detrital particulate Fe, Mn, Cu, Zn in the Humber coastal zone, U.K., *Continental Shelf Research*, 16, pp. 967-993.

Nriagu, JO. (1989). A global assessment of natural sources of atmospheric trace metals, *Nature*, 338, pp. 47-49.

Pigeot, J., Miramand, P., Guyot, T., Sauriau, P.G., Fichet, D., Le Moine, O., & Huet, V. (2006). Cadmium pathways in an exploited intertidal ecosystem with chronic cadmium inputs (Marennes-Oleron, Atlantic coast, France), *Marine Ecology-Progress Series* 307, pp. 101-114.

Ravail-Legrand, B. (1993). Incidences du débit de la Charente sur la capacité biotique du bassin ostréicole de Marennes-Oléron. Univ. Nantes, 171 pp.

Schäfer J., & Blanc, G. (2002). Relationship between ore deposits in river catchments and geochemistry of suspended particulate matter from six rivers in southwest France, *Science of the Total Environment*, 298, pp. 103-118.

Schäfer J., Blanc G., Lapaquellerie, Y., Maillet, N., Maneux, E. & Etcheber, H. (2002). Ten-Year-Observation of the Gironde Fluvial System: Fluxes of Suspended Matter, Particulate Organic Carbon and Cadmium, *Marine Chemistry*, 79, pp. 229-242.

Schäfer, J., Blanc, G., Audry, S., Cossa, D. & Bossy, C. (2006). Mercury in the Lot-Garonne River system (France): Sources, fluxes and anthropogenic component, *Applied Geochemistry*, 21, pp. 515-527.

Schleichert, U. (1975). Annual variation of the heavy metal contents of suspended sediments in the Rhine River at Koblenz, *Deutsche Gewaesserkundliche Mitteilungen* 19, pp. 150-157.

Soletchnik, P., Le Moine, O., Faury, N., Razet, D., Geairon, P. & Goulletquer, P. (1999). Mortalité de l'huître Crassostrea gigas dans le Bassin de Marennes-Oléron. Etude de la variabilité spatiale de son environnement et de sa biologie par un système d'information géographique (SIG), *Aquat. Living Resour.*, 12, pp. 131–143.

Soletchnik, P., Huvet, A., Le Moine, O., Razet, D., Geairon, P., Faury, N., Goulletquer, P. & Boudry, P. (2002). A comparative field study of growth, survival and reproduction of *Crassostrea gigas, C. angulata* and their hybrids, *Aquatic Living Resources,* 15, pp. 243-250.

Strady, E., Kervella, S., Blanc, G., Robert, S., Stanisiere, J.Y., Coynel, A. & Schäfer, J. (2011). Spatial and temporal variations in trace metal concentrations in surface sediments of the Marennes Oléron Bay. Relation to hydrodynamic forcing, *Continental Shelf Research*, 31, pp. 997-1007.

Vernier, F., Bordenave, P., Chavent, M., Leccia, O. & Petit, K. (2010). Modelling scenarios of agriculture changes on freshwater uses and water quality at a large watershed scale - the case of the Charente watershed (France), *in International congress on environmental modelling and software* - International congress on environmental modelling and software, Ottawa : Canada (2010) - http://hal.archives-ouvertes.fr/hal-00546884/en/.

Viers, J., Dupré, B. & Gaillardet, J. (2009). Chemical composition of suspended sediments in World Rivers: New insights from a new database, *Science of the Total Environment*, 407, pp. 853-868.

3

Planning and Decision Support Tools for Integrated Water Resources Management (IWRM) on River Basin Level in the Southeast-Asian Region on the Example of Vietnam – Tools for Water Quantity and Quality Risk Assessment

Björn Zindler, Andreas Borgmann, Sandra Greassidis, Sylvia Jaschinski, Christian Jolk and Harro Stolpe
Ruhr University Bochum, eE+E Environmental Engineering and Environment Germany

1. Introduction

A lot of countries worldwide suffer from emerging water related problems. Droughts, floods and water contamination are the most pressing issues. According to the United Nations (UN, 2008), in 2008 over 880 million people of the developing world's population were without access to safe drinking water and over 2.5 billion lacked adequate sanitation.

Considering a rapid global change driven by climate, land use and demographic changes, one of the most important tasks of the global community is to find means of using and protecting natural resources in a responsible way in order to support a holistic and sustainable development. The integrated approaches of the international R&D projects funded by the German Federal Ministry of Education and Research (BMBF) are helping to understand, interpret and solve problems in the water sector.

One specific program in this framework is called "Integrated Water Resources Management" (IWRM). The primary focus of the program is to establish cooperation between science, administration and economy between Germany and foreign countries. The R&D projects in the program should develop management concepts and implement action plans in the water sector in cooperation with partners of the project regions. Another task is the contribution of adapted water related technologies and the transfer of know-how.

Vietnam is one of the model regions of the funding program. Since 2007, the German-Vietnamese joint R&D project Integrated Water Resources Management in Vietnam (IWRM-Vietnam) funded by the BMBF is developing methods and technologies adapted to Vietnamese conditions.

The institute of Environmental Engineering and Ecology (eE+E) at the Ruhr University Bochum (Prof. Dr. Harro Stolpe) coordinates the R&D project IWRM-Vietnam and contributes "Planning and Decision Support Tools" in order to improve decision processes of Vietnamese decision makers in the water sector.

2. Subject for research

Water quality and quantity issues are limiting factors for the socio-economic and environmental development of Southeast-Asian societies. IWRM plays a crucial role in order to cope with the negative side effects of dynamic population and economic growth rates, expansion and intensification of agricultural land use and deforestation. The issues are further intensified by an insufficient water supply and lacking sanitation infrastructure.

Vietnam is a country with rich water resources. A dense river network provides an abundant supply of water. Despite this comfortable situation Vietnam faces numerous challenges that require large investments during the coming decades in order to implement an effective remedy. Major challenges are:

- an uneven distribution of the river network, an uneven rainfall across Vietnam and prolonged dry seasons resulting in water supply problems in some areas;
- Vietnam partly is a downstream country. Important rivers drain from bordering countries into Vietnam. The quantity and quality of the surface water depends on the water usage in the upstream countries;
- deficient water supply infrastructure, deficient wastewater management, inadequate flood protection etc;
- a fast urbanization, industrialization and the intensification of agriculture in Vietnam leads to a rapidly increasing water demand and to severe water contamination;
- an increase of conflicting water uses (e.g. agriculture vs. hydro power);
- an increase of problems stemming from untreated waste water disposal into the rivers (especially close to major cities and industrial centers);
- diffuse contamination risks from agricultural sources (pesticides, fertilizers);
- forming institutions in Vietnam that have the capacity to efficiently plan water resources usage and control water contamination.

This wide range of challenges leads to a complex problem situation that requires a holistic and sorting research approach. Research is needed to merge a multitude of approaches in an integrating manner. Developed methods within the research approach of the R&D project IWRM-Vietnam have to image the given situation as broad as possible. An important matter is to include (preferably) all institutions and stakeholders involved in the process.

The developed method should enable decision makers to plan and manage the pressing challenges in an efficient way. During the development, it has to be considered that financial resources for water management are often very limited in countries like Vietnam. Thus, the method has to be able to prioritize the challenges in order to render economical and targeted decisions possible.

Following this, the methods developed within the research have to be integrating, cost- and time-efficient and target-oriented in order to cope with the challenges of the water sector in Vietnam.

3. Background

3.1 IWRM

Since the International Conference on Water and the Environment, hosted in Dublin, and the United Nations Conference on Environment and Development, hosted in Rio de Janeiro (both in 1992), IWRM finds a stronger consideration worldwide. According to the Global Water Partnership (GWP), IWRM is "a process which promotes the coordinated development and management of water, land and related resources in order to maximize economic and social welfare in an equitable manner without compromising the sustainability of vital ecosystems and the environment" (GWP, 2000).

According to the Agenda 21 "Integrated Water Resources Management" is based on the perception of water as an integral part of an ecosystem, a natural resource and a social and economic good, whose quantity and quality determine the nature of its utilization. Water resources have to be protected, taking into account the functioning of aquatic ecosystems and the perenniality of the resource, in order to satisfy and reconcile needs for water in human activities. In developing and using water resources priority has to be given to the satisfaction of basic needs and the safeguarding of ecosystems" (UNCED, 1993).

The overall objective is "to satisfy the freshwater needs of all countries for their sustainable development". Additionally, IWRM should include "the integration of land- and water-related aspects and should be carried out at the level of the catchment basin or sub-basin" (UNCED, 1993). According to their capacity and to available resources all states worldwide should "have designed and initiated costed and targeted national action programs and [...] have put in place appropriate institutional structures and legal instruments" (UNCED, 1993).

3.2 Administration/laws

In Europe, the European Water Framework Directive (EU-WFD) is a legal instrument to implement the concept of water management on river basin scale since the year 2000. In Vietnam the legal basis of IWRM is the Prime Minister Decree 120 on River Basin Management (No: 120/2008/NĐ-CP dated 01.12.2008) and the National Target Program Water (NTP-WR 2010). The Vietnamese water law from 1999 will be amended in 2011 and will be another important legal basis of IWRM in Vietnam.

The Decree 120 on River Basin Management from 2008 regulates the principles and main tasks of river basin management as well as responsibilities for the river basin management. The decree demands an action plan for the prevention and protection of water contaminations and the restoration of contaminated water resources on river basin scale.

Essential preconditions for a successful implementation of IWRM in Vietnam will be to overcome the existing fragmentation of administrative responsibilities in the water sector and the strengthening and reorganization of the existing River Basin Organizations (RBOs). The RBOs should be authorized to raise wastewater charges and to spend these financial means on their own responsibility for required IWRM measures according to the priorities identified by the method developed in the R&D project IWRM-Vietnam.

The development in the environmental and water sector in Vietnam shows that the country already started to implement a legal framework according to the IWRM concept defined by the GWP and the Agenda 21 during the last years (cf. Zschiesche et al., 2008).

3.3 Risk assessment

Although initial steps for IWRM have been taken in Vietnam, the implementation of the entire IWRM concept, including the identification of measures for water and environmental protection, is still lacking. Within the R&D project IWRM-Vietnam and in response to this situation, the authors developed a method in order to identify areas with high priority need for action on a regional scale (Planning and Decision Support Tools).

The method follows the basic ideas of risk assessment for water quality and water quantity. It is based on concepts for the ecological risk analysis, which in Germany have been originally developed for example by Kiemstedt, Bachfischer (1978) and later completed during the further development according to the European Law of Environmental Impact Assessment from 1985 and amended in 1997, 2003 and 2009 (EC, 2011). Comparing to the European concept, the Environmental Protection Agency in the United States published their Guidelines for Ecological Risk Assessment in 1998 (EPA, 1998).

The developed method consists of a contamination risk assessment (water quality) and a water balance estimation (water quantity). Following the contamination risk assessment, the water balance estimation concludes in a risk assessment, too. Here, the risk of the occurrence of water deficit by overexploiting the water resources (or the occurrence of water surplus) are evaluated (cf. Jolk et al., 2010).

The principle idea for the contamination risk assessment is to combine contamination potentials (originating from land uses) and the sensitivity of natural resources (here of water resources), which results into a contamination risk. Two-dimensional matrices are used to aggregate the contamination potential and the sensitivity of water resources into the risk. The matrices are applied to determine the risk on a scale with the classes "low", "medium" and "high".

The principle idea of water balance risk assessment is based on the aggregation of the estimated water resources quantity on the one hand and the estimated water demands of water users on the other hand. This results in a water balance revealing deficit (and surplus). The assessment of the estimated deficit is applied to determine the risk of water deficit on a scale with the classes "low", "medium" and "high" (cf. Greassidis et al., 2011).

4. Solution oriented approach

In order to sort the method and IWRM into the given conditions in Vietnam, it is necessary to define an overall planning context. For this purpose IWRM-Vietnam designed two planning concepts.

- The IWRM planning levels provide a planning framework for the IWRM process in Vietnam and sort the method into this framework (cf. 4.1).
- Water Management Units help to show the results of the risk assessment in a structured and spatially differentiated way (cf. 4.2).

4.1 IWRM planning levels

According to the European Water Framework Directive (EU-WFD), IWRM-Vietnam designed a planning level concept suitable for IWRM in countries like Vietnam (cf. figure 1 and 2). "Planning and Decision Support Tools" were developed on the "river basin" planning level (scale approx. 1 : 300.000) to identify and prioritize areas within the river basins (Water Management Units = WMUs) with higher problem intensity and thus higher need for action through IWRM measures.

The identification and prioritization is based on an estimation of the water balance deficit (e.g. due to overexploitation) and on contamination risk assessment (e.g. due to diffuse agricultural sources or industrial point sources) and finally on a ranking according to problem intensities. This enables decision makers to distribute available financial means for the realization of IWRM measures in a practical and targeted manner.

The method was developed and discussed in close cooperation with the Vietnamese Ministry of Natural Resources and Environment, Department of Water Resources Management (MoNRE-DWRM) in Hanoi, provincial authorities (Departments of Natural Resources and Environment, DoNREs) and along with further stakeholders. Vietnamese legal demands have been considered.

Depending on its specific purpose, IWRM is being accomplished on different scales. A planning level concept based on the EU-WFD and suitable for the conditions in Vietnam has been developed within IWRM-Vietnam.

The planning level concept is a framework for IWRM in Vietnam under consideration of national and international standards. The planning level pyramid (see figure 1) structures existing and future measures and concepts of the IWRM process.

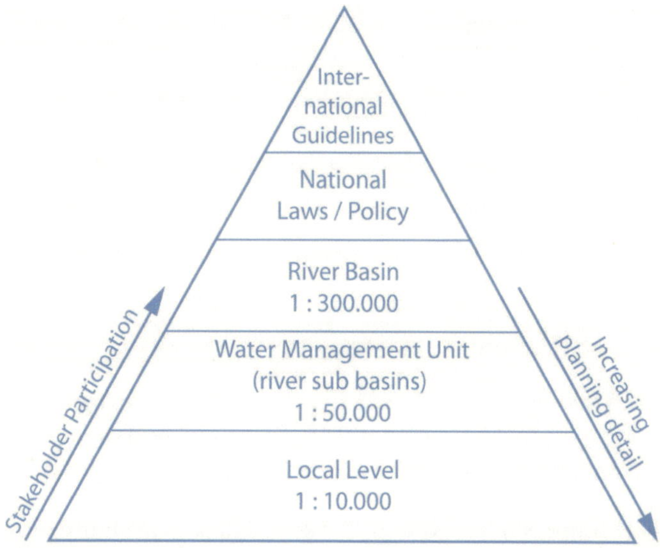

Fig. 1. Graduated IWRM planning levels (planning level 3 "River Basin": R&D project IWRM-Vietnam).

The planning levels focus on the purposes that result from the given spatial scale (see figure 2):

1. **International Level**: International scientific discussion; sharing international experience; guidelines for IWRM (e.g. Global Water Partnership)
2. **National Level in Vietnam**: Legislation (e.g. National Target Program Water); technical and water quality standards etc.; implementation of River Basin Organizations; identification of river basins with higher problem intensity and priority need for IWRM-measures (cf. Cuddihy&Frederiksen, 1996)
3. **River Basin Level:** Organization, financing and refinancing of the IWRM process by River Basin Organizations; identification of WMUs with higher problem intensity ("hot spots") and priority need for IWRM-measures (waste water treatment, water resources protection, monitoring etc.) utilizing a GIS-based evaluation of spatial and statistical information
4. **WMU Level:** Spatial planning of water management measures through the identification of most suitable locations in the areas identified on planning level 3 (river basin) by conducting detailed investigations (field investigations to confirm and complement the water balance, water quality, water demand, waste water discharge, socio-economy, water models etc.)
5. **Local Level**: Object planning of measures as identified on planning level 3 and 4 (examples for measures: monitoring, waste water treatment, water extraction, water supply etc.)

Planning Level	Planning Objectives
1. International	International guidelines for IWRM e. g. Global Water Partnership, international research projects etc.
2. National	Legislation, standards**, identification of river basins** with higher problem intensity and need for IWRM measures (scale approx. 1:1.000.000)
3. River Basin	Identification of **Water Management Units (WMUs)** with higher problem intensity and need for IWRM measures (scale approx. 1:300.000)
4. WMU	Identification of **most suitable locations for IWRM measures** (scale approx. 1:50.000)
5. Local	**Object planning:** monitoring, waste water treatment, water supply etc. (scale approx. 1:10.000)

Fig. 2. Overview of IWRM planning levels in Vietnam and their objectives (planning level 3 "River Basin": R&D project IWRM-Vietnam).

The planning level concept is adjusted to fit the scarce data availability on site. All decision relevant information for the identification of WMUs (Water Management Units) with higher problem intensity and priority need for IWRM measures on the river basin planning level could be gathered. This database consists of satellite images (land use etc.), statistical data (drinking water demand, industry demand, number of population etc.), maps (topography, administration, geology, soils, land use etc.), point data (wells, industries, monitoring stations etc.), studies, reports, laws, quality standards etc. Data that was considered to be used in the method has to meet the following requirements:

- satisfying detail grade for considered planning level and goal of planning (here: prioritize areas with need for IWRM measures)
- as detailed as needed for decision, but as easy as possible
- accessibility, homogeneity, replicability of data processing

4.2 Project areas and WMUs

The method developed within the R&D project IWRM-Vietnam has been applied in different project areas in Vietnam. The specific characteristics of these areas regarding water management, topography and land use have been taken into account during the development of the method. Thus, the Planning and Decision Support Tools can easily be assigned to other regions in Vietnam. The project areas (cf. figure 3) are:

- Red River part basin: Province Nam Dinh in the Red River Delta
- Upper Dong Nai basin: Province Lam Dong in the southern highlands
- Cuu Long part basin: City Can Tho in the Mekong Delta

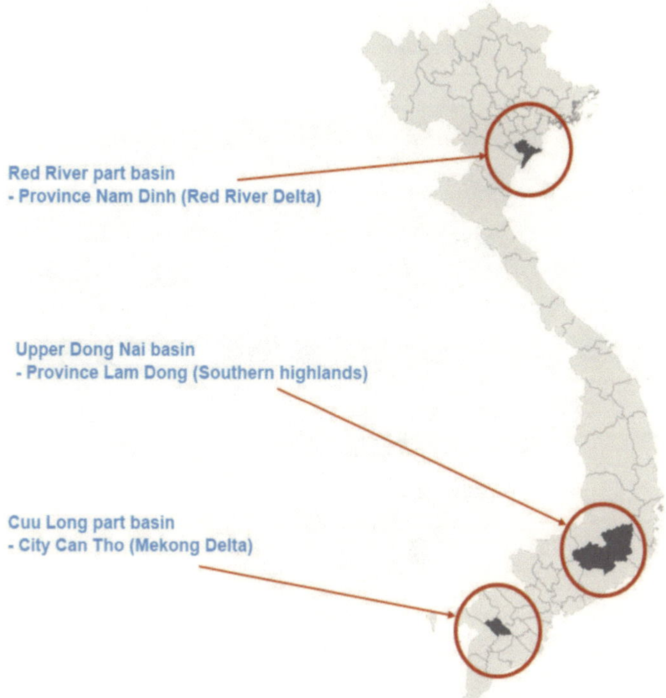

Fig. 3. Project areas in the R&D project IWRM-Vietnam. Planning and Decision Support Tools have been developed for Red River part basin, Upper Dong Nai basin and Cuu Long part basin.

In order to develop tools that fit the needs and target the right issues of the water resources situation, it is imperative to grasp all essential characteristics of the system. The result of a thorough system analysis is the basis for all further investigations and developments. Figures 4 to 6 are showing block diagrams of the water resources systems of all three project areas.

The block diagrams show information on the hydrologic system of the Red River part basin (Nam Dinh province), the Upper Dong Nai basin (Lam Dong province), and the Cuu Long part basin (Mekong Delta, Can Tho City) including elements contributing to the available water resource and all relevant water users.

The investigated Red River part basin (see fig. 4) is located in province Nam Dinh (northern Vietnam). The geological situation of this project area is dominated by alluvial and marine sedimentations. Aquifers typically are found in sand and gravel layers. Nam Dinh is a deltaic landscape with almost no elevation differences. The area is managed using polder structures. Each polder serves as a single management unit. The irrigation (and possible salt-water intrusion) is controlled by sluices in the polder dykes. The water resources are abundant but threatened by salt-water intrusion and contaminations from industries and handicraft villages within the polders (or often on the dykes). The main water user is the agriculture. Urban and rural population and the industry are further water users (as well as the watering of livestock and the tourism industry).

Fig. 4. Block diagram of Red River part basin (province Nam Dinh).

The investigated Upper Dong Nai basin (see fig. 5) is (mainly) located in province Lam Dong (southern-central Vietnam). The massif of the Lang Bian plateau and the mountain ranges of the Di Linh-Bao Loc plateau consist of basalts, granites and jurassic shales and sandstones. Quaternary deposits are located in the river valleys and the floodplains of the Dong Nai river system. Upper Dong Nai aquifers are found in the basaltic layers. The plateaus are mainly characterized by fertile soils. The south-eastern part of the province Lam Dong descends to the low-lying plains north of Ho Chi Minh City. The water resource is controlled by precipitation and groundwater abstractions. It is partly controlled by storages (major dams). The main water user is the agriculture. Urban and rural population and the industry are further water users (as well as the watering of livestock and the tourism industry).

Fig. 5. Block diagram of Upper Dong Nai basin (province Lam Dong).

The investigated Cuu Long part basin (Mekong Delta) is located within the surroundings of the city of Can Tho (southern Vietnam). The Mekong Delta is composed of alluvial sediments from the Holocene with a high amount of silt. Aquifers mostly consist of sand and are covered by thick clay layers. The Mekong Delta is a deltaic landscape with almost no elevation differences. It is characterized by a large network of rivers and artificial canals. The rivers and canals are mainly used for irrigation and for transportation of goods and people. The canal system is connected to the open sea. The water resources are abundant but threatened by salt-water intrusion. The main water user is the agriculture followed by the industry and urban population. The rural population is a further water user (as well as the watering of livestock and the tourism industry).

Fig. 6. Block diagram of Cuu Long part basin (City of Can Tho, Mekong Delta).

The Planning and Decision Support Tools originally have been developed based on the project area Upper Dong Nai river basin. After having established a basic system, the tools were adapted to fit the additional conditions of the delta regions.

After analyzing the characteristics of the project areas, basic spatial units were defined. The Planning and Decision Support Tools are being applied on planning level 3 (river basin). In order to show the results of the risk assessment in a structured and spatially differentiated way, the R&D project IWRM-Vietnam defined Water Management Units (WMUs). WMUs are sub-basins of a river basin that are defined for the purpose of IWRM in the basin.

These units should enable an effective management of both water quantity and water quality. Therefore, each water management unit combines sub basins with similar management tasks that can be joined appropriately. The following criteria are applied to delineate WMUs:

- The boundaries of the WMUs are delineated along the **natural river basins or sub-basins** and are valid for both surface- and groundwater resources
- The natural river basins are **subdivided at hydrological significant points**, i.e. confluence of a large branch river and/or the outlet of a large reservoir. Furthermore, they are **subdivided according to criteria of geomorphology, land use and administrative regions**
- If natural river basins or sub-basins do not exist (delta areas) the WMUs are divided along the main rivers, canals or polder

Each WMU is named using established local names (such as names of reservoirs, rivers or localities) and given a unique identity number. The Water Management Units are checked against the specifications of the Water Resources Allocation Plan (as defined in Decree 120/2008/ND-CP, Art. 14).

Figures 7 to 9 are showing the project areas and the respective WMUs as delineated using the above definition. The recommendations on the WMUs were subject to intense discussions with cooperating institutions and stakeholders before fixing the WMUs into their final stage.

The project area Red River part basin has been differentiated into five Water Management Units (see figure 7). Due to the special polder structure of this area, each polder has been defined to form one WMU. Areas out of polders have not been considered but influence the water management situation of each WMU.

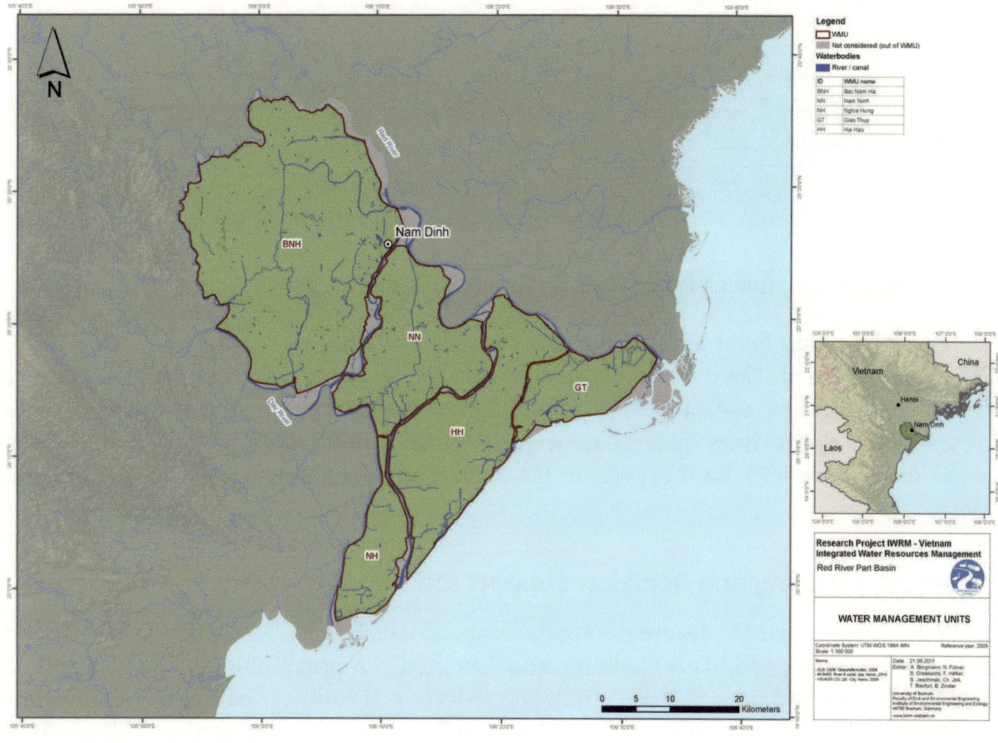

Fig. 7. WMUs in the Red River part basin (province Nam Dinh).

The project area Upper Dong Nai basin holds 22 WMUs (see figure 8). All criteria have carefully been weighed against each other. Finally, IWRM-Vietnam came up with balanced WMUs that only needed minor adjustments after the discussion with local authorities.

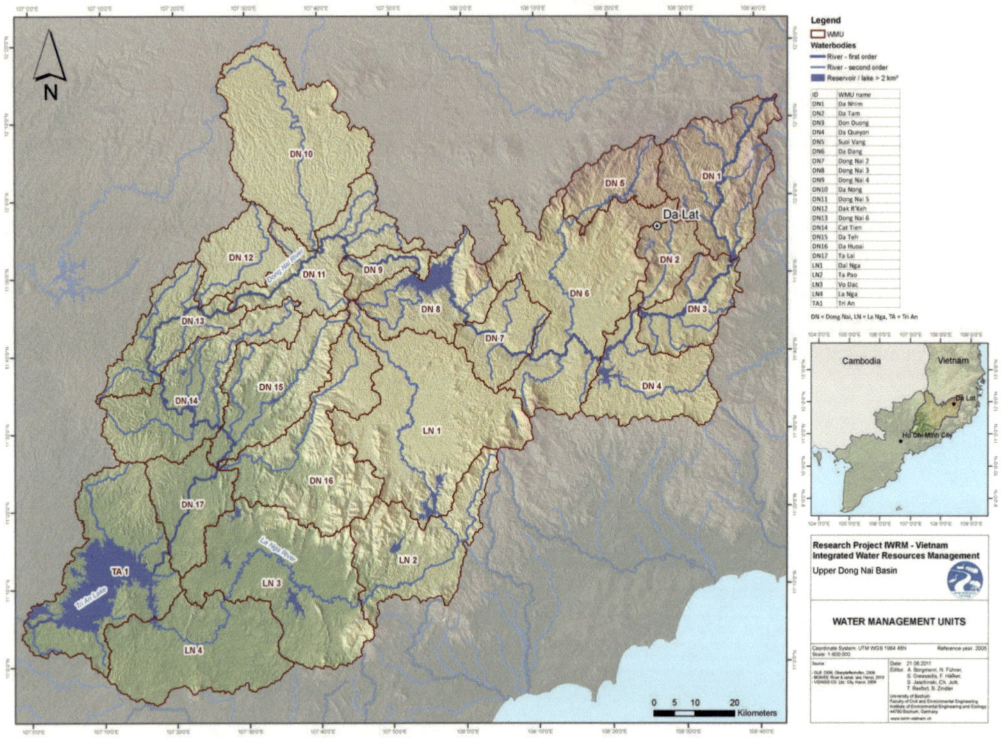

Fig. 8. WMUs in the Upper Dong Nai basin (province Lam Dong).

The project area Cuu Long part basin in the Mekong Delta has been differentiated into nine WMUs (see figure 9). The delineation of WMUs in the Mekong Delta was difficult due to a very homogenous land use and with no obvious flow directions in the rivers and canals. In a first approach, main rivers and canals were used for a delineation of WMUs. However, intense discussions with local experts in the delta were mandatory to delineate acceptable WMUs.

5. Method, planning and decision support tools

The Water Management Units are the spatial basis on which Planning and Decision Support Tools are being applied. The Planning and Decision Support Tools are instruments to spatially identify and prioritize contamination risks (e.g. diffuse agricultural contamination sources or industrial point sources) and water deficits (e.g. overexploitation). Figure 10 is showing three main tools of the developed method.

- The Contamination Risk Tool is used for risk assessment of water quality aspects (ground- and surface water).
- The Water Balance Tool is utilized to record, assess and predict the risk of water deficit (and surpluses).
- The Ranking Tool identifies WMUs with high problem intensities and priority need for IWRM measures.

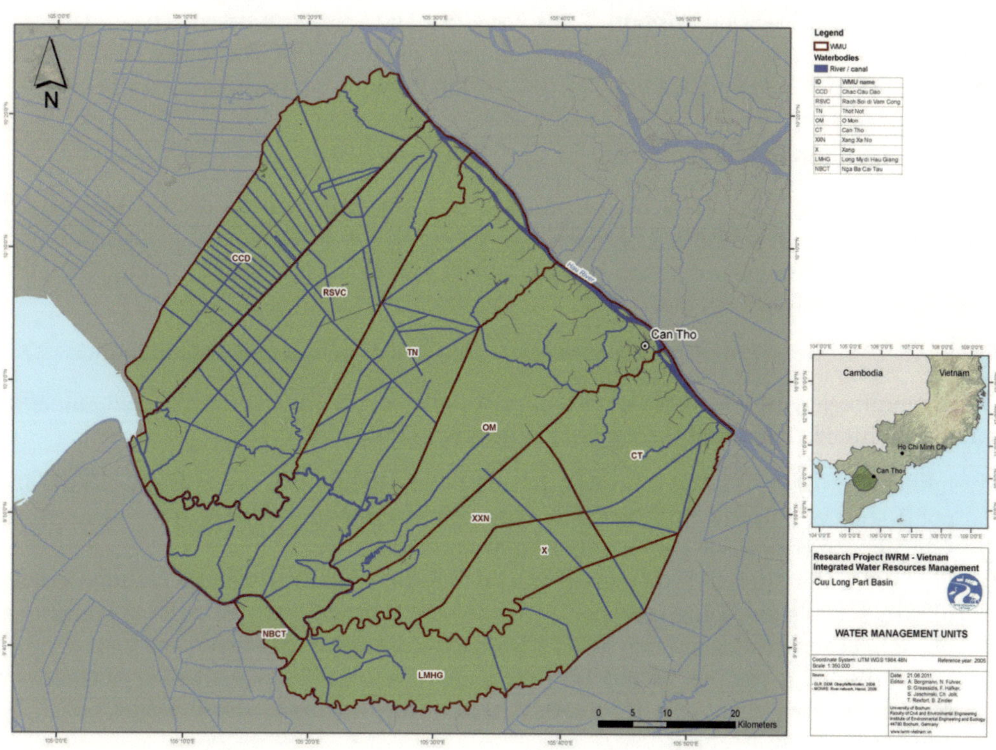

Fig. 9. WMUs in the Cuu Long part basin (city of Can Tho, Mekong Delta).

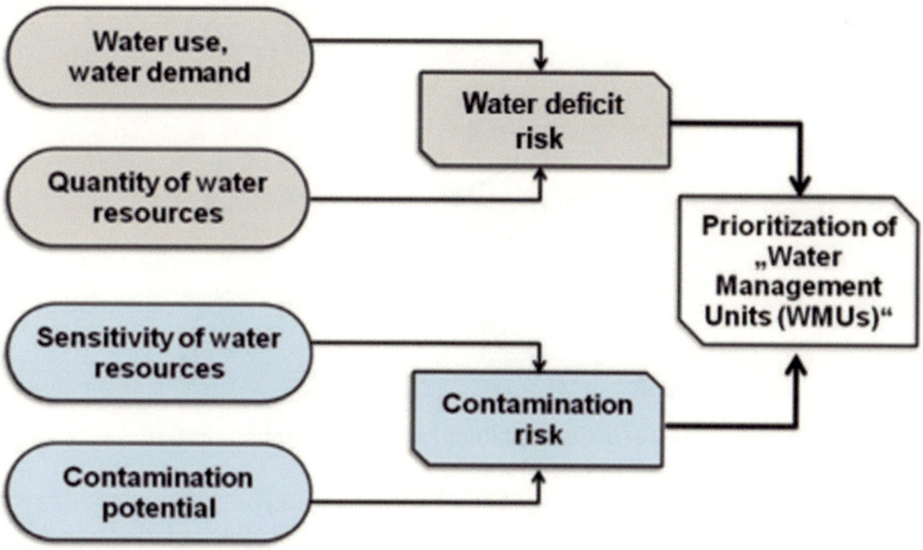

Fig. 10. Planning and Decision Support Tools (boxes) and their input elements (bubbles): Analysis and assessment of water balance (Water Balance Tool) and contamination risk (Contamination Risk Tool), prioritization of WMUs (Ranking Tool).

The Planning and Decision Support Tools are GIS[1]- and MS Excel-based. They enable the user to visualize single actual situations as well as results of the water balance, the contamination risk assessment and the prioritization (ranking) of WMUs.

5.1 Contamination risk tool

The Contamination Risk Tool is used to analyze the contamination risk of the water quality of water resources in a WMU. The method is based on the estimation of the sensitivity of water resources (groundwater and surface water) and the classification of contamination potentials from different sources.

Sensitivity of water resources + Contamination potential of polluters = Contamination risk

The contamination risk assessment is being conducted for three possible contamination paths of pollutants that affect the water resources (cf. figure 11 and 12):

1. Infiltration (figure 11 and 12) (into groundwater): Solute pollutants from diffuse and point sources directly infiltrate into the groundwater (e.g. nitrate from agricultural sources, domestic waste water, industrial waste water)
2. Erosive runoff (figure 11) / runoff (figure 12) (into surface water): Pollutants from diffuse sources are being transported by (erosive) runoff into the surface water (e.g. phosphate and pesticides from fields adsorbed to sediments or organic matter)
3. Direct discharge (figures 11 and 12) (into surface water): Pollutants from point sources are being discharged into the surface water (e.g. domestic waste water, industry, seeping landfills)

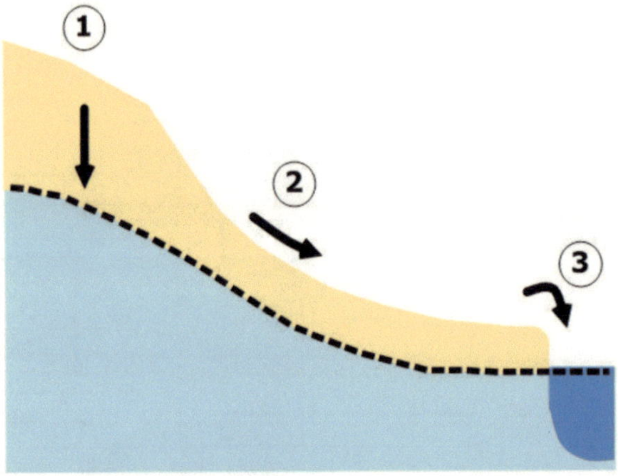

Fig. 11. Possible contamination paths of pollutants affecting ground- and surface water (Upper Dong Nai basin, province Lam Dong).

Figure 11 is showing the possible contamination paths for the Upper Dong Nai basin. The main difference compared to the delta regions (Red River Delta, Mekong Delta) is located in path 2. Path 2 in the Upper Dong Nai basin with its hilly landscape is dominated by an

[1] GIS = Geographic Information System

erosive runoff regime, where contaminants are gravitationally being transported into the surface water. In comparison, path 2 in the delta regions is ruled by inundation controlled runoff of pollutants. Pollutants are being transported extensively and into the surface water (cf. figure 12).

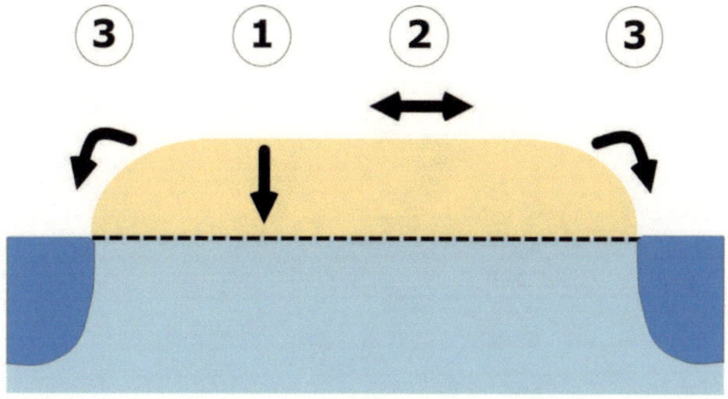

Fig. 12. Contamination paths of pollutants affecting ground- and surface water in the delta regions (Red River part basin, province Nam Dinh / Cuu Long part basin, city of Can Tho).

The contamination risk is evaluated according to these paths. It consists of two elements, the contamination potential and the water resources sensitivity. The contamination potential describes the ability of a certain polluter to negatively affect the water resources and is graded into four classes (no, low, medium, high). Only the most relevant polluters for the evaluation of contamination risks in each path have been selected for closer evaluation. Those are for path 1 (infiltration into groundwater) agricultural areas, settlements and point sources (e.g. industries), for path 2 (erosive runoff (mountainous) or runoff (deltas)) agricultural areas and for path 3 (direct discharge) settlements and point sources.

The sensitivity of water resources describes the relative ease of a contaminant applied on or near the land surface to migrate into the water resource. It is a function of different natural characteristics. The sensitivity is graded into five classes (no, low, medium, high, very high). If more than one parameter is being considered to assess a sensitivity class, matrices help to aggregate different class values into a final class. Parameters considered to assess the sensitivity of groundwater are the aquifer type, the soil type and areas with an intense use of groundwater (high density of wells). The sensitivity of surface water is assessed according to the parameters of potential soil erosion and river density. The specific regional characteristics of the different project areas are considered.

The Contamination Risk Tool and the visualization of its results are GIS-based. The tool identifies hot spot WMUs regarding the risk of contamination for water resources and helps decision makers to analyze contamination potentials from different sources. Figure 13 shows an example for a contamination risk map generated using the Contamination Risk Tool.

Table 1 gives an overview of the combinations leading to the qualitative water risk assessment and the resulting available maps. These maps are compiled in a planning atlas (together with additional maps, tables and text). A web-viewer version of the atlas is available on the internet (www.iwrm-atlas-vietnam.de). In this article, examples of maps to be published in the atlas are found in figures 7 to 9 and 13 to 15.

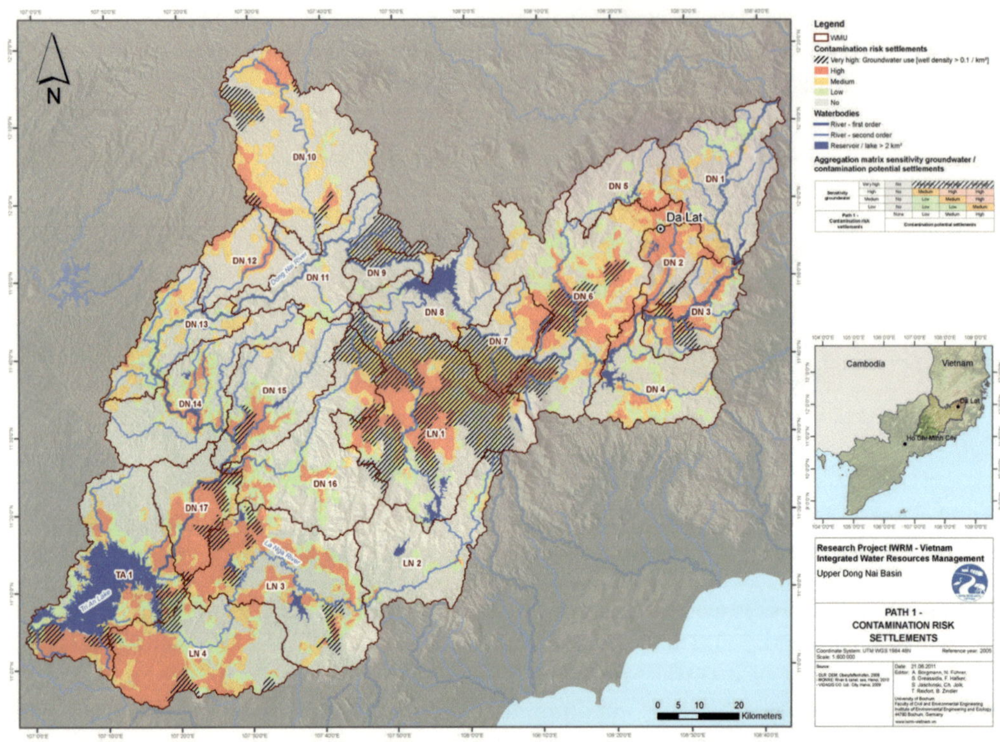

Fig. 13. Example for a resulting contamination risk map in path 1 of the Upper Dong Nai river basin. The contamination potential of settlements is combined with the sensitivity of groundwater to show the contamination risk caused by settlements (red: high risk, yellow: medium risk, green: low risk, grey: no risk).

Contamination paths	Sensitivity	+	Contamination potential	=	Contamination risk	Red River	Upper Dong Nai	Cuu Long
Path 1: Infiltration	Sensitivity of groundwater	+	Contamination potential: agriculture	=	Contamination risk: agriculture	X	X	X
		+	Contamination potential: settlements	=	Contamination risk: settlements	X	X	X
		+	Contamination potential: point sources	=	Contamination risk: point sources	X	X	X
Path 2: Erosive runoff * / runoff **	Sensitivity of surface water	+	Contamination potential: agriculture	=	Contamination risk: agriculture	X**	X*	X**
Path 3: Direct discharge	Sensitivity of surface water use	+	Contamination potential: settlements	=	Contamination risk: settlements	X	X	X
		+	Contamination potential: point sources	=	Contamination risk: point sources	X	X	X

Table 1. Aggregation of sensitivity and contamination potential into contamination risk. The "X" mark available maps for the three project areas.

5.2 Water balance tool

The Water Balance Tool is used to calculate the water quantities of the Water Management Units (WMUs). It identifies the risk of water deficit (and surplus) of a WMU by contrasting water resources and water demands of water users. Values are calculated on a monthly base and for the dry season (December to March), the rainy season (April to November) and the whole year.

Water resources + Water demand = Deficit risk (or surplus)

The tool is MS Excel-based. GIS is employed in the visualization of results (and supportively). Each WMU is represented by standardized calculation sheets in MS Excel. They include all parameters required to compile the water balance. The sheets are interconnected and conclude into an overview cascade (showing an overview of the water balance situation of all WMUs). Figure 14 characterizes the connection of the WMUs in the Upper Dong Nai basin based on their hydrological conditions and management situation. WMUs in the deltas are interconnected via a wide array of canals (Mekong Delta) or governed by a controlled inflow at the polder entrances (Red River Delta) and do not feature a directional flow as in river systems like the Upper Dong Nai basin.

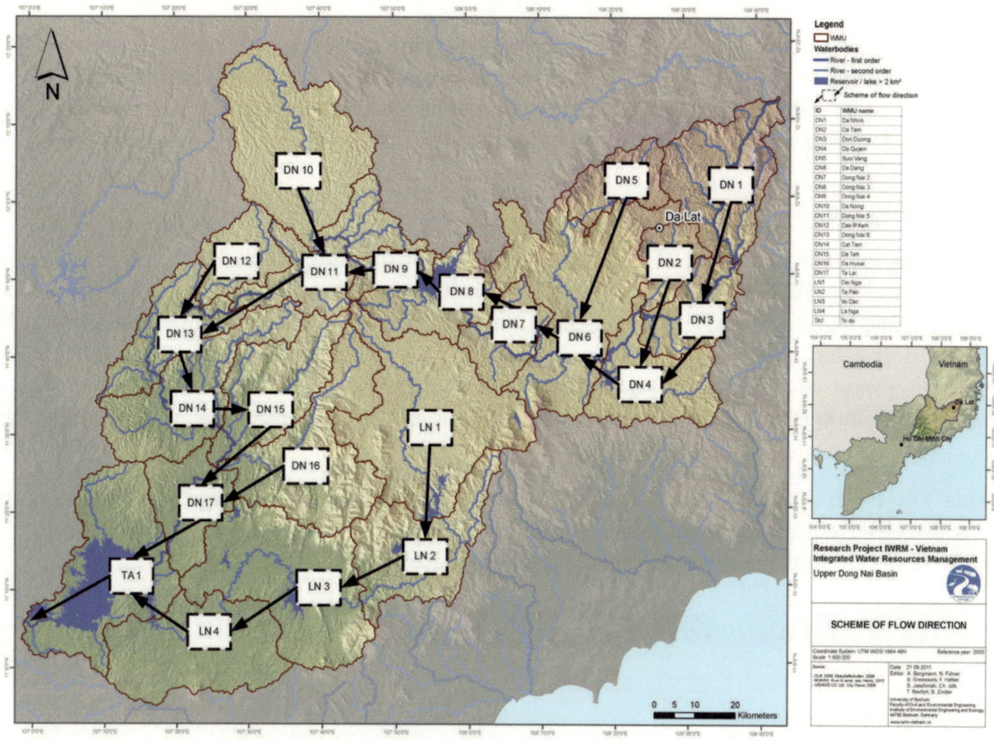

Fig. 14. Scheme of flow direction in the Upper Dong Nai basin (province Lam Dong).

Changes of the input parameters in the calculation sheets are updated automatically (scenarios). One calculation sheet bundles all components of the water balance. The water resources components consist of the discharge generated inside the WMU, the inflow coming from the upstream, the yield of reservoirs or canals, the groundwater resource and the water transfers.

The water demand components consist of the quantitatively relevant water users: rural and urban population, agriculture, industry, service sector (mainly hotels) and the minimum flow requirement ("ecowater").

The Water Balance Tool identifies hot spot WMUs regarding the available water quantities and possible shortcomings during dry seasons. The tool is a strong instrument for decision makers to analyze water resources quantities and the mixture of water users.

Table 2 gives an overview of the combinations leading to the quantitative water risk assessment and the resulting available maps. These maps are compiled in a planning atlas (together with additional maps, tables and text). A web-viewer version of the atlas is available on the internet (www.iwrm-atlas-vietnam.de).

Water balance					Red River	Upper Dong Nai	Cuu Long
Water resources	+	Water demand	=	Water deficit risk	X	X	X

Table 2. Aggregation of water resources and water demands into the risk of water deficits. The "X" mark available maps for the three project areas.

The project area Upper Dong Nai basin features a mere quantitative deficit risk. In contrast, the water resources are abundant in the delta regions (project areas Cuu Long part basin and Red River part basin). Nevertheless, the delta regions hold a certain deficit risk due to temporary high local demands (e.g. agriculture) and water quality limitations (e.g. salt water intrusion).

5.3 Ranking tool

The Ranking Tool processes the results of the Contamination Risk Tool and the Water Balance Tool (aggregated on a WMU basis) in order to prioritize the problem intensities of WMUs within a river basin regarding their water quality and quantity issues. Based on this information, decision makers can effectively prepare decisions, recommend closer inspections and coordinate measures for WMUs with priority need for IWRM measures. Figure 15 is showing an example for the ranking of a contamination risk (here ranking the contamination source: settlements).

Table 3 gives an overview of the ranking. These resulting maps are compiled in a planning atlas (together with additional maps, tables and text). A web-viewer version of the atlas is available on the internet (www.iwrm-atlas-vietnam.de).

Contamination paths	Ranking Contamination Risk	Red River	Upper Dong Nai	Cuu Long
Path 1: Infiltration	Contamination risk agriculture	X	X	X
	Contamination risk settlements	X	X	X
	Contamination risk point sources	X	X	X
Path 2: Erosive runoff * / runoff **	Contamination risk agriculture	X**	X*	X**
Path 3: Direct discharge	Contamination risk settlements	X	X	X
	Contamination risk point sources	X	X	X
Path 1, 2, 3: Aggregation	Contamination risk Path 1, 2, 3	X	X	X
Ranking Water balance		**Red River**	**Upper Dong Nai**	**Cuu Long**
Ranking Water Balance - Water deficit risk		X	X	X

Table 3. Ranking of the Contamination Risk Tool and the Water Balance Tool results. The "X" mark available maps for the three project areas.

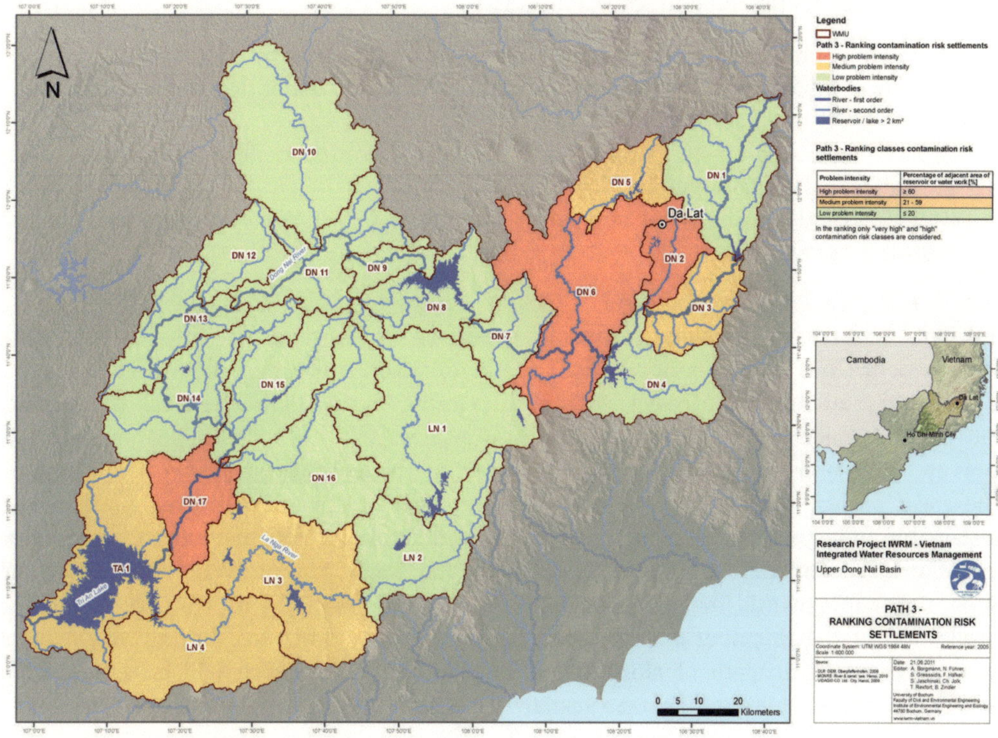

Fig. 15. Result of ranking: contamination risk for the surface water stemming from settlements (red: high risk, yellow: medium risk, green: low risk).

6. Conclusions

The developed Planning and Decision Support Tools are instruments for a sustainable water management along the principles of IWRM as defined by the Global Water Partnership and others. The integrated approach facilitates a consideration of water resources quantity and quality leading to a risk assessment. The result of the risk assessment is the identification and prioritization of areas (Water Management Units) with higher problem intensities regarding the water quantity and / or quality and thus a priority need for IWRM measures. The prioritization of WMUs helps decision makers to efficiently plan water management and funds in order to optimize the impact of IWRM measures.

In order to effectively communicate Planning and Decision Support Tools in Vietnam, it was necessary to develop a transparent, well documented and transferable method. An important aspect of the Planning and Decision Support Tools was to develop a method that fits the needs of Vietnamese water managers and decision makers. Thus, IWRM-Vietnam paid close attention to engage authorities and stakeholders into the process of developing the method in order to make the task of implementation as easy as possible.

The results of the developed methods for the analysis of the water resources on the river basin scale were documented in several reports:

- Handbook of methods: Planning and Decision Support Tools for IWRM in Vietnam
- Definition tables: methods for the estimation of water quantities and contamination risks
- GIS Manuals: step-by-step guides on how to use GIS functionalities to estimate water quantities and contamination risks
- IWRM-Atlas: Planning maps and tables for the IWRM in Vietnam (print version and web-viewer version)

The application of Planning and Decision Support Tools will enable Vietnamese stakeholders to base decisions regarding measures to improve the situation of Water Management Units on a scientific basis. The identification process enables decision makers to effectively attend to the issues with high priority ratings first. The close cooperation between the project and the Vietnamese authorities ensures a holistic implementation of the tools and a close interaction between the method and the legislative framework in Vietnam. The early participation of the responsible water agency on national level guarantees a sustainable adjustment and a nation-wide transferability of the method to Vietnamese conditions. The overall concept developed by the R&D project IWRM-Vietnam is an important step towards the implementation of IWRM principles in Vietnam. Thus, the application of the Planning and Decision Support Tools allows for the improvement of water resources management along the principles of IWRM which will be essential for further development in Vietnam on a sustainable basis

7. Outlook

The developed method (Planning and Decision Support Tools) allows for a layered, problem orientated and efficient examination of entire river basins, initially using a systematic overview examination (scale approx. 1 : 300.000) in order to establish Water Management Units (WMUs) with higher problem intensity and prioritized need for IWRM measures ("hot spots"). The next step should be the development of methods to examine these previously prioritized WMUs in more depth (scale approx. 1 : 50.000), in order to ascertain the types, extensions and locations for necessary IWRM measures.

This results in a systematic approach, rendering possible a labor efficient, targeted and cost efficient process especially for countries, such as Vietnam, which are only just starting to set up an IWRM.

The next step will be to develop methods for the closer examination of WMUs with higher need for IWRM measures on WMU planning level (planning level 4, cf. fig. 2) and to recommend practical local measures on the local planning level (planning level 5, cf. fig. 2). The inclusion of the population and the various participating stakeholders (authorities and institutions) in the planning and decision making process is a vital principle in this context.

Overall this is a holistic approach for planning and decision support for IWRM. This corresponds to the demands on Integrated Water Resources Management as formulated, for example, by the Asian Development Bank (ADB, 2006).

8. Acknowledgments

The German-Vietnamese joint R&D project IWRM-Vietnam would like to thank the German Federal Ministry of Education and Research for the funding of an important input to the Vietnamese IWRM process. IWRM-Vietnam also thanks the project managing agency Karlsruhe (PTKA-WTE) at the Karlsruhe Institute for Technology (KIT) for their support. Finally, eE+E would like to thank all partners of the R&D project for their input and cooperation.

9. References

Asian Development Bank (ADB) (2006). *Water Financing Program 2006-2010. Helping to introduce IWRM in 25 river basins in the asia-pacific region*, 09.11.2011, Available from http://www.adb.org/water/wfp/IWRM-25-River-Basins.pdf

Cuddihy, W. & Frederiksen, H. (1996). Viet Nam Water Resources Sector Review, 1st ed.; World Bank: Washington, USA, 09.11.2011, Available from http://www.vnwatersectorreview.com/detail.aspx?pid=107&r=5#4

U.S. Environmental Protection Agency (EPA) (1998). Guidelines for Ecological Risk Assessment. EPA/630/R-95/002F, Washington D.C., USA

European Commission (2011). Environmental Impact Assessement, 09.11.2011, Available from http://ec.europa.eu/environment/eia/

Government of Vietnam, Prime minister (2008). Government Decree on River Basin Management, No: 120/2008/NĐ-CP dated 01.12.2008, Hanoi, Vietnam

Greassidis, S.; Borgmann, A.; Führer, N.; Jaschinski, S.; Jolk, C.; Stolpe, H.; Zindler, B.; (2011). Überschlägige Wasserbilanz zur Planungs- und Entscheidungsunterstützung auf Einzugsgebietsebene für das Integrierte Wasserressourcen-Management in Vietnam. In: *Hydrologie und Wasserbewirtschaftung. Hydrology and Water Management*, Vol.55, No.2, (April 2011), pp. 68 – 74, Koblenz, ISSN 1439-1783

GWP (Global Water Partnership) (2000). Integrated Water Resources Management. *TAC background paper; no. 4*. Stockholm, Sweden, 09.11.2011, Available from http://www.gwptoolbox.org/images/stories/gwplibrary/background/tac_4_english.pdf

Jolk, C.; Greassidis, S.; Jaschinski, S.; Stolpe, H.; Zindler, B. (2010). Planning and Decision Support Tools for the Integrated Water Resources Management in Vietnam. In: *Water*, ISSN 2073-4441, 09.11.2011, Available from www.mdpi.com/journal/water, vol.2, pp. 711-725, Basel, Switzerland

Kiemstedt, H. & Bachfischer, R. (1977). Zur planerischen Operationalisierung der ökologischen Risikoanalyse. In: Akademie für Raumforschung und Landesplanung (Hrg.): *Arbeitsmaterial*, vol.2, ISBN 978-3-88838-357-1, Hannover, Germany

MONRE Ministry of Natural Resources and Environment (2009). *National Target Program on the improvement to the effectiveness of water resources management, protection and use* (NTP-WR), Hanoi, Vietnam

UNCED (United Nations Conference on Environment and Development) (1993). Proceedings of Earth Summit Agenda 21 - The United Nations Programme of Action from Rio, Rio de Janeiro, Brazil, June 03-14, 1992

UN (United Nations Department of Economic and Social Affairs) (2008). *The Millenium Development Goals Report 2008*, ISBN 978-92-1-101173-9, New York, USA

Zschiesche, Michael & Pham Ngoc Han (2008). *Studie zum Umweltrecht in Vietnam, Unabhängiges Institut für Umweltfragen (UfU)*, Unpublished manuscript, Berlin, Germany

4

Risk Assessment of Cyanobacteria and Cyanotoxins, the Particularities and Challenges of *Planktothrix* spp. Monitoring

Catarina Churro[1,2], Elsa Dias[1] and Elisabete Valério[2]
[1]*Laboratório de Biologia e Ecotoxicologia, Departamento de Saúde Ambiental, Instituto Nacional de Saúde Dr. Ricardo Jorge, Lisboa,*
[2]*Centro de Recursos Microbiológicos (CREM), Faculdade de Ciências e Tecnologia, Universidade Nova de Lisboa, Caparica, Portugal*

1. Introduction

Cyanobacteria are a diverse well adapted group of organisms that presents amazing morphological diversity. Cyanobacteria can be unicellular or colonial (filamentous, spherical or amorphous) (Fig. 1). Since cyanobacteria have cells larger than normal bacterial cells and behavior more similar to algae, they were classified under the microalgae for a long time and acquire the name of blue-green algae or Cyanophyta (Whitton & Potts 2000). Cyanobacteria is a phylum of bacteria that obtain their energy through photosynthesis. The name "cyanobacteria" comes from their coloration (cyano = blue). The vegetative cell wall is of Gram-negative type and in some species the peptidoglycan layer is considerably thicker than in other bacteria. Many unicellular and filamentous cyanobacteria possess an "envelope" outside the lipopolysaccharide (LPS) "outer membrane", which is called: sheath, glycocalyx, or capsule, and depending on the consistency, gel, mucilage or slime. The sheaths of cyanobacteria are predominantly polysaccharide, but a part of its weight may be polypeptides, and depending on the species, some types of sugar residues may be involved (Castenholz, 2001).

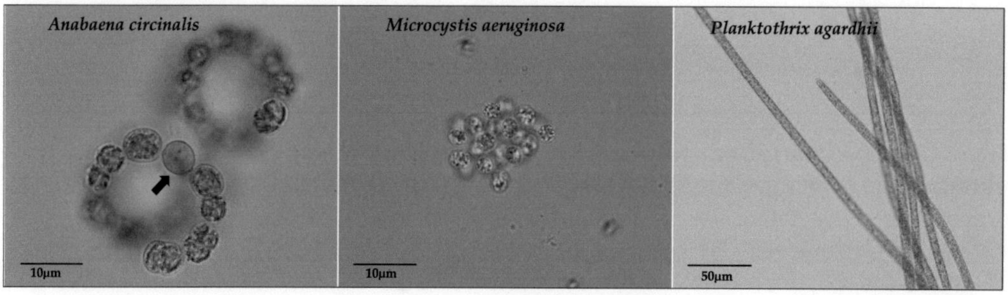

Fig. 1. Optical microscopy photographs of cyanobacteria presenting different morphologies. The arrow indicates the heterocyst cell in *Anabena circinalis*.

Cyanobacteria are autotrophs and possess all the photosynthetic pigment (chlorophyll *a*, carotenoids, allophycocyanin, phycobilins, phycoerythrins) except chlorophyll *b* (Castenholz, 2001). Prochlorophytes are also cyanobacteria that contain chlorophyll *a* and *b*, but, opposing to other cyanobacteria, lack phycobiliproteins (Castenholz, 2001). Cyanobacteria have the ability to use low light intensities effectively, since they are able to produce the accessory pigments needed to adsorb light most efficiently in the habitat in which they are present, providing them a great advantage for the colonization of a wide range of ecological niches (van den Hoek et al., 1995; WHO, 1999). Phycobiliprotein synthesis is particularly susceptible to environmental influences, especially light quality. The chromatic adaptation is largely attributable to a change in the ratio between phycocyanin and phycoerythrin in the phycobilisomes. The photosynthetic pigments are located in thylakoids that are free in the cytoplasm near the cell periphery (Fig. 2). Cell colours vary from blue-green to violet-red due to the chlorophyll *a* masking by the carotenoids and accessory pigments. The pigments are involved in phycobilisomes, which are found in rows on the outer surface of the thylakoids (Fig. 2) (WHO, 1999). Cyanobacteria are also able of storing essential nutrients and metabolites within their cytoplasm. Prominent cytoplasmic inclusions such as glycogen and cyanophycin granules (polymers of the amino acids arginine and asparagine), polyphosphate bodies, carboxysomes (containing the primary enzyme for photosynthetic CO_2 fixation, ribulose 1,5-bisphosphate carboxylase-oxygenase: RuBisCO) and gas vacuoles (Fig. 2) can be observed by electron microscopy. The occurrence of fimbriae (pili) is abundant in many cyanobacteria with varying patterns. Some filamentous forms are also able of gliding (sliding) (van den Hoek et al., 1995; WHO, 1999; Castenholz, 2001).

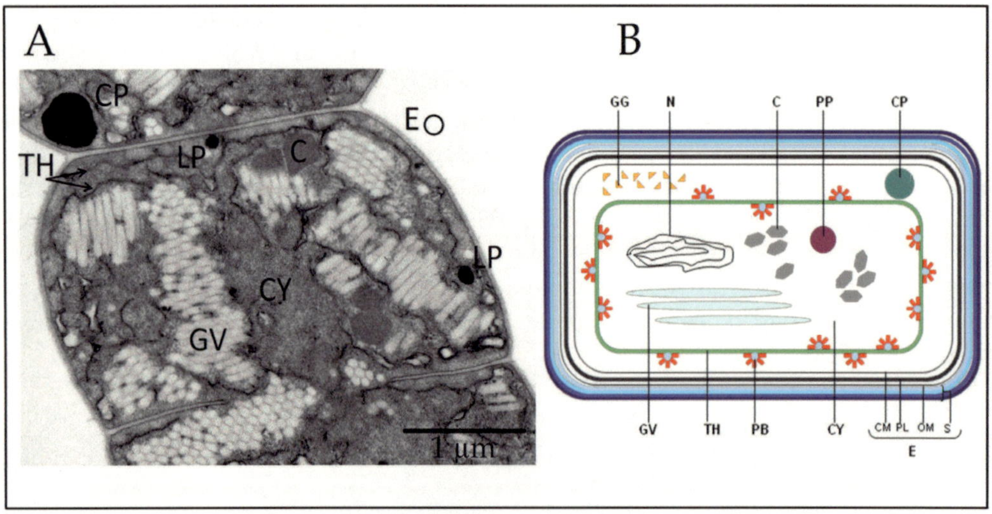

Fig. 2. Cyanobacteria cell structure. (A)Transmission electron micrographs showing the ultrastructure of an *Anabena circinalis* vegetative cell; (B) Schematic diagram of a cyanobacterial vegetative cell. S: external 4-layered cell wall; OM: outer membrane; PL: peptidoglycan layer; CM: cytoplasmic membrane; CW: cell wall; E: cell envelope; TH: thylakoid; PB: phycobilisome; CY: cytoplasm; GV: gas vesicle; GG: glycogen granules; N: nucleoplasmic region; C: carboxysome; PP: polyphosphate granule; CP: cyanophycin granule; LP: lipid droplets (adapted from van den Hoek et al., 1995; Castenholz, 2001).

Cyanobacteria can be found in the most diverse environments like hot springs, salt marshes, soils, fresh, brackish, and marine waters (Sze, 1986). In sum, cyanobacteria are ubiquitous oxygenic photosynthetic prokaryotes.

2. Why the surveillance on cyanobacteria?

Cyanobacteria are common constituents of the phytoplankton in aquatic environments. In optimal conditions these phytoplanktons can develop massively and form blooms, becoming the dominant organism in the water column and creating serious problems in water quality (Cood, 2000; Vasconcelos, 2006). The water quality deterioration produced by cyanobacterial blooms includes foul odours and tastes, deoxygenation of bottom waters (hypoxia and anoxia), fish kills, food web alterations and toxicity. Other threatening characteristic of these organisms is their ability to produce toxins that affects other living organisms and humans (Carmichael, 2001). The capacity of mass development together with the ability to produce potent toxins enlightens the importance of implementing regular monitoring programs for cyanobacteria and cyanotoxins in freshwater environments, in order to minimize potential health risks to animal and human populations that results from exposure through drinking and recreational activities. The implementation of surveillance programs on cyanobacteria involves understanding the ecophysiology of cyanobacteria, bloom dynamics, conditions that promote blooms, production of toxins and their impact in human and animal health (McPhail & Jarema, 2005).

Cyanobacteria possess some ecostrategies that allows them to overcome other organism and become dominant. In general there are four constraints on cyanobacteria growth as pre-requisites for bloom enhancement: light, nutrients, temperature and stability of the water column. Cyanobacteria requires low light intensities for growth, compared with algae, which provides competitive advantages in lakes which are turbid due to growth of other phytoplankton. They also have a higher affinity for uptake phosphorous and nitrogen than many other photosynthetic organisms and they have a substantial storage capacity for phosphorous (Mur et al., 1999). Some genera like *Anabaena, Aphanizomenon, Cylindrospermopsis, Nodularia* and *Nostoc* have specialized cells (heterocysts) (Fig. 1) for nitrogen fixation and blooms of these genera can often be related with periodic nitrogen limitation. This means that they can compete other phytoplankton under conditions of phosphorous and nitrogen limitation (Briand et al., 2003; Sunda et al., 2006).

The success of some cyanobacteria is also due to the presence of gas vacuoles that provide buoyancy regulation. During water stratification conditions cyanobacteria can migrate in the water column, accessing light in the surface layers and nutrients near the sediment. During photosynthesis, carbohydrates are accumulated which makes them heavy and sinking away from light and when the carbohydrates are respired, buoyancy is restored. As large colonies sink faster than small ones or single cells, genera like *Microcystis, Anabaena, Aphanizomenon* and *Nodularia* have scum-forming strategies (Vance, 1965; Mur et al., 1999). Cyanobacteria also produce active substances that inhibits the growth of competing algae and grazers that feed upon them, which can also promote cyanobacteria proliferation (Briand et al., 2003; Granéli & Hansen, 2006; Sunda et al., 2006; van Apeldoorn et al., 2007; Figueredo et al. 2007). As a consequence of the characteristics mentioned above the cyanobacterial cells numbers in water bodies vary seasonally. In temperate regions, seasonal successions of organisms belonging to different phytoplankton taxa are often observed. Whereas at the beginning of

Toxin	Taxon	LD_{50}* (i.p., mouse) of pure toxin	Primary target in mammals	Mechanism of action
Hepatotoxins				
Microcystins (~80 variants) (Cyclic heptapeptides)	*Microcystis* spp. *Planktothrix* spp. *Oscillatoria* *Nostoc* *Anabaena* spp. *Anabaenopsis* *Hapalosiphon* *Snowella* *Woronichinia* *Aphanocapsa*	25 to ~1000µg/kg bw	Liver	Multi-pathway process. MCs inhibit the serine/threonine protein phosphatases type 1 and type 2A (PP1/PP2A) and induces oxidative stress leading to a cascade of events responsible for the MC cytotoxic and genotoxic effects in animal cells
Nodularin (9 variants) (Cyclic pentapeptides)	*Nodularia spumigena*	30–50 µg/kg bw	Liver	Similar to MCs
Cytotoxin				
Cylindrospermopsin (3 variants) (Guanidine alkaloid)	*C. raciborskii* *Umezakia natans* *Aph. ovalisporum* *Raphidiopsis curvata* *Anabaena bergii* *Aphanizomenon* *Lyngbya*	200 - 2100 µg/kg bw/d 200 µg/kg bw/5–6 d	Liver, kidneys, lungs, heart	Inhibition of glutathione (GSH) and protein synthesis, as well as the inhibition of cytochrome P450
Dermatotoxins				
Aplysiatoxin (phenolic bislactone)	*Lyngbia* *Planktothrix* *Schizothrix*	107–117 µg/kg	Skin	Inflammatory agent, protein kinase C activator
Debromoaplysiatoxin (phenolic bislactone)	*Lyngbia*	107–117 µg/kg	Skin	Inflammatory agent, protein kinase C activator
Lyngbiatoxin-A (Alkaloid)	*Lyngbia*	250 µg/kg (?LD100)	Skin	Inflammatory agent, protein kinase C activator

Toxin	Taxon	LD_{50}* (i.p., mouse) of pure toxin	Primary target in mammals	Mechanism of action
Neurotoxins				
Anatoxin-a (5 variants) (Tropane-related alkaloids)	*Anabaena* spp. *Oscillatoria* *Cylindrospermum* *Microcystis* *Aphanizomenon* *Planktothrix*	250 μg/kg bw	Post-synaptic neuromuscular junction	Agonism of the muscular and neuronal nicotinic acetylcholine receptor
Homoanatoxin-a (alkaloid)	*Planktothrix* *Oscillatoria* *Anabaena* *Raphidiopsis* *Phormidium*	250 μg/kg bw	Post-synaptic neuromuscular junction	Same as Anatoxin-a
Anatoxin-a(s) (Guanidine methyl phosphate ester)	*Anabaena* *Aphanizomenon*	40 μg/kg bw	Post-synaptic neuromuscular junction	Irreversible inhibition of acetylcholinesterase
Saxitoxin (20 variants) (Carbamate alkaloids)	*Anabaena* *Aphanizomenon* *C. raciborskii* *Lyngbya* *Planktothrix*	10–30 μg/kg bw	Axons	Blockage of the sodium or calcium channels of the nerve axon membranes

Table 1. Cyanotoxins detected and correspondent taxa from which have been isolated, as well as their primary target in mammals. Based on the information from Chorus et al., 2000; Charmichael, 2001; Codd et al., 2005; Stewart et al., 2006; van Apeldoom et al., 2007; Bláha et al., 2009; Valério et al., 2010; Mihali et al., 2009. * - the dose needed to kill 50% of exposed animals.

the summer a great variety of microalgae and cyanobacteria usually co-exist in the same water body, towards the end of summer this diversity may drop drastically as the result of the mass development of the cyanobacterial communities (blooms) (Sze, 1986). These blooms may be formed by a consortium of cyanobacteria producing different amounts of toxins at different rates, with the same bloom-forming species having both toxigenic and non-toxigenic strains, indistinguishable by morphological examination. Cyanobacterial blooms are complex and can develop in a rather sudden and unpredictable way.

3. Cyanotoxins

Cyanobacteria are able to produce secondary metabolites that present a vast diversity of structures and variants. Most of cyanobacterial secondary metabolites are alkaloids, or possess peptidic substructures synthesised by NRPS (non-ribosomal peptide synthesis, involving peptide synthetases) or NRPS/PKS (involving peptide synthetases and polyketide synthases) hybrid pathways (Valério et al., 2010).

Cyanotoxins are usually classified according to their target in mammals, being divided in hepatotoxins (liver damaging), neurotoxins (nerve damaging), cytotoxins (cell damaging) and toxins responsible for allergenic reactions (dermatotoxins), presenting several kinds of mechanisms of action. A considerable number of these different types of toxins have been isolated from cyanobacteria, belonging to different taxa, as summarized in Table 1.

4. Cyanobacteria/cyanotoxins risk assessment

Risk assessment consists in the identification and determination of quantitative or qualitative value of risk related to the exposure to a given hazard, taking into account possible harmful effects on individuals or populations exposed to that hazard and all the possible routes of exposure. The risk assessment process includes four steps: the hazard identification, hazard characterization, exposure assessment, establishment of dose–effect and dose–response relationships in likely target individuals and populations (Duffus et al., 2007). A schematic representation of the steps involved in risk assessment of cyanotoxins is depicted in Fig. 3.

The scientific knowledge on cyanotoxins still does not enable to correctly assess the risk of human exposure to toxic cyanobacteria. Many toxicological aspects remain to clarify, epidemiological data are insufficient and the exposure assessment is a very complex task.

The human exposure to cyanobacterial cells and/or its toxins may occur through water swallowing or inhalation during recreational activities such as swimming, canoeing, sailboarding and paddling, through the intake of contaminated drinking water and through hemodialysis treatment.

Most episodes of human illness related with cyanobacteria/cyanotoxins resulted from an acute intoxication through the exposure routes mentioned above (for review see Chorus et al., 2000; Duy et al., 2000; van Apeldoom et al., 2007), such as the following examples:

Example 1 – Symptoms after exposure through recreational activity: nausea, abdominal pain, fever, dyspnea, respiratory distress, atypical pneumonia and hepatotoxicosis with a significant increase of hepatic damage biomarkers (Giannuzzi et al., 2011);

Example 2 - Symptoms after exposure during heamodialysis treatment: weakness, muscular pain, nauseas, vomiting, neurologic symptoms (head pain, vertigo, deafness, blindness and seizures), increase of hepatic damage biomarkers, hepatomegaly, hepatic failure and death (reviewed in Pouria et al, 1998).

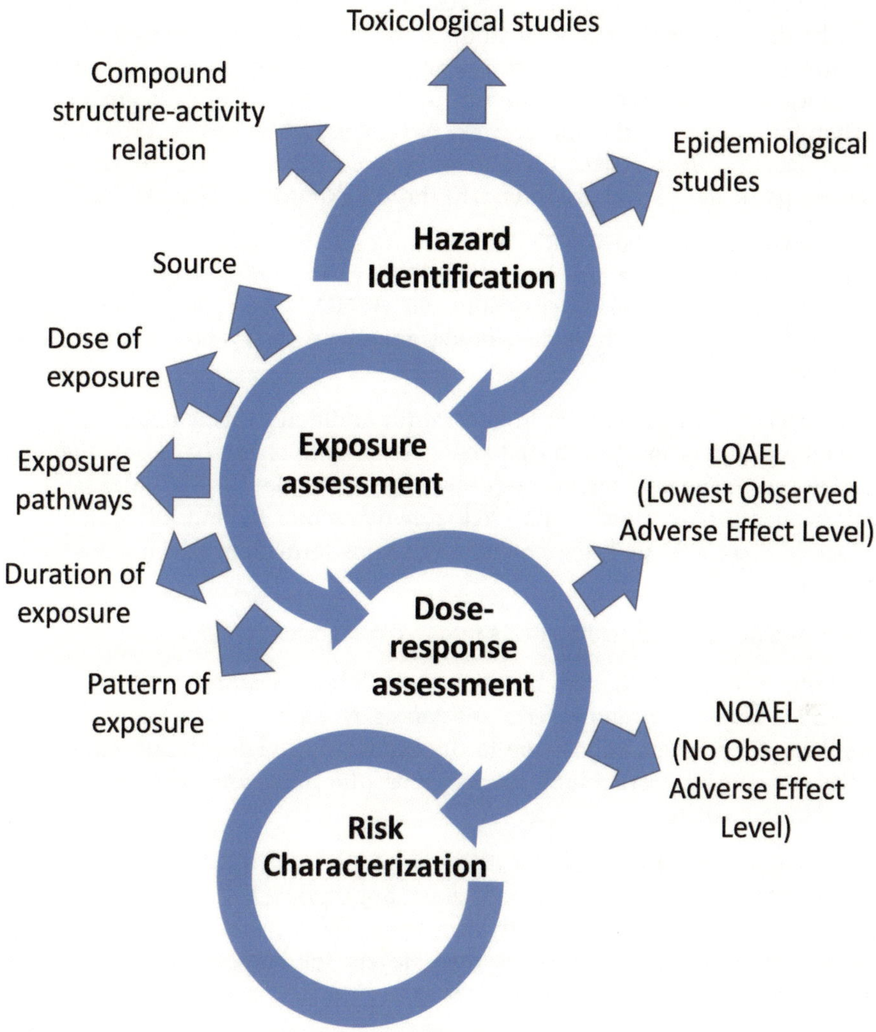

Fig. 3. Organizational chart of the steps involved in risk assessment (adapted from Dolah et al. 2001).

Besides the acute effects mentioned above, few papers reports the association between the ingestion of water contaminated with microcystins and the increase of hepatocarcinoma (Yu, 1995; Ueno et al., 1996) and colorectal cancer (Zhou et al, 2002) in human populations supplied with untreated- or ineffective-treated water.

Laboratorial studies have demonstrated that, in fact, microcystins, nodularins and cylindrospermopsin are genotoxic (reviewed in Zĕgura et al., 2011) and the carcinogenic potential of these toxins have been postulated (Gehringer, 2004; Kinnear, 2010). However, there are still many uncertainties that difficult an unequivocal conclusion about this issue.

The problem of chronic effects are particularly relevant in the case of continuous exposure to low levels of cyanotoxins, even at residual levels, that are not detected by the conventional methods employed in the monitoring procedures. Moreover, the scientific and analytical limitations hinder the complete determination of the toxicological properties of cyanotoxins, and the correct assessment of human exposure to cyanotoxins, as well as lack to provide epidemiological evidence that could confirm the chronic effects of cyanotoxins on human health. Therefore, although the surveillance programs can somehow protect against the cyanotoxins acute effects, risk assessment procedures should be developed and implemented, particularly in what concerns to chronic exposure to cyanotoxins.

During the last decade, the WHO has been regularly reviewing the public health significance of cyanobacteria occurrence in freshwater and developed guidelines for drinking and recreational water environments (WHO, 1998, 2003). This organization recommends that the approach to developing guidelines for cyanobacteria in freshwater should consider:

- the occurrence of cyanobacteria in general (in addition to their toxins) as part of the hazard, because it is not clear that all known toxic components have been identified and irritation symptoms reported may be caused by these unknown substances;
- the particular hazard caused by the well-known cyanotoxins; and
- the hazard associated with the potential of scums formation, which increase the local hazard concentration.

WHO (2003) has divided the health effects into two categories:

- Symptoms associated with skin irritation and allergic reactions resultants from dermal exposure to unknown cyanobacterial substances, and
- Potentially more severe effects due to the exposure to high concentrations of already known cyanotoxins, particularly microcystins (the most commonly found and more studied cyanotoxins).

Given the two types of severity of the symptoms, the WHO considered that the establishment of a single guideline value was not appropriate and, therefore, it has defined several guideline values associated with increasing severity and probability of impact of cyanobacteria/cyanotoxins in health at three levels for bathing waters (Table 2) and guideline values for cyanotoxins in drinking water (see 4.1).

Cyanotoxin analysis will generally be required in one of the following circumstances (WHO, 1999):

1. Action Level 1 status (i.e. > 2000 cells mL^{-1}) predominated by *Microcystis aeruginosa*, or when concentrations of other potentially toxic taxa (see Table 1) exceed 15 000 cells mL^{-1}.
2. Action Level 2 status where numbers of a cyanobacterial taxa not previously recorded as toxic exceed 100,000 cells mL^{-1} (recommended toxicity analysis by mouse bioassay or comparative method).

A brief summary of the steps that must be taken into account, when performing cyanobacteria monitoring, are presented in Fig. 4.

WHO guideline levels	Cyanobacterial cells and chlorophyll levels	Health risks	Recommended action
Low	< 20,000 of total cyanobacterial cells mL^{-1} OR < 10 µg L^{-1} chlorophyll-*a* with dominance of cyanobacteria OR < 2.5 mm^3 L^{-1} cyanobacterial biomass	Short term adverse health outcomes unlikely	Continue monitoring
Moderate	20,000 - 100,000 of total cyanobacterial cells mL^{-1} OR 10 - 50 µg L^{-1} chlorophyll-*a* with dominance of cyanobacteria OR 2.5 - 12.5 mm^3 L^{-1} cyanobacterial biomass	Short term adverse health outcomes, e.g. skin irritations, gastrointestinal illness, probably at low frequency	Add signs to indicate MODERATE alert level - increased health risk for swimming and other water contact activities
High	Cyanobacterial scum formation in contact recreation areas OR > 100,000 of total cyanobacterial cells mL^{-1} OR > 50 µg L^{-1} chlorophyll-*a* with dominance of cyanobacteria OR > 12.5 mm^3 L^{-1} cyanobacterial biomass	Short term adverse health outcomes such as skin irritations or gastrointestinal illness following contact or accidental ingestion Severe acute poisoning is possible in worst ingestion cases	Immediate action to prevent contact with scums Add signs to indicate HIGH alert level - warning of danger for swimming and other water contact activities

Table 2. WHO guideline values for safe practice in managing bathing waters that may contain cyanobacterial cells, according to the level of probability of adverse health effects (WHO, 2003).

4.1 Derivation of guideline values

Characterization of human hazards usually relies mainly on animal studies, or incidents from which quantitative estimates of the hazards to humans can be extrapolated.

Few studies in rodents and pigs enabled to estimate the tolerable daily intake (**TDI**) of some cyanobacterial toxins (Duy et al., 2000; Falconer et al., 1999; Humpage and Falconer, 2003).

Usually, studies with different quantitative animal dosing data, with follow-up over extended periods (preferably over the lifetime of the animal being tested) are necessary to estimate a no-observed-adverse-effect level (**NOAEL**), or at least a lowest-observed-adverse-effect-level (**LOAEL**).

For drinking water, the TDI for cyanotoxins can be estimated as:

$$\mathrm{TDI}=\frac{\mathrm{NOAEL\ or\ LOAEL}}{\mathrm{UF}} \tag{1}$$

Where, TDI units are mg/kg body wt/day, or µg/kg body wt/day, and UF is the product of uncertainty factors, e.g

$$UF=1000\begin{Bmatrix}10\text{ (intra-specific variations)}\\10\text{(inter-specific variations)}\\10\text{ (less-than-lifetime study)}\end{Bmatrix} \tag{2}$$

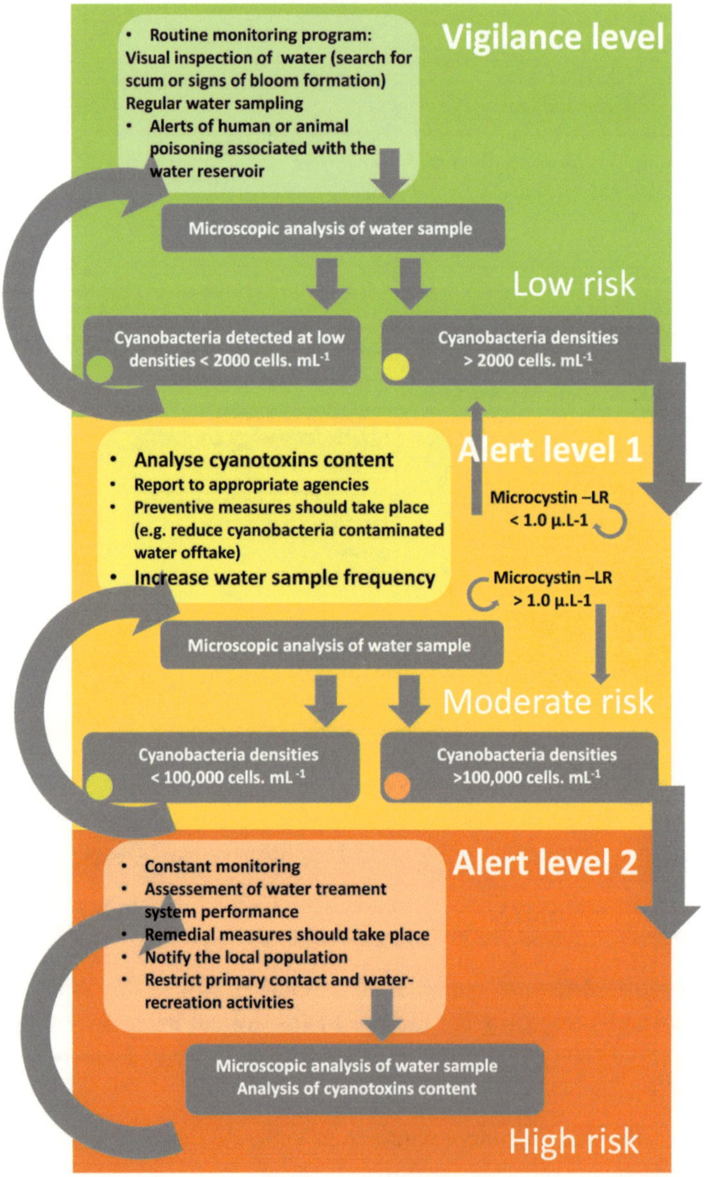

Fig. 4. Organizational chart of the steps involved in cyanobacteria risk management (adapted from Bartram et al. 1999).

Additionally, it may be also necessary consider a UF of 5 if the LOAEL is used and a UF of 3, if tumor promotion is considered (Codd et al., 2005).

The guideline value (**GV**; μg/L water) can be calculated as:

$$GV = \frac{(TDI) \times body\ wt \times AF}{C} \tag{3}$$

Where body weight is usually assumed to be 60 kg for a human adult and AF is the allocation factor, which is the proportion of daily exposure arising from drinking water ingestion. Because some oral exposure may occur via food or dietary supplements or other route, therefore, an AF of 0.8 (80% of total intake) is assumed for drinking water. Finally, C is the volume of drinking water consumption per day, assumed to be 2 L for an adult (Codd et al., 2005; van Apeldoorn et al., 2007).

4.2 Guidelines for microcystins

The drinking water guideline for microcystins was determined from a sub-cronic study (Fawell et al., 1993) with mice orally administered with microcystin-LR (since it is one of the most toxic and frequent microcystin variant and for which more information is available). In this study a NOAEL of 40 μg/kg bw was derived and a TDI of 0.040 was calculated using an uncertainly factor of 1000 (10 for intra-specific variations, 10 for inter-specific variations and 10 for limitations in the database). The resulting guideline value, using an allocation factor of 0.80 for total microcystin-LR (free plus cell bound), was aprox. 1 μg/L in drinking water.

A similar TDI for microcystins was obtained (0.067 *vs.* 0.040) from a study with pigs using freeze-thawed *Microcystis* cells containing quantified microcystins (Falconer et al., 1994). These resulted in similar GVs : 1 μg/L for mice *vs.* 1.61 μg/L for pigs.

For safety reasons, the World Health Organization (WHO) has adopted the lowest value (1 μg/L) as the GV for microcystin in drinking water for adults (WHO, 1998).

However, if tumour-promoting actions of microcystins are also considered, then an additional UF of 3 for this hazard must be used, thus originating a GV of about 0.3 μg/L (Codd et al., 2005).

The Australian guideline is 1.3 μg/L for total microcystin. This slightly differs from the WHO provisional guideline of 1 μg/L microcystin-LR due to the use of a different average body weight for an adult (70 kg *vs.* 60 kg) and different Alocation Factor (0.9 *vs.* 0.8).

4.3 Guidelines for nodularin

No NOAEL can be derived for nodularin(s) due to the absence of suitable toxicological data. However, since nodularin(s) and microcystin-LR have identical mechanisms of action, the guideline value determined for MC-LR (1 μg/L) can also be used for nodularin(s).

4.4 Guidelines for anatoxin-a

A NOAEL of 98 μg/kg has been derived from a 28-day gavage study using mice (Fawell et al., 1999). If a uncertainly factor (UF) of 1000 (10 for intra-specific variations, 10 for inter-specific variations and 10 for limitations in the database) is used, a TDI of 0.1 µg/kg bw can be reached. Svrcek & Smith (2004) have suggested a guideline limit of 3.0 µg/L.

4.5 Guidelines for anatoxin-a(S)

There are no sufficient data to derive an NOAEL or LOAEL and, consequently, insufficient data to determine a TDI for anatoxin-a(S). However, in the Guidelines for Drinking-Water Quality Management for New Zealand 2005, a Maximum Acceptable Values (MAVs) for anatoxin-a(S) of 1.0 µg/L is suggested (Chorus, 2005).

4.6 Guidelines for cylindrospermopsin

According to the 90-day study of Shaw et al. (2000) using drinking water in mice a NOAEL of 150 µg/kg bw was obtained. A second study with mice administered by gavage with cylindrospermopsin for 11-weeks from Humpage and Falconer (2003) resulted on a NOAEL of 30 µg/kg bw. If a uncertainly factor (UF) of 1000 (10 for intra-specific variations, 10 for inter-specific variations and 10 for limitations in the database) is used, a TDI of 0.03 µg/kg bw can be calculated. Considering the "standard" adult body wt of 60 kg and a 0.9 AF, a GV of 0.81 is obtained, leading the authors to propose a Guideline Value of 1 µg/L (Humpage and Falconer, 2003).

4.7 Guidelines for saxitoxin

There are no attempts to determine a NOAEL or LOAEL and thus calculate a TDI for saxitoxin, because the range of lowest concentration where adverse effects were observed varies greatly. Given the different susceptibilities of person, it has been difficult to decide which uncertainty factor should be also used (van Appeldoorn et al., 2007).

Although there are no official guidelines, Australia considers a GV of 3 µg STX eq/L of drinking water, which was based on the data from marine shellfish toxicity (van Appeldoorn et al., 2007).

4.8 Guidelines for aplysiatoxin and lyngbyatoxins

There are no sufficient data to derive an NOAEL or LOAEL and thus calculate a TDI for these toxins.

The members of the population presenting greatest risk when exposed to cyanotoxins are children because of their water intake: body weight ratio, which is higher than that of adults (Falconer, 1999). Also the people having already certain pathologies may be more susceptible to the intake of the toxins (Falconer, 1999).

Ideally, the guidelines values established should protect against acute and chronic effects derived from the contact with cyanobacteria and their toxins, although, such it was stated above, the knowledge on the chronic effects of cyanotoxins still presents many gaps. The guideline values determined/suggested for each known cyanotoxin are summarized in Table 3.

Toxin	Drinking water guideline values	Countries using the GV	References
MC-LR	1.0 µg/L (most generally accepted)	Brazil Czech Republic Denmark France Great Britain Greece Italy New Zealand Poland Portugal South Africa Spain U.S.A.	Chorus, 2005; Codd et al., 2005; van Apeldoorn et al., 2007
MC-LR	1.3 µg/L	Australia Canada	Chorus, 2005; van Apeldoorn et al., 2007
Nodularin	No guideline, however, hazard assessment can be guided by that for microcystins		Fitzgerald et al., 1999; Chorus, 2005; van Apeldoorn et al., 2007
	1.0 µg/L	New Zealand	
Anatoxin-a	3.0 µg/L (no official guideline)		Codd et al., 2005; Svrcek & Smith, 2004; Chorus, 2005
	6.0 µg/L	New Zealand	
Homoanatoxin-a	2.0 µg/L	New Zealand	Chorus, 2005
Anatoxin-a(S)	Nd		Chorus, 2005
	1.0 µg/L	New Zealand	
Cylindrospermo psin	1.0 µg/L (suggested)	Canada, New Zealand	Humpage &Falconer, 2003; Svrcek & Smith, 2004; Chorus, 2005
	15.0 µg/L	Brazil	
STX	3.0 µg STX eq/L	Australia Brazil New Zealand	Svrcek & Smith, 2004; Chorus, 2005; Codd et al., 2005
Aplysiatoxins	nd		
Lyngbyatoxins	nd		

Table 3. Guideline values (GV) estimated for cyanobacterial toxins in drinking water. Nd - not determined.

5. Challenges and gaps in *Planktothrix* spp. risk assessment and management

Usually, *Microcystis* is the genus that occurs more frequently and is usually considered the main responsible for the production of microcystins. However, there is another emergent genus that has also the ability to produce microcystins, which is *Planktothrix*.

The cyanobacteria of the genus *Planktothrix* have a planktonic life style, occur in solitary filaments and lack sheaths, heterocysts and akinetes. Formerly classified into to the genus *Oscillatoria*, *Planktothrix* now represents a well-defined independent genus based in phylogenetic and morphologic characteristics and comprises 13 species (Komárek & Komárkova 2004). Similar to other cyanobacteria, *Planktothrix* can achieve high cellular densities in water forming blooms that unbalance the ecosystem and it can also produce several types of cyanotoxins, namely microcystins, homoanatoxin-a, anatoxin-a, aplysiatoxins, saxitoxins, anabaenopeptins, (Luukkainen et al., 1993; Erhard et al., 1999; Kouzminov, 2001; Viaggiu et al., 2004; Wood, 2005; Kosol et al., 2009), thus threatening humans and animals.

From the 13 species described, *Planktothrix rubescens* and *Planktothrix agardhii* are the most studied and common species reported to cause water related problems. A summary of *Planktothrix* occurrence in European lakes where they form recurrent blooms and the associated toxicity found is presented in Table 4.

Unlike other cyanobacteria, *P. agardhii* and *P. rubescens* are well adapted to very low light intensities and this characteristic provides to them several advantages. For *P. agardhii* it allows them to grow in waters with high turbidity, in which it can be homogeneously dispersed throughout the epilimnion in eutrophic waters having a competitive advantage upon other phytoplankton species. For *P. rubescens* the low light intensity requirements together with the high content of the red pigment phycoerythrin enables it to growth in the metalimnetic layer in thermally stratified waters away from the photic surface zone (Mur et al., 1999; Bright & Walsby, 2000). Furthermore, these two species have different irradiance tolerances; *P. agardhii* is more tolerant to high irradiance than *P. rubescens*, what is related with their occurrence in different ecological niches in the water column and inhabit in different types of water systems (Oberhaus et al., 2007). Therefore *P. agardhii* grows well in the upper part of the water column of shallow eutrophic lakes, however it can also grow at several depths along the water column (Halstvedt et al., 2007). On the other hand, *P. rubescens* is well adapted in forming metalimnic populations of deep stratified lakes in spring and summer and when the lake loses its thermal stratification in the winter, it can be dispersed through the entire water column (Bright & Walsby, 2000; Briand et al., 2005). *Planktothrix* also has different water temperature tolerances when compared to other cyanobateria, making them organisms that can be easily found in subalpine lakes or in temperate regions during winter, so *Planktothrix* blooms may persist all year around and not only during summer or spring where temperatures and light irradiance are higher. *P. agardhii* has been found viable under ice covers (Sivonen & Jones 1999; Oberhaus et al., 2007). Since both species occupies different water niches they can coexist in the same water body forming surface and deep layer blooms, althougth this coexistence is rare it has been reported (Davis & Walsby, 2002; Halstvedt et al., 2007).

Regarding the risk management measures that are usually followed to overcome the presence of cyanobacteria and cyanotoxins in the freshwater, *Planktothrix* has some particularities that need to be taken into account. One of them is *Planktothrix*´s ability to

Country	Lake	Maximum depth (m)	Trophic status	Species	Toxins detected	References
Austria	Mondsee	68	Mesotrophic	*P. rubescens*	---	Kurmayer et al. 2004; Kurmayer & Gumpenberger 2006
	Irrsee	32	Mesotrophic	*P. rubescens*	---	Kurmayer et al. 2004; Kurmayer & Gumpenberger 2006
	Afritzer	2.5	Mesotrophic	*P. rubescens*	---	Kurmayer & Gumpenberger 2006
	Wörthersee	85.2	Mesotrophic	*P. rubescens*	---	Kurmayer & Gumpenberger 2006
Czech Republic	Bílá Lhota	1	Hypertrophic	*P. agardhii*	---	PoulíČková et al. 2004
England	Blelham Tarn	14.5	Eutrophic	*P. rubescens, P. agardhii*	---	Davis & Walsby 2002; Davis et al. 2003
	Elmhirst	Shalow	---	*P. agardhii*	3.2 μ.L^{-1} MC-LR [a,c]	Akcaalan et al. 2006
	Bassenthwaite	21	Mesotrophic	*P. agardhii*	27- 41 pg.filament^{-1} MC-LR [b,c]	Akcaalan et al. 2006
Estonia	Verevi	11	Hypertrophic	*P. agardhii*	---	Nõges & Kangro 2005; Kangro et al. 2005
Finland	Vargsundet	32	Oligotrophic	*P. agardhii*	11.1 μg.L^{-1} MC-RR [a,d]	Lindholm et al. 1999
France	Bourget	145	Mesotrophic	*P. rubescens*	Max. 6.7 μ.L^{-1} MC-LR [a,d]	Briand et al. 2005, Jacquet et al. 2005
	Viry-Châtillon	5.3	Hypertrophic	*P. agardhii*	Máx. 7.4 μg.L^{-1} MC-LR [a,e] Máx. 34.5 μg.L^{-1} MC-LR [a,d]	Catherine et al. 2008, Yéprémian et al. 2007
Germany	Ammersee	81.1	Mesotrophic	*P. rubescens*	Mean 0.43 ± 0.06 μg.L^{-1} MC-LR [a,c]	Ernst et al. 2009
	Stolpsee	Shalow	Eutrophic	*P. agardhii*	1.81 μg.L^{-1} MC-LR [a,d]	Mbedi et al 2005
	Breiter Luzin	58.3	Mesotrophic	*P. rubescens*	2.31 μ.L^{-1} MC-LR [a,d]	Mbedi et al 2005
	Langer See	3.8	---	*P. agardhii*	----	Rücker et al. 1997
	Wolziger See	13	---	*P. agardhii*	----	Rücker et al. 1997
	Lebbiner See	4	---	*P. agardhii*	----	Rücker et al. 1997
Greece	Ziros	56	Oligotrophic	*P. rubescens*	Máx. 199 μ.L^{-1} MC-LR [a,c]	Vareli et al. 2009
Hungary	Balaton	3.2[f]	Eutrophic	*P. agardhii*	---	Honti et al. 2007

Country	Lake	Maximum depth (m)	Trophic status	Species	Toxins detected	References
Italy	Pusiano	24.3	Eutrophic	*P. rubescens*	---	Legnani et al. 2005
	Como	410	Mesotrophic	*P. rubescens*	---	Bettinetti et al. 2000; Buzzi 2002
	Maggiore	370	Oligo-mesotrophic	*P. rubescens*	---	Morabito et al. 2002
	Fiastrone			*P. rubescens*	---	Viaggiu et al. 2003
	Spino	8		*P. rubescens*	12.13 ng.mg Anatoxin-a (a,d)	Viaggiu et al. 2003, Viaggiu et al. 2004
	Gerosa	50	Oligotrophic	*P. rubescens*	Máx. 1.94 µg.L^{-1} MC-LR (a,c)	Manganelli et al. 2010
	Pozzillo	50	Meso-eutrophic	*P. rubescens*	34 mg.L^{-1} MC-LR (a,d)	Naselli-Flores et al. 2007
	Prizzi	44	Meso-eutrophic	*P. rubescens*	7 µg L^{-1} MC-LR (a,d)	Naselli-Flores et al. 2007
	Nicoletti	36	Mesotrophic	*P. rubescens*	---	Naselli-Flores et al. 2007
	Garcia	43	Meso-eutrophic	*P. rubescens*	---	Naselli-Flores et al. 2007
	Garda	350	Oligo-mesotrophy	*Planktothrix* sp. *P. rubescens*	---	Salmaso 2000; Salmaso 2010
Netherlands	Vechten	11.9	---	*P. agardhii*	---	Montealegre et al. 1995
Norway	Steinsfjorden	24	Mesotrophic	*P. rubescens* *P. agardhii*	---	Halstved et al. 2007; Rudi et al. 2005
Poland	Bytynskie	3.5(f)	Eutrophic	*P. agardhii*	15.8 µ.L^{-1} MC-LR (a,e)	Mankiewicz-Boczek et al. 2009
	Lubosinskie	2.6(f)	Eutrophic	*P. agardhii*	21.9 µ.L^{-1} MC-LR (a,e)	Mankiewicz-Boczek et al. 2009
	Laskownickie	7.4	Hypertrophic	*P. agardhii*	---	Stefaniak et al. 2005
	Grylewskie	6.5	Hypertrophic	*P. agardhii*	---	Stefaniak et al. 2005
Portugal	Beliche	52	Eutrophic	*P. rubescens*	1.1 µg.L^{-1} MC-LR (a,c)	Paulino et al. 2009a
Spain	El Atazar	100	Oligo-mesotrophic	*P. rubescens*	---	Almodóvar et al. 2004
Switzerland	Zürich	143	Mesotrophic	*P. rubescens*	---	Bright & Walsby 2000; Kurmayer & Gumpenberger 2006
	Hallwilersee	48	Mesotrophic	*P. rubescens*	---	Kurmayer & Gumpenberger 2006
Turkey	Spanca	55	Oligo-Mesotrophic	*P. rubescens*	18.4-66.1 pg.$filament^{-1}$ MC-LR (b,c)	Akcaalan et al. 2006
	Iznik	70	Mesotrophic	*P. rubescens*	29.2-114 pg.$filament^{-1}$ MC-LR (b,c)	Akcaalan et al. 2006
	Küçükçekmece	20	---	*P. agardhii*	Nontoxic (a)	Al-Tebrineh et al. 2011
	Sapanca	52	---	*P. rubecens*	Mean 8.4 ± 0.5µL^{-1} MC-LR (a,d)	Al-Tebrineh et al. 2011

Table 4. Lakes were *Planktothrix* spp. has been reported to form recurrent blooms. **(a)** Bloom Sample/Environmental sample, **(b)** Filaments isolated from bloom samples, **(c)** anti-Adda ELISA Kit, **(d)** HPLC, **(e)** protein phosphatase 2A inhibition assay (PP2A), **(f)** Mean depth., (---) Information not available.

establish populations at several depts. in the water column that allows them to access nutrients located near the bottom and still have enough light for photosynthesis, making them able to form blooms away from the surface. This unique characteristic of *Planktothrix* may possess a problem for the water monitoring authorities, since their bloom may be overlook by surface monitoring inspection (Sivonen & Jones 1999). Furthermore, *Planktothrix* blooms may co-occur with other cyanobacterial surface blooms what can also be misleading in water monitoring. Generally cyanobacteria blooms are expected to occur in highly nutrient rich waters during summer or spring months (Chorus et al., 2000). The responsible agencies for the reservoirs monitoring often restricts or increases to normal level the water inspection and water sampling frequency. *Plantothrix* species such as *P. rubescens* occurs in low nutrient oligotrophic waters forming perennially blooms that can prevail for many years. Furthermore since nutrients are not a limiting factor for *P. rubescens* it has been reported the lodging and development of population of this species after restoration lake activities and decrease in nutrient input since it improves trophic level and increases water transparency (Jacquet et al., 2005; Legnani et al., 2005; Ernst et al., 2009). So, in lakes were *Planktothrix* species occur the surveillance must be during all year (Utkilen et al., 1999; Naselli-Flores et al., 2007). Other important feature is that *Planktothrix* may contain higher microcystins content per cell, when compared with other microcystins producers; and that the proportion of toxic strains is higher in *Planktothrix* blooms than for example *Microcystis* blooms, this may result in the occurrence of high toxin concentrations in water without scum formation (Falconer et al., 1999; Briand et al., 2008; Ernst et al., 2009).

6. *Planktothrix* spp. occurrence in Portugal

Planktothrix species can be commonly found in Portuguese freshwater reservoirs. Some of the species reported are *P. mougeotii/P. isothrix* from a wastewater treatment plant in the north of Portugal (Vasconcelos & Pereira 2001, Martins et al. 2010), *P. rubescens* from Beliche reservoir in the South of Portugal (Paulino et al. 2009a) and *P. agardhii* and *P. pseudoagardhii* isolated from several reservoirs in the center and south of Portugal that are maintained in laboratory cultures (Paulino et al. 2009b). However, their occurrence is more pronounced in the center and south of Portugal where it has been increasing and causing problems in some water reservoirs over the last years, such as the deep layer *P. rubescens* bloom with associated microcystin production reported by Paulino et al. 2009. Another example is the particular case of a drinking water reservoir located in the center of Portugal that has been monitored over the last eight years and where a continuous *Planktothrix* spp. bloom persists since 2006 (Fig. 5).

As it can be depicted from Fig. 5 high *Planktothrix* cell concentrations started to appear in the reservoir in 2006 and microcystin concentration increased significantly since 2007. Furthermore, the microcystin concentrations in raw water does not correlate will *Planktothrix* cell numbers, since a high cell concentration does not indicate the presence of high microcystin concentrations and high concentrations of microcystins are not directly associated with high cell densities. This is probably because distinct strains/species of this genus with distinct ability to produce microcystins may occur together. In fact, a natural cyanobacterial population is usually a consortium of toxic and nontoxic strains, and this is believed to be the reason why the population toxicity can vary over time and between samples (WHO, 1999).

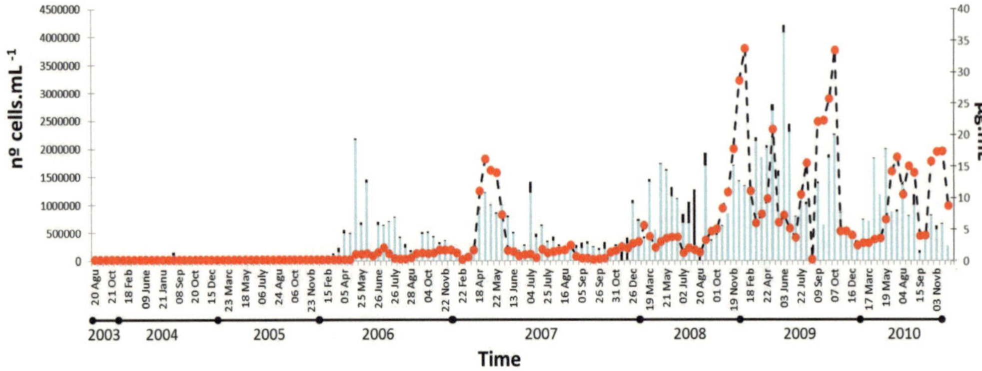

Fig. 5. *Planktothrix* occurrence in a freshwater reservoir located in the center of Portugal and microcystin-LR concentration in raw water over the same sampling period (--•-- microcystin concentration in µg.mL $^{-1}$, light blue bars represent *Planktothrix* spp. cell concentration in n°cells.mL^{-1}, black bars represent total phytoplankton cell concentration in n°cells.mL^{-1}).

As it can be seen by this monitoring data, *Planktothrix* can suddenly reach high cell densities and dominate the phytoplankton community presenting cell densities values close to total phytoplankton concentration. The figure also shows that *Planktothrix* can form perennial blooms but during this time no visible scum formation was observed within the reservoir. It is still unknown why this bloom of toxic *Planktothrix* persists for 5 years in this reservoir and the answer to this issue will be certainly an important contribution to the knowledge of cyanobacteria ecotoxicology. Since the begining of this P*lanktothrix* bloom this reservoir has been under strict vigilance: monitoring sampling is regular, cellular composition/densities and microcystin content in the samples are always screened and the water treatment plant efficiency analysed to avoid any possible harmfull effect on the population. Nevertheless, due to the persistence of high cell densities and high toxin contents occasionally observed, the reservoir represents a potential risk for human and wild life. Therefore, studies must be performed in order to understand the factors underlying the bloom appearance, persistence and toxicity and to access the risk that this reservoir represents to human health, in order to apply measures to prevent and manage the risk of *Planktothrix* occurrence in the reservoir and to restore the quality of this water-supply.

7. Conclusion

The risk of human exposure to toxic cyanobacteria is very difficult to assess because many scientific issues remain to be clarified, such as the toxicological properties of cyanotoxins and their real impact on human health. Nevertheless, the establishment of several guidelines for the most common toxins and the establishment of surveillance programs have contributed to minimize the human exposure to toxic cyanobacteria. However, particular attention should be taken for those species, such as *Planktothrix,* that develop particular strategies to adapt, survive and proliferate in freshwater environments. Therefore, the monitoring programs in water reservoirs where *Planktothrix* species occur must have into account that samples should be taken at several depths, microcystin concentration should be accessed constantly and the water system should be monitored regularly throughout the years since perennial persistence of *Planktothrix* may occur. In water capture for potable water treatment plants the selection of water off-take depth is important and the infrastructures must be equipped with multiple off-takes. In water reservoirs where *Planktothrix* species occurs, certain particularities must be taken into account (Fig. 6) in order to implement the most adequate risk assessment procedures, monitoring programs and preventive measures to protect public health from cyanotoxin occurrence in freshwater supplies.

Fig. 6. Schematic representation of the steps involved in *Planktothrix* risk management.

8. Acknowledgments

We acknowledge the Ph.D research grant SFRH/BD65706/2009 to Catarina Churro from Fundação para a Ciência e a Tecnologia (Portugal) and the research grant BIC/04/DSA/2008 attributed to Elsa Dias by Instituto Nacional de Saúde Dr. Ricardo Jorge (Portugal).

9. References

Akcaalan, R.; Young, F.M.; Metcalf J.S.; Morrison, L.F.; Albay, M. & Codd, G.A. (2006) Microcystin analysis in single filaments of *Planktothrix* spp. in laboratory cultures and environmental blooms. *Water Research*, Vol.40, No.8, pp. 1583-1590.

Almodóvar, A.; Nicola, G. G. & Nuevo, M. (2004) Effects of a bloom of *Planktothrix rubescens* on the fish community of a Spanish reservoir. *Limnetica*, Vol.23, No.1-2, pp. 167-178.

Al-Tebrineh, J.; Gehringer, M.M.; Akcaalan, R. & Neilan, B.A. (2011) A new quantitative PCR assay for the detection of hepatotoxigenic cyanobacteria. *Toxicon* Vol.57, No.4, pp. 546-554.

Bartram, J.; Burch, M.; Falconer, I.R.; Jones, G. & Kuiper-Goodman, T. (1999) Chapter 6. *Situation Assessment, Planning and Management*. In: *Toxic cyanobacteria in water: A guide to their public health consequences, monitoring and management*. I. Chorus, Bartram, J. (Ed), WHO.

Bettinetti, R.; Morabito, G. & Provini, A. (2000) Phytoplankton assemblage structure and dynamics as indicator of the recent trophic and biological evolution of the western basin of Lake Como (N. Italy). *Hydrobiologia*, Vol.435, pp. 177–190.

Bláha, L.; Babica, P. & Maršálek, B. (2009) Toxins produced in cyanobacterial water blooms - toxicity and risks. *Interdisciplinary Toxicology*, Vol.2, No.2, pp. 36-41.

Briand, E.; Gugger, M.; Francois, J.C.; Bernard, C.; Humbert, J.F. & Quiblier, C. (2008) Temporal variations in the dynamics of potentially microcystin-producing strains in a bloom-forming Planktothrix agardhii (Cyanobacterium) population. *Applied and Environmental Microbiology*, Vol.74, No.12, pp. 3839-3848.

Briand, J.; Jacquet, S.; Bernard, C. & Humbert J. (2003) Health hazards for terrestrial vertebrates from toxic cyanobacteria in surface water ecosystems. *Veterinary Research* Vol.34, No.4, pp. 361-377.

Briand, J.F.; Jacquet, S.; Flinois, C.; Avois-Jacquet, C.; Maisonnette, C.; Leberre, B. & Humbert, J.F. (2005) Variations in the microcystin production of Planktothrix rubescens (cyanobacteria) assessed from a four-year survey of Lac du Bourget (France) and from laboratory experiments. *Microbial Ecology*, Vol.50, No.3, pp. 418-28.

Bright, D.I. & Walsby A.E. (2000) The daily integral of growth by Planktothrix rubescens calculated from growth rate in culture and irradiance in Lake Zürich. *New Phytologist*, Vol. 146, pp. 301-316.

Buzzi, F. (2002). Phytoplankton assemblages in two sub-basins of Lake Como. *Journal of Limnology*, Vol.61, No.1, pp. 117-128.

Carmichael, W.W. (2001) Health effects of toxin-producing cyanobacteria: "The CyanoHABs". *Human and Ecological Risk Assessment*, Vol.7, No.5, pp. 1393-1407.

Castenholz, R. W. (2001) Phylum BX. *Cyanobacteria.* In *Bergey´s Manual of Systematic Bacteriology.* D. R. Boone, Castenholz, R.W. (Ed). Springer, New York, 2001, pp. 473-599.

Catherine, A.; Quiblier, C.; Yepremian, C.; Got, P.; Groleau, A.; Vincon-Leite, B.; Bernard, C. & Troussellier, M. (2008) Collapse of a *Planktothrix agardhii* perennial bloom and microcystin dynamics in response to reduced phosphate concentrations in a temperate lake. *FEMS Microbiology Ecology,* Vol. 65, No.1, pp. 61-73.

Chorus, I.; Falconer, I.R.; Salas, H.J. & Bartram, J. (2000) Health risks caused by freshwatercyanobacteria in recreational waters. *Journal of Toxicology and Environmental Health, Part B,* Vol.3, No.4, pp. 323-347.

Codd, G.A. (2000) Cyanobacterial toxins, the perception of water quality, and the priorization of eutrophication control. *Ecological Engineering,* Vol.16, No.1, pp. 51-60.

Codd, G.A.; Morrison, L.F. & Metcalf, J.S. (2005) Cyanobacterial toxins: risk management for health protection. *Toxicology and Applied Pharmacology,* Vol.203, No.3, pp. 264–272.

Davis, P.A. & Walsby, A.E. (2002) Comparison of measured growth rates with those calculated from rates of photosynthesis in *Planktothrix* spp. isolated from Blelham Tarn, English Lake District. *New Phytologist,* Vol. 156, pp. 225–239.

Davis, P.A.; Dent, M.; Parker, J.; Reynolds, C.S., & Walsby, A.E. (2003) The annual cycle of growth rate and biomass change in *Planktothrix* spp. in Blelham Tarn, English Lake District. *Freshwater Biology,* Vol.48, pp. 852–867.

Dolah, F. M. V.; Roelke, D. & Greene, R. M. (2001) Health and eological impacts of Harmful Algal Blooms: Risk assessment needs. *Human and Ecological Risk Assessment,* Vol.7, No.5, pp. 1329-1345.

Duffus, J.H.; Nordberg, M. & Templeton, D.M. (2007) Chemistry and Human Health Division. Glossary of terms used in toxicology, 2nd edition (IUPAC Recommendations 2007). *Pure and Applied Chemistry,* Vol.79, No.7, pp. 1153–1344.

Duy, T.N.; Lam, P.K.S.; Shaw, G.R. & Connell, D.W. (2000) Toxicology and risk assessment of freshwater cyanobacterial (blue-green algal) toxins in water. *Reviews of Environmental Contamination & Toxicology,* Vol.163, pp. 113-186.

Erhard, M.; von Döhren, H.; Jungblut, P. R. (1999). Rapid identification of the new anabaenopeptin G from *Planktothrix agardhii* HUB 011 using matrix-assisted laser desorption/ionization time-of-flight mass spectrometry. *Rapid Communications in Mass Spectrometry,* Vol. 13, No.5, pp. 337–343.

Ernst, B.; Hoeger, S.J.; O'Brien, E. & Dietrich, D.R. (2009) Abundance and toxicity of Planktothrix rubescens in the pre-alpine Lake Ammersee, Germany. *Harmful Algae,* Vol.8, No.2, pp. 329-342.

Falconer, I. (1999) An overview of problems caused by toxic blue-green algae (cyanobacteria) in drinking and recreational water. *Environmental Toxicology,* Vol.14, No.1, pp. 5 - 12.

Falconer, I.; Bartram, J.; Chorus, I.; Kuiper-Goodman, T.; Utkilen, H.; Burch M.; Codd G.A.(1999) Chapter 5. Safe levels and safe practices In: *Toxic cyanobacteria in water: A guide to their public health consequences, monitoring and management.* I. Chorus, Bartram, J. (Ed), WHO.

Falconer, I.R.; Burch, M.D.; Steffensen, D.A.; Choice, M. & Coverdale, O.R. (1994) Toxicity of the blue-green alga (cyanobacterium) *Microcystis aeruginosa* in drinking water to growing pigs, as an animal model for human injury and risk assessment. *Environmental Toxicology and Water Quality*, Vol.9, No.2, pp. 131- 139.

Fawell, J.K.; James, C.P. & James, H.A. (1993) *Toxins from Blue-Green Algae: Toxicological Assessment of Microcystin-LR and a Method for its Determination in Water*. Foundation for Water Research, Marlow, England.

Fawell, J.K.; Mitchell, R.E.; Everett, D.J. & Hill, R.E. (1999) The toxicity of cyanobacterial toxins in the mouse: I Microcystin-LR. *Human & Experimental Toxicology*, Vol.18, No.3, pp. 162-167.

Figueredo, C.; Giani, A. & Bird, D. (2007) Does allelopathy cntribute to *Cylindrospermopsis raciborskii* (Cyanobacteria) bloom occurrence and geographic expansion? *Journal of Phycology* Vol.43, pp. 256-265.

Fitzgerald, D.J.; Cunliffe, D.A. & Burch, M. D. (1999) Development of health alerts for cyanobacteria and related toxins in drinking water in South Australia. *Environmental Toxicology*, Vol.14, No.1, pp. 203-209.

Gehringer, M.M. (2004) Microcystin-LR and okadaic acid-induced cellular effects: a dualistic response. *FEBS Letters*, Vol.557, No.1-3, pp. 1–8.

Giannuzzi, L.; Sedan, D., Echenique, R. & Andrinolo, D. (2011) An acute case of intoxication with cyanobacteria and cyanotoxins in recreational water in Salto Grande Dam, Argentina. *Marine Drugs*, Vol.9, pp. 2164-2175; doi:10.3390/md9112164.

Granéli, E. & Hansen, J. (2006) Allelopathy in Harmful Algae: A mechanism to compete for resours? In: *Ecology of Harmful Algae*. E. Granéli & Turner J. (Ed.). Ecological studies Vol. 189. pp. 189-201.

Halstvedt, C.B., Rohrlack, T.; Andersen, T.; Skulberg, O. & Edvardsen, B. (2007) Seasonal dynamics and depth distribution of *Planktothrix* spp. in Lake Steinsfjorden (Norway) related to environmental factors. *Journal of Plankton Research*, Vol.29, No.5, pp. 471-482.

Honti, M.; Istvánovics, V.; Osztoics, A (2007) Stability and change of phytoplankton communities in a highly dynamic environment—the case of large, shallow Lake Balaton (Hungary). *Hydrobiologia*, Vol. 581(1), pp. 225-240.

Humpage, A.R. & Falconer, I.R. (2003). Oral toxicity of the cyanobacterial toxin cylindrospermopsin in male Swiss albino mice: Determination of no observed adverse effect level for deriving a drinking water guideline value. *Environmental Toxicology*, Vol.18, No.2, pp. 94-103.

Jacquet, S.; Briand, J.-F.; Leboulanger, C.; Avois-Jacquet, C.; Oberhaus, L.; Tassin, B.; Vinçon-Leite, B.; Paolini, G.; Druart, J.-C.; Anneville, O. & Humbert, J.-F. (2005) The proliferation of the toxic cyanobacterium *Planktothrix rubescens* following restoration of the largest natural French lake (Lac du Bourget). *Harmful Algae*, Vol.4, No.4, pp. 651-672.

Kangro, K.; Laugaste, R.; Nõges, P.; Ott, I. (2005). Long-term changes and seasonal development of phytoplankton in a strongly stratified, hypertrophic lake. *Hydrobiologia* Vol. 547, pp. 91-103.

Kinnear, S. (2010) Cylindrospermopsin: a decade of progress on bioaccumulation research. *Marine Drugs*, Vol.8, No.3, pp. 542-564.

Komárek, J. & Komárková, J. (2004) Taxonomic review of the cyanoprokaryotic genera Planktothrix and Planktothricoides. *Czech Phycology*, Vol.4, pp. 1-18.

Kosol, S.; Schmidt, J. & Kurmayer, R. (2009) Variation in peptide net production and growth among strains of the toxic cyanobacterium *Planktothrix* spp. *European Journal of Phycology*, Vol.44, No.1, pp. 49-62.

Kurmayer, R. & Gumpenberger, M. (2006) Diversity of microcystin genotypes among populations of the filamentous cyanobacteria *Planktothrix rubescens* and *Planktothrix agardhii*. *Molecular Ecology*, Vol.15, No.12, pp. 3849-3861.

Kurmayer, R.; Christiansen, G.; Fastner, J. & Borner, T. (2004) Abundance of active and inactive microcystin genotypes in populations of the toxic cyanobacterium *Planktothrix* spp. *Environmental Microbiology*, Vol.6, No.8, pp. 831-841.

Legnani, E.; Copetti, D.; Oggioni, A.; Tartari, G.; Palumbo, M. T. & Morabito, G. (2005). *Planktothrix rubescens'* seasonal dynamics and vertical distribution in Lake Pusiano (North Italy). *Journal of Limnology*, Vol.64, No.1, pp. 61-73.

Lindholm, T.; Öhman, P.; Kurki-Helasmo, K.; Kincaid, B. & Meriluoto, J. (1999) Toxic algae and fish mortality in a brackish-water lake in Åland, SW Finland. *Hydrobiologia*, Vol.397, pp. 109–120.

Luukkainen, R.; Sivonen, K.; Namikoshi, M.; Fardig, M.; Rinehart, K.L.; Niemela, S.I. (1993) Isolation and Identification of eight Microcystins from thirteen *Oscillatoria agardhii* strains and structure of a new Microcystin. *Applied and Environmental Microbiology*, Vol. 59, No. 7, pp. 2204-2209.

MacPhail, R.C. & Jarema, K.A. (2005) Prospects on behavioral studies of marine and freshwater toxins. *Neurotoxocology and Teratology*, Vol. 27, No.5, pp. 695-699.

Manganelli, M.; Scardala, S.; Stefanelli, M.; Vichi, S.; Mattei, D.; Bogialli, S.; Ceccarelli, P.; Corradetti, E.; Petrucci, I.; Gemma, S.; Testai, E. & Funari, E. (2010) Health risk evaluation associated to *Planktothrix rubescens*: An integrated approach to design tailored monitoring programs for human exposure to cyanotoxins. *Water Research*, Vol.44, No.5, 1297-1306.

Mankiewicz-Boczek, J.; Gagala, I.; Kokocinski, M.; Jurczak, T. & Stefaniak, K. (2011) Perennial toxigenic *Planktothrix agardhii* bloom in selected lakes of Western Poland. *Environmental Toxicology*, Vol.26, No.1, pp. 10-20.

Martins, J.; Peixe, L.; Vasconcelos, V. (2010) Cyanobacteria and bacteria co-occurrence in a wastewater treatment plant: absence of allelopathic effects. Water Science Technology, Vol. 62I(8), pp. 1954-1962.

Mbedi, S.; Welker, M.; Fastner, J. & Wiedner, C. (2005) Variability of the microcystin synthetase gene cluster in the genus *Planktothrix* (Oscillatoriales, Cyanobacteria). *FEMS Microbiology Letters*, Vol.245, No.2, pp. 299-306.

Mihali, T.K.; Kellmann, R. & Neilan, B.A. (2009) Characterisation of the paralytic shellfish toxin biosynthesis gene clusters in *Anabaena circinalis* AWQC131C and *Aphanizomenon* sp. NH-5. *BMC Biochemistry*, Vol.10, No.8, DOI:10.1186/1471-2091-10-8.

Montealegre, R. J.; Verreth, Steenbergen, K.; Moed, J.; Machiels, M. (1995) A dynamic simulation model for the blooming of *Oscillatoria agardhii* in a monomictic lake. *Ecological Modelling* Vol. 78, pp. 17-24.

Morabito, G.; Ruggiu, D. & Panzani, P. (2002) Recent dynamics (1995-1999) of the phytoplankton assemblages in Lago Maggiore as a basic tool for defining association patterns in the Italian deep lakes. *Journal of Limnology*, Vol.61, No.1, pp. 129-145.

Mur, L.; Skulberg, O. & Utkilen, H. (1999) Chapter 2. Cyanobacteria in the environment. In: *Toxic cyanobacteria in water: A guide to their public health consequences, monitoring and management*. I. Chorus, Bartram, J. (Ed), WHO.

Naselli-Flores, L.; Barone, R.; Chorus, I.; Kurmayer, R. (2007). Toxic Cyanobacterial Blooms in Reservoirs Under a Semiarid Mediterranean Climate: The Magnification of a Problem. *Environmental Toxicology*, Vol. 22, pp. 399–404.

Nõges, T. & Kangro, K. (2005). Primary production of phytoplankton in a strongly stratified temperate lake. *Hydrobiologia* Vol.547, pp. 105-122.

Oberhaus, L.; Briand, J.F.; Leboulanger, C.; Jacquet, S. & Humbert, J. F. (2007) Comparative effects of the quality and quantity of light and temperature on the growth of *Planktothrix agardhii* and *P. rubescens*. *Journal of Phycology*, Vol.43, No.6, pp. 1191-1199.

Paulino, S.; Sam-Bento, F.; Churro, C.; Alverca, E.; Dias, E.; Valério, E. & Pereira, P. (2009) The Estela Sousa e Silva Algal Culture Collection: a resource of biological and toxicological interest. *Hydrobiologia*, Vol.636, No.1, pp. 489-492.

Paulino, S.; Valério, E.; Faria, N.; Fastner, J.; Welker, M.; Tenreiro, R. & Pereira, P. (2009a) Detection of *Planktothrix rubescens* (Cyanobacteria) associated with microcystin production in a freshwater reservoir. *Hydrobiologia,* Vol.621, No.1, pp. 207-211.

Pouliskova, A.; Hasler, P. & Kitner, M. (2004) Annual cycle of *Planktothrix agardhii*(Gom.) Anagn. & Kom. Nature Population. *International Review of Hydrobiology*, Vol.89, No.3, pp. 278-288.

Pouria, S.; Andrade, A.; Barbosa, J.; Cavalcanti, R.L.; Barreto, V.T.S.; Ward, C.J.; Preiser, W. & Poon, G.K. (1998) Fatal microcystin intoxication in haemodialysis unit in Caruaru, Brazil. *The Lancet*, Vol.352,No.9121, pp. 21-26.

Rücker, J.; Wiedner, C.; Zippel, P. (1997) Factors controlling the dominance of *Planktothrix agardhii* and *Limnothrix redekei* in eutrophic shallow lakes. *Hydrobiologia,* Vol. 342/343, pp. 107–115.

Rudi, K.; Skulberg, O.M. & Jakobsen, K.S. (2005) 16S rDNA analyses of the cyanobacterial microbiota through the water-column in a boreal lake with a metalimnic *Planktothrix* population. *Preparative Biochemistry & Biotechnology*, Vol.35, No.4, pp. 301-312.

Salmaso, N. (2000). Factors affecting the seasonality and distribution of cyanobacteria and chlorophytes: a case study from the large lakes south of the Alps, with special reference to Lake Garda. *Hydrobiologia* Vol. 438, pp. 43–63.

Salmaso, N. (2002). Ecological patterns of phytoplankton assemblages in Lake Garda: seasonal, spatial and historical features. *Journal of Limnology*, Vol.61, No.1, pp. 95-115.

Salmaso, N. (2010) Long-term phytoplankton community changes in a deep subalpine lake: responses to nutrient availability and climatic fluctuations. *Freshwater Biology*, Vol.55, No.4, pp. 825-846.

Shaw, G.R.; Seawright, A.A.; Moore, M.R.& Lam, P.K.S. (2000) Cylindrospermopsin, a cyanobacterial alkaloid: evaluation of its toxicologic activity. *Therapeutic Drug Monitoring*, Vol.22, No.1, pp. 89-92.

Sivonen, K. & Jones, G. (1999) Chapter 3. Cyanobacterial Toxins. In: *Toxic cyanobacteria in water: A guide to their public health consequences, monitoring and management*. I. Chorus, Bartram, J. (Ed), WHO.

Stefaniak, K.; Kokocinski, M. & Messyasz, B. (2005) Dynamics of *Planktothrix agardhii* (Gom.) Anagn. et Kom. blooms in polimictic lake Laskownickie and Grylewskie (Wielkopolska region) Poland. *Oceanological and Hydrobiological Studies*, Vol.34, No.3, pp. 125-136.

Stewart, I.; Webb, P.; Schluter, P. & Shaw, G. (2006) Recreational and occupational field exposure to freshwater cyanobacteria - a review of anecdotal and case reports, epidemiological studies and the challenges for epidemiologic assessment. *Environmental Health: A Global Access Science Source*, Vol.5, No.1, pp. 6. doi:10.1186/1476-1069X-1185-1186.

Sunda, W.; Graneli, E. & Gobler, C. (2006) Positive feedback and the development and persistence of ecosystem disruptive algal blooms. *Journal of Phycology* Vol.42, pp. 963-974.

Svrcek, C. & Smith, D.W. (2004) Cyanobacteria toxins and the current state of knowledge on water treatment options: a review. *Journal of Environmental Engineering and Science*, Vol.3, No.3, pp. 155-185.

Sze, P. (1986) Prokaryotic Algae (Cyanophyta, Prochlorophyta). In *A Biology of the Algae*. WCB Publishers, pp. 19-34.

Ueno, Y.; Nagata, S.; Tsutsumi, T.; Hasegawa, A.; Watanabe, M.F.; Park, H.D.; Chen, G.C.; Chen, G. & Yu, S.Z. (1996) Detection of microcystins, a blue-green algal hepatotoxins, in drinking water sampled in Haimen and Fusui, endemic areas of primary liver cancer in China, by highly sensitive immunoassay. *Carcinogenesis*, Vol.17, No.6, pp. 1317-1321.

Utkilen, H.; Fastner J.; Bartram J. (1999) Chapter 11. Fieldwork: site inspection and sampling. In: *Toxic cyanobacteria in water: A guide to their public health consequences, monitoring and management*. I. Chorus, Bartram, J. (Ed), WHO.

Valério, E.; Chaves, S. & Tenreiro, R. (2010) Diversity and Impact of Prokaryotic Toxins on Aquatic Environments: A Review. *Toxins* Vol.2, No.10, pp. 2359-2410.

van Apeldoorn, M.E.; Egmond, H.P.; Speijers, G.J.A. & Bakker, G.J.I. (2007) Toxins of cyanobacteria. *Molecular Nutrition & Food Research*, Vol.51, No.1, pp. 7-60.

van den Hoek, C.; Mann, D.G. & Jahns, H.M. (1995) *Algae. An introduction to phycology*. Cambridge University Press.

Vance, B. (1965) Composition and succession of cyanophycean water blooms. *Journal of Phycology* Vol. 1, pp. 81-86.

Vareli, K.; Briasoulis, E.; Pilidis, G. & Sainis, I. (2009) Molecular confirmation of *Planktothrix rubescens* as the cause of intense, microcystin—Synthesizing cyanobacterial bloom in Lake Ziros, Greece. *Harmful Algae*, Vol.8, No.3, pp. 447-453.

Vasconcelos, V. (2006) Eutrophication, toxic cyanobacteria and cyanotoxins: when ecosystems cry for help. *Limnetica* Vol.25, No.1-2, pp. 425-432.

Vasconcelos, V.M. & Pereira, E. (2001) Cyanobacteria diversity and toxicity in a wastewater treatment plant (Portugal). *Water Research*, Vol. 35, No.5, pp. 1354-1357.

Viaggiu, E.; Calvanella, S.; Mattioli, P.; Albertano, P.; Melchiorre, S. & Bruno, M. (2003) Toxic blooms of *Planktothrix rubescens* (Cyanobacteria/Phormidiaceae) in three waterbodies in Italy. *Algological Studies*, Vol.109, No.1, pp. 569-577.

Viaggiu, E.; Melchiorre, S.; Volpi, F.; Di Corcia, A.; Mancini, R.; Garibaldi, L.; Crichigno, G. & Bruno, M. (2004) Anatoxin-a toxin in the cyanobacterium *Planktothrix rubescens* from a fishing pond in northern Italy. *Environmental Toxicology*, Vol.19, No.3, pp. 191-197.

Whitton, B.A. & Potts, M. (2000) Chapter 1. Introduction to the cyanobacteria. In *The ecology of cyanobacteria*. B.A Whitton, Potts, M. (Ed). Kluer Academic Publishers, Netherlands, pp. 1-11.

WHO (1998) *Guidelines for drinking water quality*, 2nd ed., Addendum to Vol. 2, Health criteria and other supporting information. World Health Organization: Geneva, Switzerland.

WHO (1999) Chapter 2. Cyanobacteria in the environment. In *Toxic Cyanobacteria in Water: Aguide to their public health consequences, monitoring and management.* I. Chorus, Bartram, J. (Ed). World Health Organization.

WHO (2003) *Guidelines for Safe Recreational Water Environments. Coastal and Freshwaters.* World Health Organization: Geneva, Switzerland, Vol.1, pp. 136–158.

Yepremian, C.; Gugger, M.F.; Briand, E.; Catherine, A.; Berger, C.; Quiblier, C. & Bernard, C. (2007) Microcystin ecotypes in a perennial *Planktothrix agardhii* bloom. *Water Research*, Vol.41, No.19, pp. 4446-4456.

Yu, S.Z. (1995) Primary prevention of hepatocellular carcinoma. *Journal of Gastroenterology and Hepatology*, Vol.10, No.6, pp. 674-82.

Zĕgura, B.; Štraser, A. & Filipič, M. (2011) Genotoxicity and potential carcinogenicity of cyanobacterial toxins - a review. *Mutation Research*, Vol.727, No.1-2, pp. 16–41.

Zhou, L.; Yu, H. & Chen K. (2002) Relationship between microcystin in drinking water and colorectal cancer. *Biomedical and Environmental Sciences*, Vol.15, No.2, pp. 166-71.

5

Generalized Additive Models in Environmental Health: A Literature Review

Jalila Jbilou* and Salaheddine El Adlouni
Université de Moncton, Moncton, Canada

1. Introduction

Time series regression models are especially suitable in epidemiology for evaluating short-term effects of time-varying exposures. Typically, a single population is assessed with reference to its change over the time in the rate of any health outcome and the corresponding changes in the exposure factors during the same period. In time series regression dependent and independent variables are measured over time, and the purpose is to model the existing relationship between these variables through regression methods. Various applications of these models have been reported in literature exploring relationship between mortality and air pollution (Katsouyanni et al. 2009; Wong et al. 2010; Balakrishnan et al. 2011); hospital admissions and air pollution (Peng et al. 2008; Zanobetti et Schwartz 2009; Lall et al. 2011); pollution plumes and breast cancer (Vieira et a. 2005); diet and cancer (Harnack et al. 1997); and mortality and drinking water (Braga et al. 2001). Different time series methods have been used in these studies, i.e. the linear models (Hatzakis et al. 1986) the log-linear models (Mackenbach et al. 1992), the Poisson regression models (Schwartz et al. 2004), and Generalized Additive Models (Dominici 2002; Wood, 2006). The Generalized Additive Models represent a method of fitting a smooth relationship between two or more variables and are useful for complex correlations, that not easily fitted by standard linear or non-linear models.

The present chapter reviews The Genralized Additive Model (GAM), a class of statistical models which have commonly been used in time series regression, specially allowing for serial correlations, which make them potentially useful for environmental epidemiology.

2. Generalized additive models

The classic multiple linear regression model has the form:

$$Y = X\beta + \varepsilon \tag{1}$$

where Y is the response variable, X is the matrix (n×p) of the independent p variables $X_1,\dots,X_p$, β is the vector of the parameters and ε is the vector of errors normally

---

* Corresponding Author

distributed with average 0 and variance σ^2 . Consequently, the variable Y is also Normal distribution with $E[Y]=\mu=X\beta$ and the covariance matrix $\sigma^2 I$ (I is the identity matrix). The linear models are central in applied statistics, mainly because of their simple structure and their interpretative ease. However, they present certain limits and are inadequate when the assumption of normality of the response variable is no longer justified. The linear model is extended to the Generalized Linear Model (GLM) to include a large class of the response variable distribution which belongs to the exponential family of distribution. The distribution Y is related to the linear combination of the covariables, $\eta = X\beta$, via the link function g(.), such as $g(\mu)=g(E[Y])=\eta$.

To introduce more flexibility in the dependence structure between the response variables and covariables, the Generalized Additive Models (GAM), an extension of the GLM, replace the linear dependence functions by more flexible non-linear functions (Hastie and Tibshirani, 1990). The dependences are generally presented by non-parametric smoothing functions. The statistical inference consists on the estimation of the non-linear functions $f(X_j), j=1,\ldots,p$, for each explicative variable X_j . This allows the identification of the specific form of the effect of each explicative variable on the dependant variable Y.

In practice, the objective is to model the dependence between the response variable, Y, and the explicative variables $X_1,\ldots,X_p$, for three main reasons: the description, the inference and the prediction. The goal is to find an explicit form of the effect $f(X_j)$ of each variable X_j on the variability of Y. The Generalized Additive Model (GAM) can be summarized by the flowing three components:

1. The random component: Y that follows a distribution of the exponential family and the mean and the variance are, respectively, $E[Y]=\mu$ and $\mathrm{var}[Y]=\sigma^2$.
2. The systematic component: the explicative variables $X_1,\ldots,X_p$ that compose the regressor, defined by

$$\eta = \alpha + \sum_{j=1}^{p} f(X_j) \quad (2)$$

3. The link function g(.) is such that $g(\mu)=g(E[Y])=\eta$, which implies that $E[Y]=\mu=g^{-1}(\eta)$.

The exponential family of distributions contains the Normal, Binominal, Gamma, Poisson, Geometric, Negative Binominal, and Exponential.

The non-linear functions $f(.)$ are usually represented by non-parametric dependence functions based on smoothing. The smoothing consists on creating a polynomial function that summarizes the data's tendencies. Some types of smoothing designed to express non-linear relations between the Y variable and the covariates Xj, j = 1… p, of the GAM models are the following: smoothing by scatter plot, parametric regression, mobile average, kernel smoothing and spline smoothing. A spline is a combination of polynomial functions. The knots are the points that mark the transition between the pieces of the polynomials (Eilers and Marx, 1996). The constraints allowing the joining of the polynomial pieces are defined by the number of continuous derivatives from the polynomial to the knots. The most popular choice of the spline function is the natural cubic spline. It is a polynomial of the 3rd degree whose second derivative is zero at the limits. It offers less flexibility at the limits but this constitutes an advantage since the fit given by the regression spline presents a large variance around the limits (Hastie and Tibshirani, 1990). A smoothing B-spline basis is independent of the response variable Y and depends only on the following information: (i) the extent of the explicative variable; (ii) the number and position of the knots, and (iii) the degree of the B-spline. The properties of the B-splines are:

- It is formed of q + 1 polynomial pieces, each of q degree;
- The polynomial pieces are joined at q knots;
- The derivatives of order one to q at the joining points of the polynomial pieces are continuous;
- The B-spline function is positive on the q + 2 nodes extent and neutral elsewhere;
- It straddles 2q surrounding polynomial pieces (except at the edges);
- For all x, the q + 1 B-spline function are non-null.

One of the main advantages of the generalized additive model (GAM) is that it offers a great flexibility in order to represent the relations between the dependant variable and the explicative variables. Berger et al. (2004) present advantages related to the GAM to describe the relation between the use of the fluoroquinolone antibiotic and the resistance of the Staphylococcus aureus bacteria collected on the adult patients hospitalized for at least 48 hours. The dependant variable Y(t) of the model was the monthly number of cases in which the bacteria collected from the infected patient resisted to the fluoroquinolone and the explicative variables (Xm(t), m=1...p) were the monthly indicators of the antibiotics doses daily administered. The variable Y(t) follows the Poisson distribution P(λ), where the parameter λ corresponds to the average number of the cases per month and is function of the covariates. The link function is the logarithmic function and the regressor has the form $\lambda(t) = a + \sum_{m=1}^{p} f_m(X_m(t))$ in which $f_m(.)$ is a spline function. The results have shown the existence of a significant relation between the use of fluoroquinolone and the resistance of the bacteria.

The GAM models are used in the prognostic analyses of diseases. For example, Gehrmann et al. (2003) explored multiple sclerosis disease in order to identify the variables that have significant effects on the supported progression of the disease, to determine the intensity and the form of these effects and to estimate the survival curves. The use of Generalized Additive Models helped identify that among the available explicative variables; only the level of initial severity and the number of relapses during the twelve months preceding the study had significant effects on the hazard rate. The hazard rate h(t) means the probability of death after the time t, given that the patient has survived up to the time t.

In a study on the failure rate h(t) of patients with breast cancer (Hastie et al., 1992), the GAM model has been considered to identify among the prognostic factors those which presented significant non-linear relations with h(t). These prognostic factors are: the presence or absence of necrosis of the tumor, the size of the tumor, the number of samples examined, the patient's age, the body mass index and the number of days between the surgical intervention and the beginning of the study. Among these variables, the non-significant relation has been identified with the age and the body mass index. The authors stated that the non-linear modeling had the advantage, firstly, of preventing against the false definition of the model which would lead to incorrect conclusions with regards to the effectiveness of a treatment, and also of provide information on the relation between the prognostic factors and the risk of disease which the standard (linear regression, normal distribution) models do not provide.

The GAM models are also employed in the analyses on the impact of climate and environmental variables on the public's health. In Quebec, a study of the impact of climate variables on mortality was conducted by Doyon et al. (2006). The number of daily deaths was modeled by the Poisson regression with a linking logarithmic function and the explicative climate variables selected were the humidity, the heat threshold and the functions of the average daily temperatures. A similar project carried out on European cities characterized by diverse climatic conditions arrived at the same conclusion of the existence of a significant relation between mortality and the temperature in several cities in Europe (Michelozzi P et al. 2007). The number of deaths and the number of hospital admissions were classified by age groups (15-64 years, 65-74 years, 75 and above years) and by cause (all causes - except death due to external causes -, cardiovascular diseases, cerebrovascular diseases, respiratory diseases, influenza). Considered climate variables are: temperature, dew point, wind speed, wind direction, pressure, total coverage of clouds, solar radiation, precipitations, and visibility. The variables of pollution were SO2, TSP (black smoke), PM10, NO2, and CO. The analysis was done separately for the warm season (April-September) and the cold season (October-March). This provides flexibility for the analysis, allowing the use of different model structures for each season (Terzi and Cengiz, 2009). Recently, Bayentin et al. (2010) used the GAM model to study the association between climate variables and circulatory diseases. The short term effect of climate conditions on the incidence of ischemic heart disease (IHD) over the 1989-2006 period was examined for Quebec's 18 health regions, with control for seasonality and socio-demographic conditions.

3. Parameter estimation

3.1 Local scoring procedure algorithm

The algorithm (presented in Appendix C.1) is summarized as an iterative and weighted process which allows the adjustment of a function f_j, $j = 1\dots p$, while keeping the other p-1 dimensions in their actual state. GAM models, in which the iterative algorithm is incorporated in S-Plus, became a popular analytical tool in epidemiology, especially in studies on the effects of environmental variables on public health (Dominici et al,. 2002). However, estimation by this algorithm presents problems of convergence and validity when the weighting matrix W (Appendix C.1) is not diagonal and if the independence hypothesis is not respected. Even if augmenting the number of iterations improves the estimations, the typical estimation errors remain difficult to evaluate and the model's effective dimension is statistically demanding (Wood, 2006). Many authors have suggested more direct approaches to remedy these problems.

3.2 Simultaneous estimation

The most effective way to estimate parameters is the use of a parametric GLM model with a limited number of regression splines or smoothing splines. This reduces the parameter estimation problem in both cases to that of a GLM model with all its advantages related to the linear dependence functions. Despite the simplicity of the penalized GLM model, in the case of smoothing splines (Hastie and Tibshirani, 1990), the problem of the large system of equations remains. In the case of regression splines, each spline function is the function of the sum of the basis B-spline functions. This situation features the ease of B-spline construction, but the problem of the optimum choice in the position and number of B-spline nodes arises (Hastie and Tibshirani, 1990). Eilers and Marx (1996) have shown that this problem could be avoided by combining the B-splines to a differential penalty. In fact, the penalty is applied directly to the parameters in order to control the roughness of the spline functions. Criterion can be employed for the number of knots and the value of the penalty parameter.

When the P-spline are considered, the GAM has the form $g(\mu) = E(Y) = \alpha + \sum_{j=1}^{p} f_j(X_j)$ with $f_j(X_j) = B_j A_j$ and a response variable distribution belongs to the exponential family. In this section, B_j, $j = 1\ldots p$ is the B-spline matrix (with n_j knots) of $N \times n_j$ dimension, A_j is the n_j-vector of the basic B-spline function coefficients and then represents the part of the the variability of Y explained by X_j. The model can be rewritten as follows:

$$E[Y] = g(\mu) = BA \tag{3}$$

where $B = \left[1\, B_1\, B_2 \ldots B_p\right]$ and $A = \left(\alpha, A_1, \ldots, A_p\right)$. We are left with a GLM model and the estimation of the parameters α by maximization of the penalized log-likelihood is done by the penalized GLM Fisher scoring, below, until the desired convergence criterion is obtained.

$$\hat{A}_{t+1} = (B'\hat{W}_t B + P)^{-1} B'\hat{W}_t \hat{z}_t \tag{4}$$

where

$\hat{W} = diag\left\{\dfrac{[h'(\hat{\eta}_i)]^2}{Var(Y_i)}\right\}$, $\hat{z}_i = \hat{\eta}_i + \dfrac{(y_i - \hat{\mu}_i)}{h'(\hat{\eta}_i)}$ and $\boldsymbol{P} = \text{blockdiag}(0, \lambda_1 P_1, \ldots \lambda_p P_p)$.

$\boldsymbol{P}$ is the component which summarizes the penalty on the B-spline coefficients of the p covariates and h is the opposite of the linking function g.

The approach assumes that the effect functions f_j of a covariate X_j can be approximated by a polynomial spline written in terms of a linear combination of B-spline basis functions. The crucial problem with such regression splines is the choice of the number and the position of the knots. A small number of knots may lead to a function space which is not flexible enough to capture the variability of the data. A large number of knots may lead to a serious overfiting. Similarly, the position of the knots may potentially have a strong influence on the estimation. A remedy can be based on a roughness penalty approach as proposed by Eilers and Marx (1996).

Smoothing parameters are used to balance the goodness-of-fit and smoothness. A performance measure is used to find the optimum values of the penalties. The number and location of knots are no longer crucial as long as the minimum number of knots is reached. In practice, this approach poses problems to get a solution when the number of the model's smoothing functions is high (Lang and Brezger, 2004). The P-spline approach is easy to conceive and has the advantage of the explicit formula of the estimation matrix and standard errors estimations (Marx and Eilers, 1998). However, the simplicity if reduced if the knots are at unequal distances (Wood, 2006). Thus, despite the advantages of the P-spline approach in the GAM models, the problems of the estimation of the parameters with the *penalized GLM Fisher scoring* algorithm, remains important (Zhao et al., 2006, Wood, 2006, Binder and Tutz, 2006).

3.3 Bayesian method

The Bayesian approach is essentially based on the concept that the parameters to be estimated are not constants but are considered as random variables. Bayesian statistical inference is based on the posterior distributions of the parameters, which combine the prior information and observed one from the sample. In the case of the GAM models, we wish to estimate the parameter α and the functions $f_1,\ldots,f_p$. One of the advantages of the Bayesian approach compared to the penalized GLM Fisher scoring algorithm is the fact that the uncertainty related to the variance of the components is taken into account trough the posterior distribution of the parameters (Fahrmeir and Lang, 2001, Zhao et al., 2006). In practice, the analytical form of the posteriori distribution is rarely available and then it s difficult to extract their characteristic for risk assessment purposes. The Markov chain Monte Carlo procedure (MCMC), allows to obtain all these characteristics by simulating samples from the posterior distribution and thus to deduce parameter estimators, the quantiles and associated risk as well as estimator uncertainty. More details on the MCMC approach and their convergence diagnostics are studied in El Adlouni et al. (2006).

In the case of the P-spline functions, the parameters ***a*** of the GAM model, in equation (2), are a random variables. The penalties based on the finite differences of the B-spine coefficients are replaced by their stochastic equivalent which correspond to a random walks of order one or two, defined by

$$a_{j\rho} = a_{j,\rho-1} + u_{j\rho} \text{, or } a_{j\rho} = 2a_{j,\rho-1} - a_{j,\rho-2} + u_{j\rho} \tag{5}$$

with $u_{j\rho} \sim N(0,\tau_j^2)$ and the initial values a_{j1}, a_{j2} are constants. The level of smoothing is thus controlled by the variance parameter τ_2, which must also be estimated. Lang and Brezger (2004) suggest a prior distribution of the parameters a_j of the form:

$$a_j \left| \tau_j^2 \right. \propto \frac{1}{(\tau_j^2)^{rk(K_j)/2}} \exp\left(-\frac{1}{2} a_j' K_j a_j \right) \tag{6}$$

where K is the penalty and depends on the smoothing function f_j and on the nature of the X_j variable. The prior distribution of the parameter τ_2 is an Inverse Gamma distribution

$IG(c_j, d_j)$, where c_j, d_j are the hyper-parameters and are usually given by prior knowledge on the variables. It is however necessary to perform a sensitivity analysis on the prior choice.

The posteriori distribution of the model has the following form:

$$p(\alpha, a_1, \tau_1^2, \ldots, a_p, \tau_p^2 | y) \propto L(y, \alpha, a_1, \tau_1^2, \ldots, a_p, \tau_p^2)$$

$$\propto \prod_{j=1}^{p} \frac{1}{(\tau_j^2)^{rk(K_j)/2}} \exp\left(-\frac{1}{2\tau_j^2} a_j' K_j a_j\right) \prod_{j=1}^{p} (\tau_j^2)^{-a_j-1} \exp\left(-\frac{b_j}{\tau_j^2}\right) \quad (7)$$

All the inference is based on the posterior distribution. The MCMC algorithm can be performed to estimate the empirical posterior distribution and the predictive distribution of the quantile to deduce the risk values.

The assumptions of the Bayesian estimation model are completed by the following conditional independence assumptions:

a. For all explicative variables and f_j parameters, the observations Y_i are conditionally independent.
b. The prior distributions of the parameters are conditionally independent.
c. The priori distribution of the fixed effects and variances τ_j^2, $j = 1, \ldots, p$ are mutually independent.

4. Performance measure

In order to select the smoothing penalty and the number of knots that leads to the most adequate fit some performance measures are used. The most used performance measures are the Akaïke information criterion (*AIC*) and the generalized cross-validation (*GCV*). They are based on the deviance statistic (or the statistical likelihood ratio) that, for a counting GAM model (the case of the Poisson distribution), is obtained by the following formula:

$$D(y;\hat{\mu}) = 2\sum_{i=1}^{n}[y_i \ln(y_i / \hat{\mu}_i) - (y_i - \hat{\mu}_i)] \quad (8)$$

The Akaïke information criterion developed by Akaike (1973) measures the quality of the model fit to observed data series. It is the function of the deviance function $D(y; \mu)$ and is obtained by the following formula:

$$AIC = \frac{1}{n}[D(y;\hat{\mu}) + tr(R)\phi] \quad (9)$$

where $tr(R)$ the sum of the diagonal elements of the matrix R of the weighted additive-fit operator of the last iterations in the estimation process, and ϕ the scale parameter.

The generalized cross-validation for the smoothing penalty is obtained by the following formula:

$$GCV(\lambda) = \frac{1}{n}\sum_{i=1}^{n}\left\{\frac{y_i - \hat{f}_\lambda(x_i)}{1 - tr(S_\lambda)/n}\right\}^2 \tag{10}$$

Where S is the smoother. For the GCV of the model, the corresponding criterion is based on :

$$GCV = \frac{\frac{1}{n}D(y;\hat{\mu})}{[1 - tr(R)/n]} \tag{11}$$

R is the weighted additive-fit operator of the last iteration in the estimation of the model.

5. Confounding variables, concurvity, and interaction

5.1 Confounding variables

Confounding is potentially present in all observational studies. A confounding factor in the field of environmental health refers to a situation in which an association between an exposure (i.e. air pollution) and a health outcome (i.e. morbidity or mortality) is distorted because it is mixed with the effect of a third variable - the confounding variable (i.e. humidity). The confounding variable is related to both the exposure and the outcome. The distortion introduced by a confounder can lead to an overestimation (positive confounding, affecting the outcomes in the same direction as the exposure under study) or underestimation (negative confounding, affecting the outcomes in the opposite direction of the exposure under study) of the association between exposure and outcome. Confounding variables can be controlled for by using of one or more of a variety of techniques that eliminate the differential influence of the confounder. For example, if one group is mostly females and the other group is mostly males, then the gender may have a differentially effect on the outcome. As a result, we will not know whether the outcome is due to the treatment or due to the effect of gender. If the comparison groups are the same on all extraneous variables at the start of the experiment, then differential influence is unlikely to occur. The control techniques are essentially attempts to make the groups similar or equivalent. Confounding variables are to be differentiated from intermediating or latent variables that are part of the causal pathway between the exposure and the outcome (Budtz-Jorgensen et al., 2007).

Peng et al. (2006) identified two types of confounding variables: those that are measured and are already included in the model, and those that are not. They propose as an adjustment to this problem the inclusion of a non-linear function of actual and future data in the model. In the study of the relation between air pollution and mortality, the non-measured confounders are the factors that influence the mortality in the same way as the air pollution variables (Peng et al., 2006). These factors produce seasonal effects and long-term tendencies on the mortality which deforms the relation between the mortality and the air pollution (i.e. Influenza epidemics and pulmonary infections). In these situations, the inclusion of the variable "time" helps to reduce the bias caused by these factors.

Other processes for managing the confounding effects are the methods of sampling: specification and matching. Specification is the scheme that specifies the value of the potential confounding and excludes other values (i.e. non smokers only in the study). This method of sampling allows focusing solely on the subjects of the study in question but does not enable the generalization of results. The matching consists on grouping the subjects with similar values of the confounding variable. It has the advantage of eliminating the influences of the confounding with important effects and of improving the precision (strength) by balancing the number of cases and controlling each layer.

5.2 Concurvity

The non-linear dependence that remains between the covariates is referred to as the concurvity in the GAM models by analogy to the co-linearity in GLM models. Researchers (Ramsay et al., 2003) insist that a certain degree of concurvity exists in every epidemiological time series, especially when the time is included in the model as a confounding variable. The main problem caused by concurvity in a GAM model is the presence of a bias in the model, more specifically the overestimation of the parameters and the underestimation of the standard errors. The use of asymptotically unbiased estimator of standard errors introduced by Hastie and Tibshirani (1990) and demonstrated by Dominici et al. (2003) does not solve the bias problem. The consequence of this is the inflation of type I errors in the signification tests, resulting in the conclusion of the presence of significant effect (Ramsay et al., 2003).

Several approaches have been proposed to control the problem of concurvity in time series. One method of estimation of the variation, based on the bootstrap parametric, has also produced biased results based on simulations by Ramsay et al. (2003). These recommend instead the use of parametric models such as the GLM model with natural splines (Dominici et al., 2002). He (2004) suggests the use of a non parametric model GAM to explore data in a primary level of analysis and when the appropriate variables are retained, to pursue the analysis with a parametric model GLM with natural splines, all while keeping the same degree of smoothing.

Figueras et al. (2005) developed the conditional bootstrap method in order to control the effect of the concurvity. In this type of bootstrap, B bootstrap replicates are generated. In each of these, the values of the independent variables are the same as those of the observed data, with only the values of the response variable being varied from replicate to replicate. The value assumed by the outcome in each observation is conditional (hence the technique's name) upon the values of the set of independent variables in said observation. The conditional Bootstrap approach has been tested on simulated data and leads to good results.

5.3 Interactions in the GAM model

The interaction within a statistical model denotes the effect of two or more variables, which is not simply additive. In other words, the effect is due to the combination of two or more variables in the model. A consequence of the interaction between two variables is that the effect of a variable depends on the value observed for the other one. A form of interaction often found in bibliography is the modification of the effect. The modification of the effect happens when the statistical measure of the association between the explicative variable X_1 and the response variable Y depends on the level of another variable X_2, known as the effect modifier. The extent of the relationship depending on the value of the effect modifier contributes to the improvement of the model fit. In the field of environmental health, this allows us to identify the most vulnerable groups to a particular condition (Wood, 2006; Bates and Maechler, 2009).

6. Conclusion

Environmental health research is becoming a cornerstone for supporting evidence-based (informed) decision making in healthcare services and management. Providing evidence through robust and relevant epidemiologic studies in environmental health research may be improved through an adequate utilization of statistics methods. In this chapter, we reviewed the Generalized Additive Models and the most used estimating methods and presented their advantages and limits. Knowing this, researchers should take into account these aspects when it is time to define exposures and outcomes, to map spatial variations, to design epidemiologic studies' conceptual frameworks and to select suitable estimating models. These critical aspects are of central importance for developing clinical and public health decision making to reduce the burden of environment impacts on individual and population health. Moreover, using accurate and relevant methods, i.e. GAM, in environmental epidemiology studies is a cornerstone for developing effective actions that may help save cost and improve decision making performance.

Improvements will be seen also in clinical practices through a better understanding and the integration in medical decisional algorithms of the effects of long term exposition to specific environmental factors. These effects are translated into risks of occurrence and prognosis of sensitive diseases (i.e. breast cancer, lung cancers). Spreading GAM method utilization in environmental epidemiology through a clinical perspective is highly recommended to develop effective decisional tools that may greatly improve personalized medicine. Moreover, GAM method may help to better manage follow-up of patients exposed to long term medications and reduce side-effects and complications. This review highlights the utility of Generalized Additive Model (GAM) for risk assessment (such as breast cancer) related to environmental factors and explored the use of the GAM for risk assessment in the presence of multiple non-linear effects. The selection and the estimation of the parameters and non-linear functions (B-Splines and P-splines) are essential for an adequate estimate of the risk. Next research should explore how GAM models may help the development of relevant risk assessment tools that may be integrated in personalized medical decision making algorithms. The GAM will allow the integration of environmental factors and others health determinants in clinical algorithms that may help improve the personalization of healthcare delivery. These algorithms will be implemented in public health programs (i.e. personalization of breast cancer screening based on women individual risk) and clinical algorithms (i.e. for patients with a diagnosis of breast cancer the personalization of follow-up will be based on the surveillance of relevant factors such as the biomarkers, the clinical signs and the exposition to environmental factors).

This chapter presented the potential of the Generalized Additive Model (GAM) for environmental studies. Generalized additive models (GAMs) are a generalization of generalized linear models (GLMs) and constitute a powerful technique to capture nonlinear relationships between explanatory variables and a response variable. Selection of the best parameter estimation methods, control for confounding variables and concurvity aims to reduce bias and improve the use of the GAM model. Moreover, when using the GAM model in environmental health, and for an adequate interpretation of the outputs, socio-economic and demographic parameters should be considered.

7. Acknowledgement

The authors would like to thank the CIHR Team on Familial breast cancer at Université Laval (QC) leaded by Dr Jacques Simard; and also the Consortium national de formation en santé-Université de Moncton (NB) for the financial support they provided to prepare and publish this chapter.

8. References

Akaike H. (1973). Information theory as an extension of the maximum likelihood principle. Second International Symposium on Information Theory (B. N. Petrov, et F. Csaki), pp. 267-281, Akademiai Kiado, Budapest.

Balakrishnan K, Ganguli B, Ghosh S, Sankar S, Thanasekaraan V, Rayudu VN, Caussy H; HEI Health Review Committee. Short-term effects of air pollution on mortality: results from a time-series analysis in Chennai, India. *Res Rep Health Eff Inst.* 2011 Mar;(157):7-44.

Bates D. and M. Maechler (2009). lme4: Linear mixed-effects models using S4 classes. URL http://CRAN.R-project.org/package=lme4. R package version 0.999375-31.

Bayentin, L., S. El Adlouni, T.B.M.J. Ouarda, P. Gosselin, B. Doyon and F. Chebana (2010). Spatial variability of climate effects on ischemic heart disease hospitalization rates for the period 1989-2006 in Quebec, Canada. International Journal of Health Geographics, 9:5. doi:10.1186/1476-072X-9-5.

Ballester F., Rodríguez P., Iñíguez C., Saez M., Daponte A., Galán I., Taracido M., Arribas F., Bellido J., Cirarda F.B., Cañada A., Guillén J.J., Guillén-Grima F., López E., Pérez-Hoyos S., Lertxundi A. and Toro S. (2006). Air Pollution and cardiovascular admissions association in Spain: results within the EMECAS Project. Epidemiol. Community Health 60: 328-336.

Berger P., L. Pascal, C. Sartor , J. Delorme, P. Monge, C. P. Ragon, M. Charbit, R. Sambuc and M. Drancourt (2004). Generalized Additive Model demonstrates fluoroquinolone use/resistance relationships for Staphylococcus aureus. European Journal of Epidemiology, 19, pp. 453-460.

Binder H. and G. Tutz (2006). Fitting Generalized Additive Models: A comparison of Methods. Universität Freiburg i. Br., Nr. 93.

Braga, A.L., Zanobetti, A. and Schwartz J (2001), The time course of weather related deaths, Epidemiology, 12, 662-667.

Budtz-Jorgensen E., N. Keiding, P. Grandjean and P. Weiher (2007). Confounder selection in environmental epidemiology: Assessment of health effects of prenatal mercury exposure. Ann. Epidemiol. 17(1), pp 27-35.

Dominici F., A. McDermott and T.J. Hastie (2003). Issues in Semi-parametric Regression with Applications in Time Series Models for Air pollution and Mortality. Rapport de recherche. http://biosun01.biostat.jhsph.edu/~fdominic/trevorpaper.pdf

Dominici F., McDermott A., Zeger S. L. and J.M. Samet (2002). On the use of generalized additive models in time-series studies of air pollution and health. *American Journal of Epidemiology*. 156(3), pp 193-203.

Doyon B., D. Bélanger and P. Gosselin (2006). Effets du climat sur la mortalité au Québec méridional de 1981 à 1999 et simulations pour des scénarios climatiques futurs. *Rapport de recherche*, INSPQ.

Eilers P. H. C. and B. D. Marx (1996). Flexible Smoothing with B-splines and Penalties. Statistical Science, Vol. 11, No. 2, pp. 89-121.

El Adlouni, S., Favre, A.C. and Bobée, B. (2006). Comparison of methodologies to assess the convergence of Markov Chain Monte Carlo methods. Computational Statistics and Data Analysis 50(10): 2685-2701.

Fahrmeir L. and S. Lang (2001). Bayesian inference for generalized additive mixed models based on Markov random field priors. *Appl. Statist.*, 50, Part 2, pp. 201-220.

Figueiras A., J. Roca-Pardiñas and C. Cadarso-Suárez (2005). A bootstrap method to avoid the effect of concurvity in generalized additive models in time series of air pollution. Journal of Epidemiology and Community Health. 59, pp. 881-884.

Gehrmanu U., B. Hellriegel, A. Neiss and L. Fahrmeir (2003). Analysis of the time to sustained progression in Multiple Sclerosis using generalized linear and additive models. Discussion paper 354-SFB 386- LMU Munich.

Hastie T., L. Sleeper and R. Tibshirani (1992). Flexible covariate effects in the proportional hazard model. Breast Cancer Research and *Treatment*, 22, pp. 241-250.

Hastie T. and R. Tibshirani (1990). Generalized Additive Models. *Chapman & Hall*, New-York.

Hastie T. and R. Tibshirani (1993). Varying-coefficient models. *Journal of the Royal Statistical Society B*, 55, pp. 757-796.

Harnack, L., G. Block, et al. (1997). "Association of Cancer Prevention-Related Nutrition Knowledge, Beliefs, and Attitudes to Cancer Prevention Dietary Behavior." *Journal of the American Dietetic Association* 97(9): 957-965.

Hatzakis, A., Katsouyanni, K., Kalandidi, A., Day, N. and Trichopoulos, D. (1986) Short-term effects of air pollution on mortality in Athens. Int. J. Epidemiol., 15, 73-81.

He S.(2004). Generalized additive models for data with concurvity: statistical issues and a novel fitting approach. Rapport de recherche. *http://etd.library.pitt.edu/ETD/available/etd-12022004-103805/unrestricted/ShuiHe.pdf*

Katsouyanni K, Samet JM, Anderson HR, Atkinson R, Le Tertre A, Medina S, Samoli E, Touloumi G, Burnett RT, Krewski D, Ramsay T, Dominici F, Peng RD, Schwartz J, Zanobetti A; HEI Health Review Committee. Air pollution and health: a European and North American approach (APHENA). *Res Rep Health Eff Inst*. 2009 Oct;(142):5-90.

Lall R., Ito K. and G.D. Thurston (2011). Distributed Lag Analyses of Daily Hospital Admissions and Source-Apportioned Fine Particle Air Pollution. *Environ Health Perspect.* April; 119(4): 455–460.

Lang, S. and A. Brezger (2004): Bayesian P-Splines. Journal of Computational and Graphical Statistics, 13, 183-212.

Liang K.Y. and S. L. Zeger (1986). Longitudinal data analysis using general linear models. *Biometrika*, 73, 13-22.

Mackenbach, J., A. Kunst, and C. Looman (1992). Seasonal variation in mortality in The Netherlands. Journal of Epidemiology and Community Health 46, 261–265.

Marx B.D. and P. H. C. Eilers (1998). Direct generalized additive modeling with penalized likelihood. *Computational Statistics & Data Analysis*, 28, pp.193-209.

McCullagh P. and J. A. Nedler (1989). Genralised linear models. *Chapman&Hall.*

Michelozzi P et al. (2007) Assessment and prevention of acute health effects of weather conditions in Europe, the PHEWE project: background, objectives, design. Environ Health; 6:12.

Peng R.D., F. Dominici and A. L. Thomas (2006). Model choice in time series studies of air pollution and mortality. *J. R. Statist. Soc. A*, 169, *Part 2*, pp. 179-203.

Peng R.D., Chang H.H., Bell M.L., McDermott A., Zeger S.L., Samet J.M. and F. Dominici (2008) Coarse Particulate Matter Air Pollution and Hospital Admissions for Cardiovascular and Respiratory Diseases Among Medicare Patients *JAMA*;299(18):2172-2179.

Ramsay T.O., R.T. Burnett and D. Krewski (2003). Exploring bias in generalized additive models for spatial air pollution data. *Environmental Health Perspectives*, 111(10), 1283-1288.

Ruppert, D. (2002). Selecting the number of knots for penalized splines. *Journal of Computational and Graphical Statistics*,11(4), pp 735-757.

Schwartz J., J. M. Samet and J. A. Patz (2004). Hospital admissions for heart disease: the effects of temperature and humidity. *Epidemiology*, 15(6), pp. 755-761.

Terzi Y. and M. A. Cengiz (2009). Using of generalized additive model for model selection in multiple poisson regression for air pollution data. Scientific Research and Essay Vol.4 (9), pp. 867-871.

Vieira V., Webster Thomas., Weinberg J., Aschengrau A. and Ozonoff D. (2005) Spatial analysis of lung, colorectal, and breast cancer on Cape Cod: An application of generalized additive models to case-control data. *Environmental Health: A Global Access Science Source.* 4:11

Wood, S. N. (2006). Generalized Additive Models, An introduction with R. Chapman & Hall/ CRC.

Wong CM, Thach TQ, Chau PY, Chan EK, Chung RY, Ou CQ, Yang L, Peiris JS, Thomas GN, Lam TH, Wong TW, Hedley AJ; HEI Health Review Committee. Part 4. Interaction between air pollution and respiratory viruses: time-series study of daily mortality and hospital admissions in Hong Kong. *Resp Rep Health Eff Inst.* 2010 Nov;(154):283-362.

Zanobetti, A. and J. Schwartz. (2009). A Novel Approach to Estimate Distributed Lag Model Between Hospital Admissions and Ozone: A Multi-City Time Series Analysis. *Epidemiology*: November - Volume 20 - Issue 6 - p S62

Zhao Y., J. Staudenmayer, B.A. Coull and M. P. Wand (2006). General design bayesian generalized linear mixed models. *Statistical Science*. 21(1), pp 35-51.

6

The Science and Opportunity of Wildfire Risk Assessment

Matthew P. Thompson, Alan A. Ager,
Mark A. Finney, Dave E. Calkin and Nicole M. Vaillant
US Forest Service
USA

1. Introduction

Wildfire management within the United States continues to increase in complexity, as the converging drivers of (1) increased development into fire-prone areas, (2) accumulated fuels from historic management practices, and (3) climate change potentially magnify threats to social and ecological values (Bruins et al., 2010; Gude et al., 2008; Littell et al., 2009). The need for wildfire risk assessment tools continues to grow, as land management agencies attempt to map wildfire risk and develop strategies for mitigation. Developing and employing wildfire risk assessment models can aid management decision-making, and can facilitate prioritization of investments in mitigating losses and restoring fire on fire prone landscapes. Further, assessment models can be used for monitoring trends in wildfire risk over space and across time.

The term risk is generally used to measure the chance of loss, as determined from estimates of likelihood and magnitude of particular outcomes. Probabilistic approaches to risk assessment estimate the expected value of the conditional probability of the even occurring and the consequence of the event given that it has occurred. Risk assessments are conducted when predicted outcomes are uncertain, but possible outcomes can be described and their likelihoods can be estimated. Wildfire risk assessment entails projecting wildfire extent and intensity, and the consequences of fires interacting with values-at-risk.

We begin by introducing a conceptual model of wildfire management (Figure 1) that considers the major drivers of wildfire risk and strategic options for mitigation. Ignition processes influence the spatiotemporal pattern of wildfire occurrence (natural and human-caused), and strategic prevention efforts can reduce the number of wildfires and associated damage (Prestemon et al., 2010). Given an ignition that escapes suppression, fuel, weather, and topography jointly drive wildfire behavior. Of these, only fuel conditions (loading, structure, continuity) can be altered to induce desirable changes in fire behavior (Agee & Skinner 2005). Suppression efforts are intended to slow the growth of active wildfires and reduce the chance of loss. Collectively these factors influence wildfire extent and intensity, which in turn determine the consequences (detrimental and beneficial) to social and ecological values. Wildfire losses can also be prevented or reduced by activities that lessen the consequences of an interaction with fire, for instance the use of fire-resistant materials in home construction.

The challenge of wildfire management is to find efficient combinations of investments in mitigation options, recognizing heterogeneity in the environmental and socioeconomic factors contributing to wildfire risk. Assessing wildfire risk and evaluating mitigation options are highly complex tasks that integrate multiple interacting components including fire simulation modeling, mapping valued resources and assets, characterizing first- and second-order fire effects, quantifying social and managerial preferences and priorities, and exploring feasible management opportunities. Wildfire risk analysis is therefore fundamentally interdisciplinary, requiring the pairing of substantive expertise (fire behavior modeling, silviculture, fire ecology, etc.) with methodological expertise (statistics, engineering, decision analysis, etc.). Improved assessment of wildfire risk in turn ideally leads to improved strategic risk management across planning scales, and ultimately to enhanced resource protection and ecosystem resiliency.

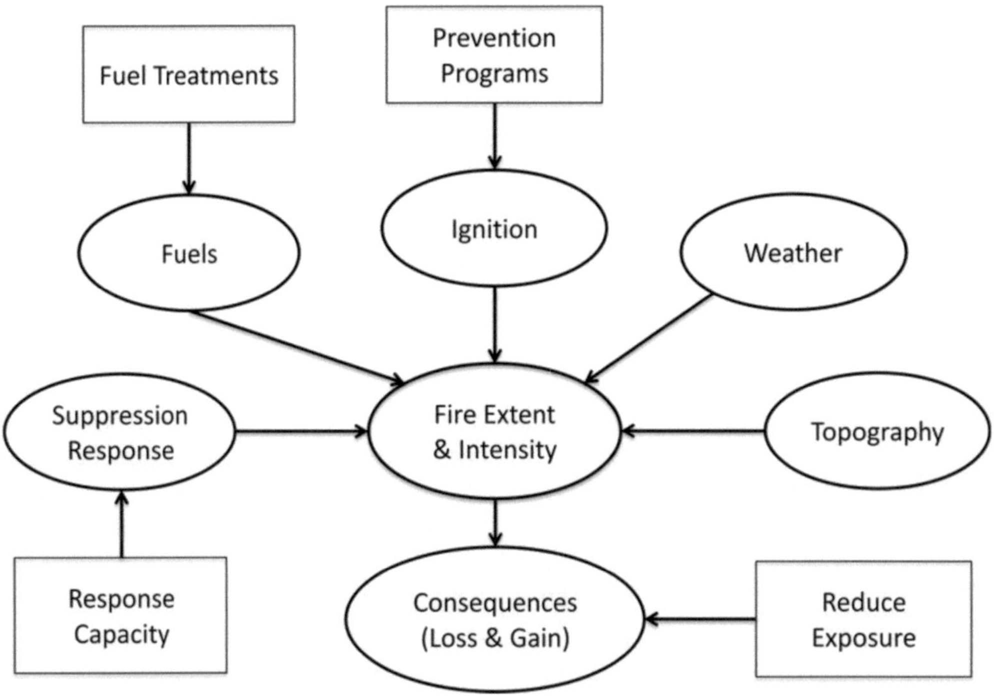

Fig. 1. Conceptual Model of Wildfire Management (Modified from Calkin et al., 2011).

The major drivers of fire extent and intensity are represented as ovals, and the major strategic options for mitigating risk are represented as rectangles.

In this chapter we review the state of wildfire hazard and risk analysis, in particular highlighting a risk assessment framework that is geospatial, quantitative, and considers multiple social and ecological values. Contextually our focus is federal wildfire management in the United States, although the framework we present has broader applicability across geographic locations and ownerships. First we review concepts of hazard and risk in the wildfire management context. Second, we describe newer developments in the application of burn probability modeling for exposure analysis, and illustrate how this modeling approach can inform fuel management and wildfire suppression efforts. Third, we discuss challenges in quantifying risk for the array of non-market values that are the primary management concern on federal lands, and how expert judgment can be used to advance wildfire effects analysis. We use examples from recent and ongoing broad scale risk assessments and describe their use for informing strategic policy. Lastly we conclude by discussing potential benefits to wildfire management and policy from embracing risk management principles.

2. Wildfire hazard and risk assessment

In is important to recognize the difference between wildfire hazard and wildfire risk, since these terms are often used interchangeably in the literature. Wildfire hazard characterizes the potential for wildfire to harm human life and safety or damage highly valued resources and assets (HVRAs) (Keane et al., 2010). Wildfire risk, by contrast, includes quantification of the magnitude of fire outcomes (beneficial and detrimental) as they relate to fire hazard (Finney, 2005). From this perspective, mapping fire hazard can reveal patterns of one component of risk, but offers less complete information to decision-makers faced with questions of how to understand and mitigate potential impacts to HVRAs.

2.1 Wildfire hazard

A variety of approaches have been adopted to characterize wildfire hazard. Typically hazard is described in relation to factors affecting the fire environment and likely fire behavior, including fuel and vegetation properties, topography, climate and weather variables, and ignition characteristics (Hessburg et al., 2007; Vadrevu et al., 2010). Conceptually, probabilistic, spatially-explicit models of wildfire hazard are most relevant for risk assessment. For instance, hazard can be described with a probability distribution for a given fire characteristic at a given location, such as fire occurrence or behavior. Fire occurrence likelihood is often estimated using logistic regression models (Brillinger et al., 2009; Finney et al., 2011a; Martínez et al., 2009; Prasad et al., 2007; Priesler & Westerling 2007). Some approaches have considered likelihood of wildfire occurrence as a separate component, and characterized hazard instead as the potential to cause harm given a wildfire occurs (i.e., hazard is measure of conditional fire behavior). Here we include wildfire likelihood in our definition of hazard, which incorporates not only the likelihood of ignition for any particular area on the landscape but also the likelihood of burning due to fire spread from remote ignitions.

Modeling fire behavior given fire occurrence typically entails estimating spread rate, flame length, fireline intensity, and crown fire activity, and involves the integration of multiple sub-models (Ager et al., 2011; Cruz & Alexander 2010). Modeling fire spread allows the computation of fire travel pathways and fire size distributions, and a robust characterization of the spatial process. Simulating fire growth across heterogeneous landscapes can identify emergent behavioral properties that may not be predictable and may not be captured with localized estimates of fire behavior (Carmel et al., 2009; Parisien et al., 2007). Modeling fire spread also allows for estimates of fireline intensity as a function of fire spread direction (flanking, heading, or backing).

Rapid advancements in geospatial data management, fire behavior modeling, and computing power have vastly improved the spatial assessment of fire impacts on HVRAs. In particular, estimation of burn probability (BP), an estimate of the likelihood of a point burning under a predefined set of assumptions about ignition and fire behavior, is now feasible for large landscapes. Explicit consideration of fire spread from remote ignitions is particularly important in parts of the western United States, where large lightning-caused fires typically spread over large distances. In other locations and in different planning environments ignition likelihood may be much more of a driver.

Simulation modeling can further produce burn probabilities for fire intensity (BP_i) as a function of the number of times a pixel burned at a given intensity level. The intensity with which a fire burns is an important variable for predicting fire effects. Fire intensity (KW/m) is typically converted to flame length to measure fire effects. Fire intensity is relative to the spread direction, and thus quantifying intensity for a particular point needs to consider all possible arrival directions and their probabilities. The conditional flame length (CFL), or the probability weighted flame length given a fire occurs (Scott, 2006; Equation 1) is used for this purpose, and is a statistical expectation, summing over burn intensity probabilities multiplied by the midpoints of the corresponding flame length category (Ager et al., 2010).

$$CFL = \Sigma BP_i FL_i \qquad (1)$$

Figure 2 displays burn probability maps (a) and conditional flame lengths (b) for National Forests in the states of Oregon and Washington, in the Pacific Northwest of the United States. These estimates were derived from the large fire simulation model FSim (Finney et al., 2011a). Maps of BP and CFL differentiate regions and forests with higher relative wildfire hazard, for instance the eastern-most National Forests. Hazard is lower in the western portion of the region, where forests are generally moister and where annual rainfall is much higher.

2.2 Wildfire risk

A widely accepted ecological risk assessment framework was developed by the U.S. Environmental Protection Agency that entails four primary steps: (1) problem formulation, (2) exposure analysis, (3) effects analysis, and (4) risk characterization (U.S. Environmental Protection Agency, 1998). The two primary analytical components are exposure analysis, which explores the predicted scale and spatiotemporal relationships of causative risk factors, and effects analysis, which explores the response of HVRAs to varying levels of the risk factors (Fairbrother & Turnley, 2005). Risk characterization integrates information from exposure analysis and effects analysis to formulate a conclusion about risk. The ability to characterize risk in a common metric facilitates the integration of multiple HVRAs and allows for economic analysis of management alternatives on the basis of cost-effectiveness, although challenges exist especially for non-market resources (Chuvieco et al., 2010; Venn & Calkin, 2011).

Assessing wildfire risk requires an understanding of the likelihood of wildfire interacting with valued resources, and the magnitude of potential beneficial and negative effects to resources from fire (Finney, 2005). In the above formulation, the components required to estimate wildfire risk are wildfire hazard maps generated from wildfire simulation models, HVRA maps, and characterization of fire effects to HVRAs. Exposure analysis intersects mapped HVRAs with spatially-explicit measures of wildfire hazard (burn probability and conditional fire intensity). Effects analysis quantitatively defines the response of the HVRA to wildfire hazard, in this case using response functions. Collectively exposure and effects analysis characterize risk to the HVRAs in question, which can be analyzed separately or aggregated using valuation techniques and/or multi-criteria decision analysis (Thompson & Calkin, 2011).

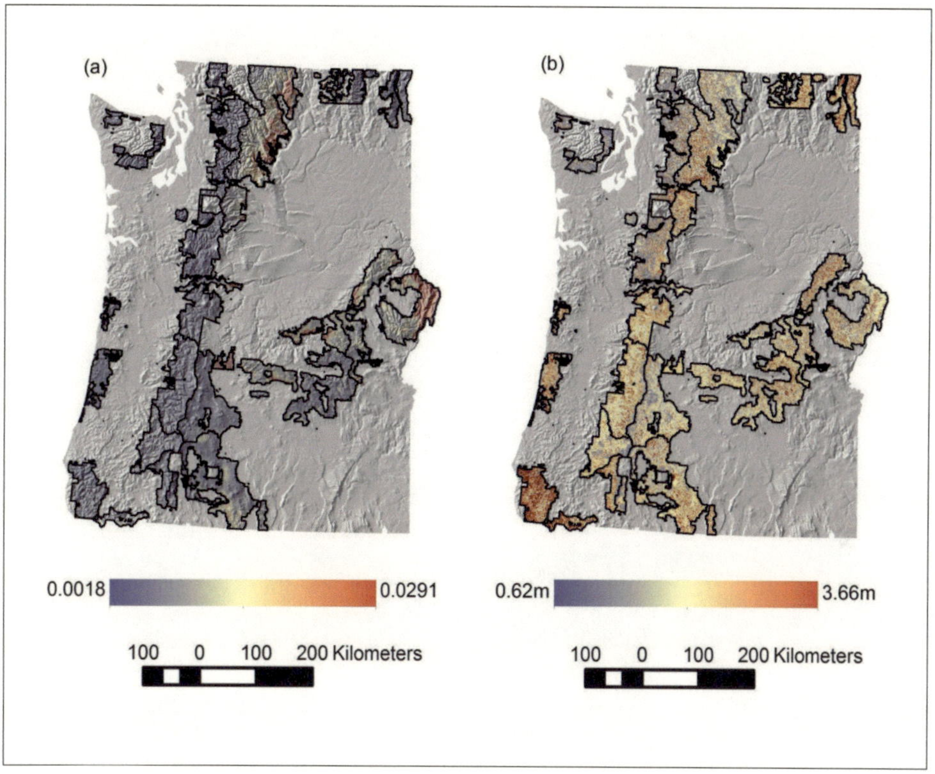

Fig. 2. Burn probability (a) and conditional flame length (b) for National Forests in the states of Oregon and Washington, in the Pacific Northwest of the United States. Figure from (Ager et al., submitted).

Figure 3 presents our conceptual model for assessing wildfire risk combining exposure and effects analysis. Here an integrated assessment is illustrated, using a representative set of HVRAs (air quality, wildlife habitat, municipal watersheds, and human communities) for which federal agencies manage. Equation 2 presents the mathematical formulation for calculating risk (Finney, 2005), where $E(NVC_i)$ is the expected net value change to resource j, and RF_i and is a "response function" for resource j as a function of fire intensity i and a vector of geospatial variables X_i that influence fire effects to resource j.

$$E(NVC_j) = \Sigma BP_i RF_j(i,X_j) \quad (2)$$

This framework quantifies risk in terms of relative net value change (NVC), or the percentage change in initial value resulting from interaction with fire. That is, response functions address relative rather than absolute change in resource or asset value. Response functions translate fire effects into NVC to the described HVRA. In response functions illustrated in Figure 3 NVC is based on fire intensity, which is a robust fire characteristic that integrates fuel consumption and spread rate, and is often used to estimate fire effects (Thompson et al. 2011a; Ager et al. 2007). In Figure 3 the response function varies according to categorical fire intensity levels, although the framework is perfectly amenable to definition of multivariate response functions incorporating additional geospatial information that influence response to fire (see Section 4).

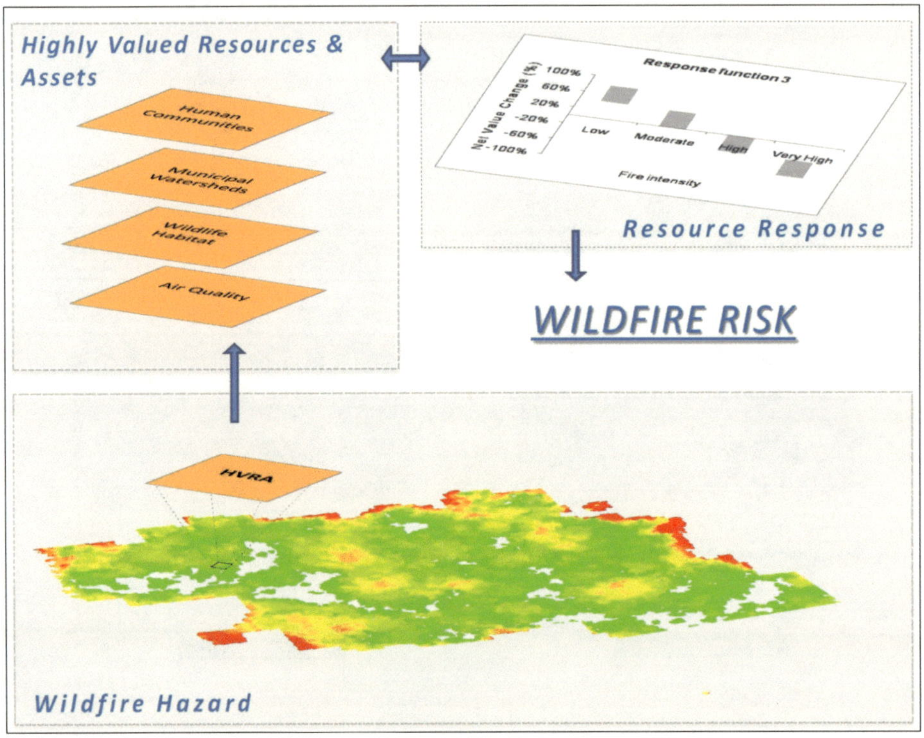

Fig. 3. Conceptual model for calculating wildfire risk (Modified from Calkin et al., 2010).

Characterizing fire effects has presented a major challenge to risk assessment, due to limited understanding of the spatiotemporal dynamics of ecological changes wrought by wildfire. Many past analyses focused on wildfire risk to commercial values, such as commercial timber (Konoshima et al. 2008), with a much more limited set focusing on broader non-market resource values and public infrastructure (Venn & Calkin 2011). There exist a variety of models that can estimate first-order fire effects such as tree mortality, soil heating, fuel consumption, and smoke production, although managers are generally more concerned with second- and third-order effects such as air quality, water quality, and habitat degradation (Reinhardt & Dickinson 2010). The management context, availability of appropriate models, and quality of spatial data will inform selection of the appropriate fire effects modeling approach (Reinhardt et al. 2001). In the absence of fire effects models, a reliance on local knowledge by resource managers is a common substitute for formal effects analyses.

3. Applications of burn probability modeling & exposure analysis

The design and functionality of simulation-based approaches span a range of intended applications, from modeling a specific fire event given an ignition to projecting wildfire likelihood and intensity at landscape scales across multiple fire seasons. Advances in burn probability modeling have enabled increasing sophistication and analytical rigor across a variety of wildfire management applications. Researchers and practitioners are able to, for instance, project near-term fire behavior using real-time weather information (Andrews et al., 2007) or to project wildfire behavioral changes in response to fuel treatments (Kim et al., 2009). In this section we focus on application of burn probability modeling and exposure analysis to support management of wildfire incidents and to support proactive hazardous fuels reduction treatments.

3.1 Incident management

Development of suppression strategies for escaped wildland fires is subject to considerable uncertainty and complexity. Factors to balance include likely weather and fire behavior, topography, firefighter safety, and the availability and productivity of firefighting resources (ground crews, fire engines, air tankers, etc.). Of particular importance is the ability to project where and under what conditions fire is likely to interact with HVRAs. This information can help fire managers decide where aggressive fire suppression may be effectives to protect HVRAs, and where fires may have a positive impact in fire-prone ecosystems.

In the United States, all wildfires occurring on federal lands are cataloged within the Wildland Fire Decision Support System (WFDSS). WFDSS provides decision documentation and analysis functionality to describe the fire incident, create objectives and requirements, develop a course of action, validate key dependencies and evaluate risks (Noonan-Wright et al., 2011). The system combines a suite of fire behavior predictions with identification and quantification of values at risk to inform incident management considering safety, complexity, economics, and risk (Calkin et al., 2011).

The two primary risk-based analytical components within WFDSS are the Fire Spread Probability model (FSPro) and the Rapid Assessment of Values at Risk (RAVAR). FSPro calculates the probability of fire spread from a current fire perimeter or ignition point, for a specified time period. Burn probability maps are derived from simulating fire growth for thousands of statistically generated weather scenarios (Finney et al., 2011b). As implemented in WFDSS burn probabilities are mapped as probability zones, or contours, of similar burn probability; exterior contours have lower probability of fire occurrence than interior contours.

Figure 4 displays an FSPro analysis for the SQF Canyon Fire, a human-caused fire that ignited on September 20, 2010 in California in the Sequoia National Forest. The figure provides a 7-day projection of fire growth as of September 14, 2010. The fire spread probability contours, moving outward from the red center, correspond to intervals of >80%,

60-80%, 40-60%, 20-40%, 5-20%, 1-5%, and <1% of likely fire spread given the current fire location and perimeter.

The RAVAR analytic model produces two distinct map products and associated reports, inventorying mapped Critical Infrastructure (CI) and Natural and Cultural Resources (NCR). HVRAs identified in CI reports include private structures, recreation facilities, water supply systems, major power lines, pipelines, communication towers, and hazardous waste sites. NCR products focus on regionally identified natural resources and wildland management priorities, such as sensitive wildlife habitat and restoration priority areas. Table 1 provides example tabular RAVAR output quantifying the number and value of structures at risk according to FSPro Fire Spread Zones, using county tax records.

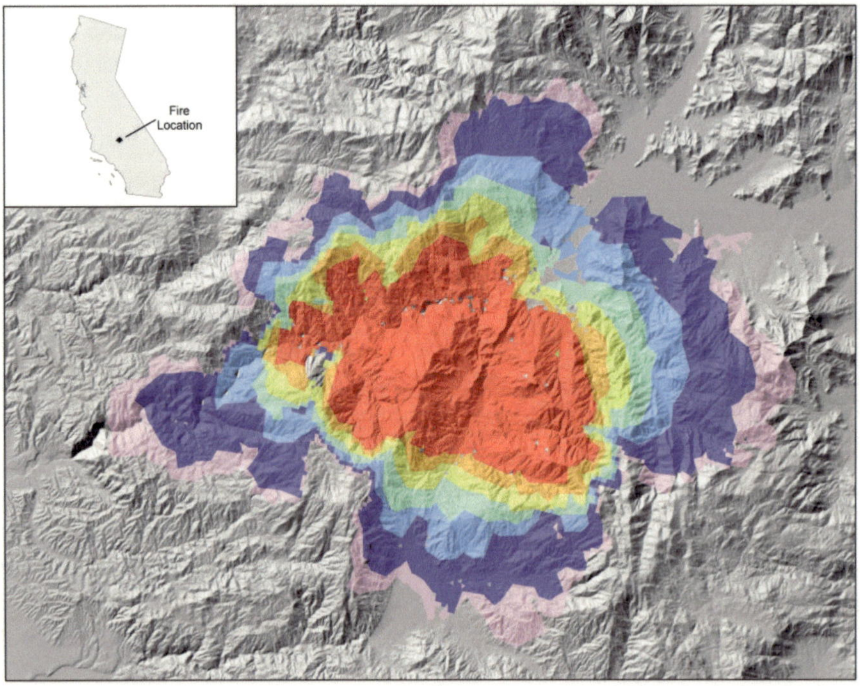

Fig. 4. FSPro run for the Canyon Fire in the Sequoia National Forest, California.

Figure 5 displays a close-up view of an FSPro-RAVAR analysis for the SQF Canyon Fire, which overlays geospatial identification of Critical Infrastructure on top of probability contours. (RAVAR maps are approximately 4′ x 3′ and are intended for poster display, generally making detailed displays on computer screens difficult.) The current fire perimeter is outlined in red, overlaid on top of associated probability contours of likely spread (see Figure 4). Threatened resources include private structures (black triangles), federal structures (green triangle), power transmission lines (inverted "T", dashed connector), and mine sites (pick and shovel). The green line demarcates the National Forest boundary, and yellow/red dots identify "hot" points from satellite images.

Together FSPro and RAVAR provide state-of-the-art exposure analysis, linking near real time probabilistic fire spread predictions with values at risk These analytical products inform managers regarding the likelihood of fire impacting HVRAS and assist in developing target fire containment perimeters. WFDSS supports risk-informed decision making by analyzing HVRA exposure to fire, allowing local managers to evaluate the likely impacts and prioritize suppression efforts accordingly.

	Acres Threatened		Kern County			
Fire Spread Probability Zone	**Acres by Zone**	**Cumulative Acres**	**Count by Zone**	**Cumulative Zone**	**Value by Zone**	**Cumulative Value**
> 80 %	47,894	47,894	290	290	$43,399,080	$43,399,080
60 - 80 %	12,029	59,923	215	505	$32,175,180	$75,574,260
40 - 60 %	14,062	73,985	289	794	$43,249,428	$118,823,688
20 - 40 %	15,602	89,586	208	1,002	$31,127,616	$149,951,304
5 - 20 %	24,995	114,582	297	1,299	$44,446,644	$194,397,948
1 - 5 %	53,989	168,571	794	2,093	$118,823,688	$313,221,636
Expected Value (without suppression)		67,980		679		$101,665,338

Table 1. Estimates of Structure Values at Risk, as output by WFDSS-RAVAR, using data from Kern County, California.

3.2 Hazardous fuels management

Fuel management seeks to alter the quantity, spatial arrangement, structure, and continuity of fuels so as to induce desirable changes in fire behavior. Broadly speaking, fuel management activities are designed to reduce the risk of catastrophic fire, protect human communities, reduce the extent and cost of wildfires, and restore fire-adapted ecosystems. For a fuel treatment to function effectively it must first spatially interact with an actual wildfire, and second mitigate fire behavior according to design objectives (Syphard et al., 2011).

Recognized principles for fuels management planning (Agee & Skinner, 2005) largely relate to individual treatments and their effects on localized fire behavior. Less understood is how in aggregate fuel treatments can affect landscape-scale processes of fire spread (Hudak et al., 2011). Prospective evaluation the influence of fuel treatments requires the estimation of altered fire behavior both within and outside of treated areas (Finney et al., 2007). Spatial fire growth models and burn probability modeling have emerged as useful tools for analyzing the influence of fuel treatments on topological fire spread, and to enable risk-based analysis of fuel treatment effectiveness.

A workflow for fuel treatment planning includes identifying the purpose and need for treatments, simulating wildfire behavior across the current, untreated landscape to characterize hazard and risk, developing treatment strategies, and iteratively simulating and evaluating changes to wildfire hazard and risk stemming from the treatment. Primary variables comprising a treatment strategy include the size of individual treatment units, the placement/pattern of the treatments, the proportion of the landscape treated, and treatment longevity (Collins et al., 2010).

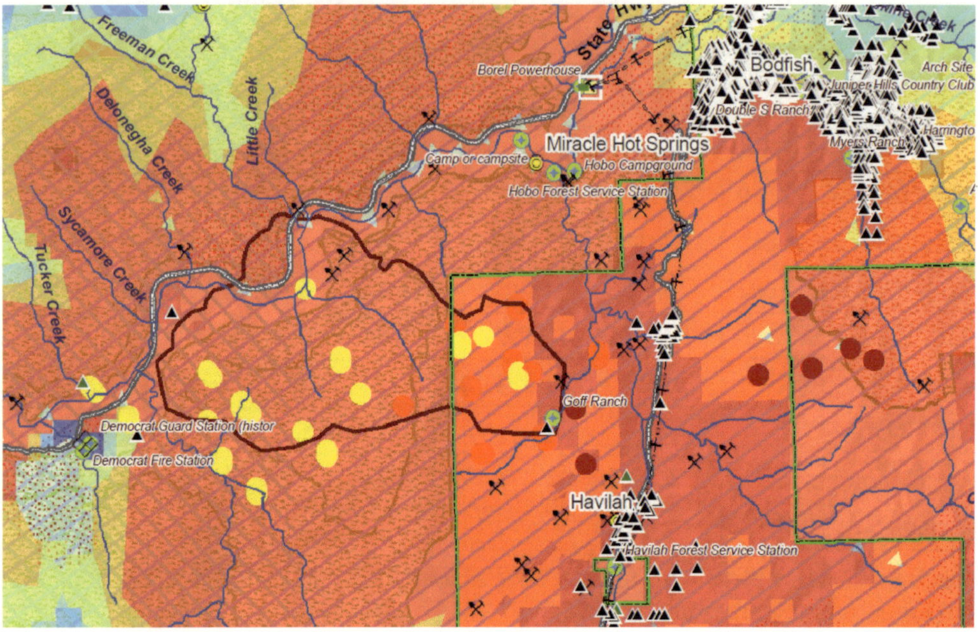

Fig. 5. Detail of RAVAR analysis for the Canyon Fire in Sequoia National Forest, California.

Ager et al. (2011) reviewed the development and use of ArcFuels, an integrated system of tools to design and test fuel treatment programs within a risk assessment framework. A number of fuel treatment case studies have employed the same basic analytical approach of comparative burn probability and intensity modeling across untreated/treated landscapes (Ager et al., 2010; Parisien et al., 2007). Figure 6 illustrates such a case study that investigated the influence of different treatment strategies on burn probability. Four scenarios, representing treating 0%, 10%, 20%, and 50% of the landscape were fed into wildfire simulation models to estimate impacts to burn and intensity probabilities.

In addition to evaluating prospective fuel treatments and informing treatment design, burn probability modeling can also be used to evaluate the effectiveness of previously implemented treatments. Field-based evaluations of fuel treatments have relied on the relatively rare occurrence of wildfires interacting with treatments. Of these treatments that have engaged wildfire, few have been subject to rigorous review to characterize treatment effectiveness (Hudak et al., 2011). Only recently has it been possible to estimate the spatial probabilities of landscape burning as a function of extant fuels treatments for real wildland fire-affected landscapes (Cochrane et al., in press). Figure 7 displays an example of burn probability modeling to analyze the impact of implemented treatments and their engagement with the School Fire. Rather than simulating the impacts of hypothetically implemented treatments, as in Figure 6, here the analysis simulates hypothetical fuel conditions had treatments not been implemented. The actual fire perimeter is outlined in red, and probability zones reflect contours of likely fire spread as output from wildfire simulations, had the treatments not been implemented. Areas of positive probability (yellow, orange, red) reflect that the treatments were effective in preventing spread. Exposure analysis intersects probability zones with mapped HVRAs including US Forest Service structures, improved structures (identified from county tax records), and bull trout (*Salvelinus confluentus*) critical habitat. Quantification of reduced exposure can inform estimates of fuel treatment effectiveness.

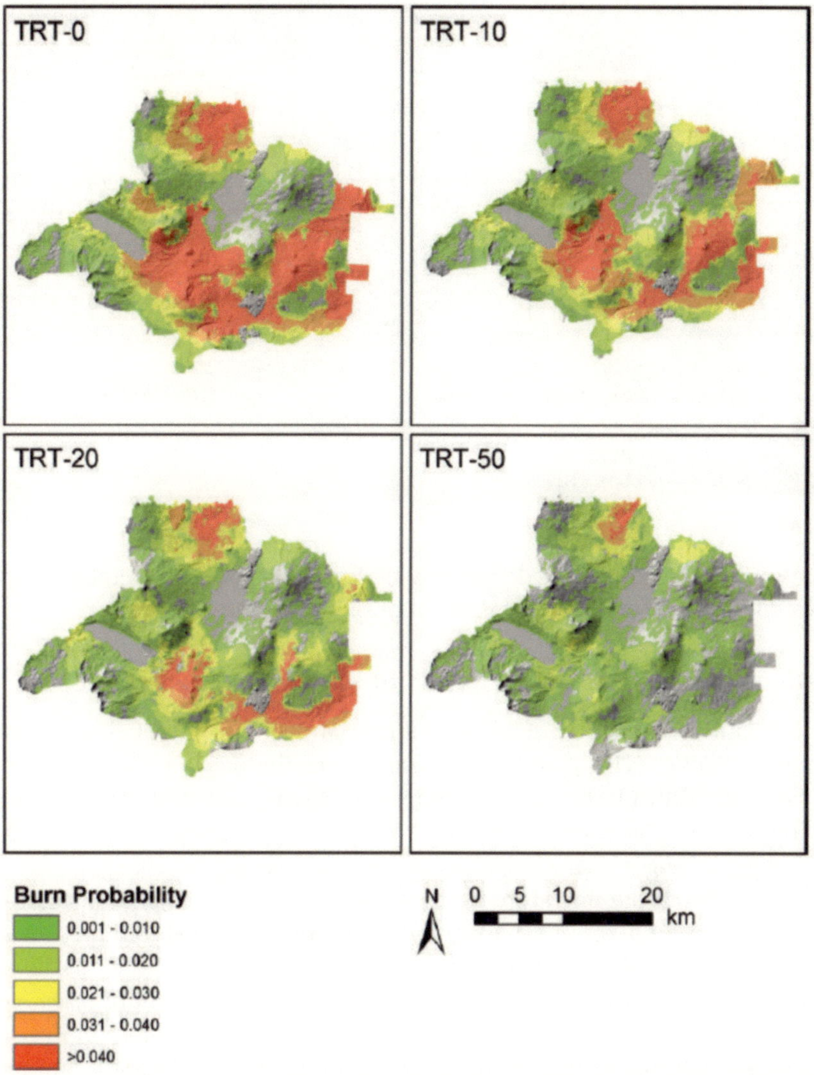

Fig. 6. Illustration of reductions in burn probability as a function of percent of the landscape treated (Ager et al., 2007). "TRT-X" refers to different modeled scenarios in which X percent of the landscape is treated.

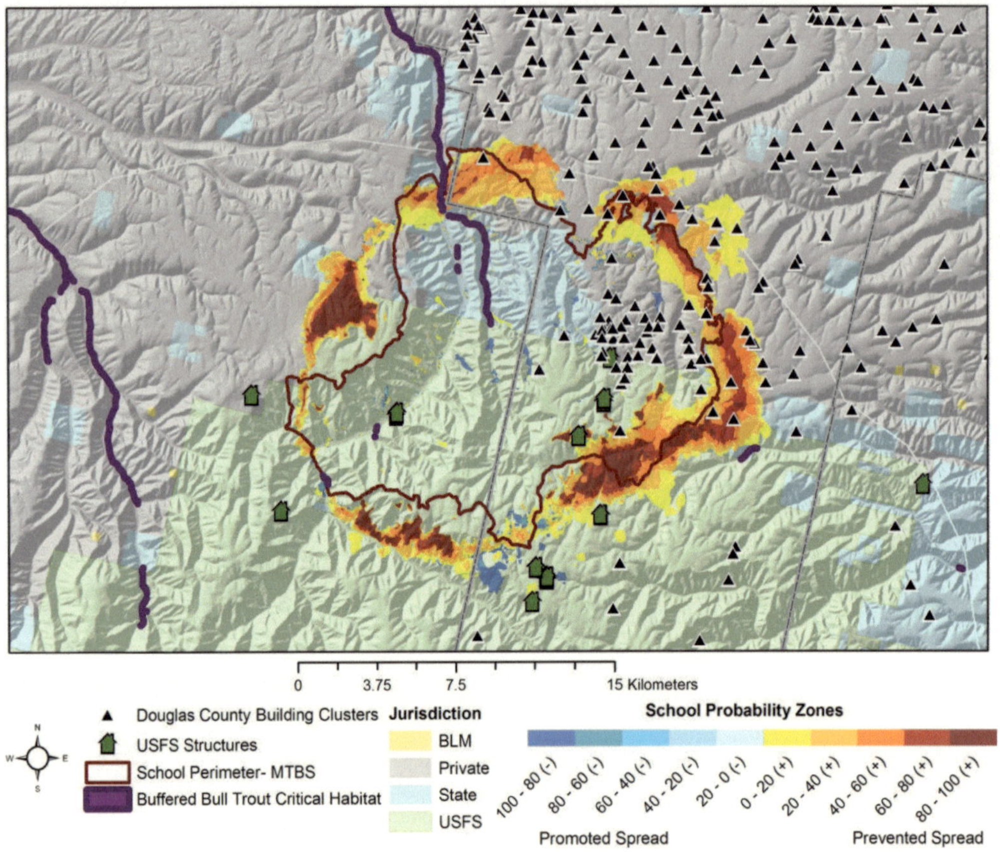

Fig. 7. Analysis for the School Fire demonstrating the impact of implemented fuel treatments. Modified from (Cochrane et al., in press).

4. Fire effects analysis & incorporation of expert judgment

Estimating resource response to wildfire is a crucial step for quantitative risk assessment (Fairbrother & Turnley, 2005), and yet is also one of the most challenging steps. Effects analysis is made difficult by the scientific uncertainty and lack of data/information surrounding wildfire effects on non-market resources; specifically in that limited scientific understanding challenges characterization of marginal ecological changes, and further in that economic methods are immature for broad scale monetization of such changes (Keane & Karau 2010; Venn & Calkin, 2011). Expert systems are commonly used in natural resource management decision-making (González et al., 2007; Hirsch et al., 2004; Kaloudis et al., 2005; Vadrevu et al., 2010), and rely on the unique expertise and judgment of professionals as a proxy for empirical data. Increasingly in a variety of natural resource management applications researchers and practitioners are adopting structured approaches for eliciting and using expert knowledge (Kuhnert et al., 2010; Martin et al., 2009). Elicitation of expert judgment is particularly useful where decisions are time-sensitive and management or policy cannot wait for improved scientific knowledge (Knol et al., 2010).

Structured Elicitation Process

1. Articulate research question
2. Identify and characterize uncertainties
3. Resolve scope and format of elicitation
4. Select experts
5. Design elicitation protocol
6. Prepare elicitation protocol
7. Elicit expert judgment
8. Feedback

Fig. 8. Eight major steps in organizing and implementing a structured elicitation of expert judgment. Modified from (Knol et al., 2010; Kuhnert et al., 2010)

Figure 8 presents a structured process for eliciting expert judgment. In the first step, a clear articulation of the research question(s) will inform the design and implementation of the study as well as the larger structure of the modeling process. This is followed by identification and characterization of uncertainties, which will influence choices regarding the type of experts and elicitation format. Resolving the scope and format of the elicitation entails identifying the number of experts to engage and the nature of the engagement (interview, group workshop, survey distribution, etc.), while considering available resources and other constraints. Selection of experts includes choices between generalists, subject matter experts, and normative experts (those with experience to support elicitation itself). Design of the protocol considers the type of information to be elicited, the most appropriate metric(s), the most appropriate elicitation mechanism, and how to clearly communicate information needs to avoid issues of linguistic uncertainty (Regan et al., 2002). Preparation of the elicitation protocol includes providing selected experts with sufficient information on the nature of the research problem and associated uncertainties, the scope and purpose of the elicitation, and the nature of the elicitation procedure itself. Lastly, the elicitation procedure is implemented, with opportunities for post-elicitation feedback and revision.

In terms of the wildfire management context, the research question generally involves assessing wildfire risk to HVRAs, potentially in a comparative risk framework to evaluate the effectiveness of alternative management actions (Figure 1). In the second step, wildfire management is subject to myriad sources of uncertainty, not all of which are necessarily best handled with expert judgment (burn probability modeling to capture environmental stochasticity, e.g.). Thompson & Calkin (2011) present a typology of uncertainties faced in wildfire management, and identify that with regard to knowledge uncertainty surrounding fire effects, expert systems are perhaps the most appropriate approach. In our past experience we have adopted group workshops, and assembled resource scientists (hydrologists, soil scientists, wildfire biologists, fire ecologists, etc.) as appropriate given the HVRAs being assessed (Thompson et al., 2011b). The elicitation protocol identifies response functions that quantitatively characterize resource-specific response functions as a function of fire intensity, and response functions are iteratively explored, justified, and updated until expert consensus is reached.

Figure 9 illustrates expert-based response functions for two HVRAs with contrasting response to fire, mapped across six fire intensity level (FIL) classes. These response functions were assigned in a group workshop format as part of a broader wildfire risk assessment conducted for the Lewis and Clark National Forest in Montana, United States. Aspen stands are highly valued because they provide habitat for a broad diversity of wildlife, and due to their relative rarity across the landscape. Aspens are reliant on wildfire for natural regeneration, and so are modeled as experiencing substantial benefit from fire at low to moderate intensities, with minor loss expected high intensity fires. For high investment infrastructure (e.g., developed campgrounds, cabins, ranger stations), damages are expected from any interaction with fire, and are expected to increase in severity as fire intensity increases.

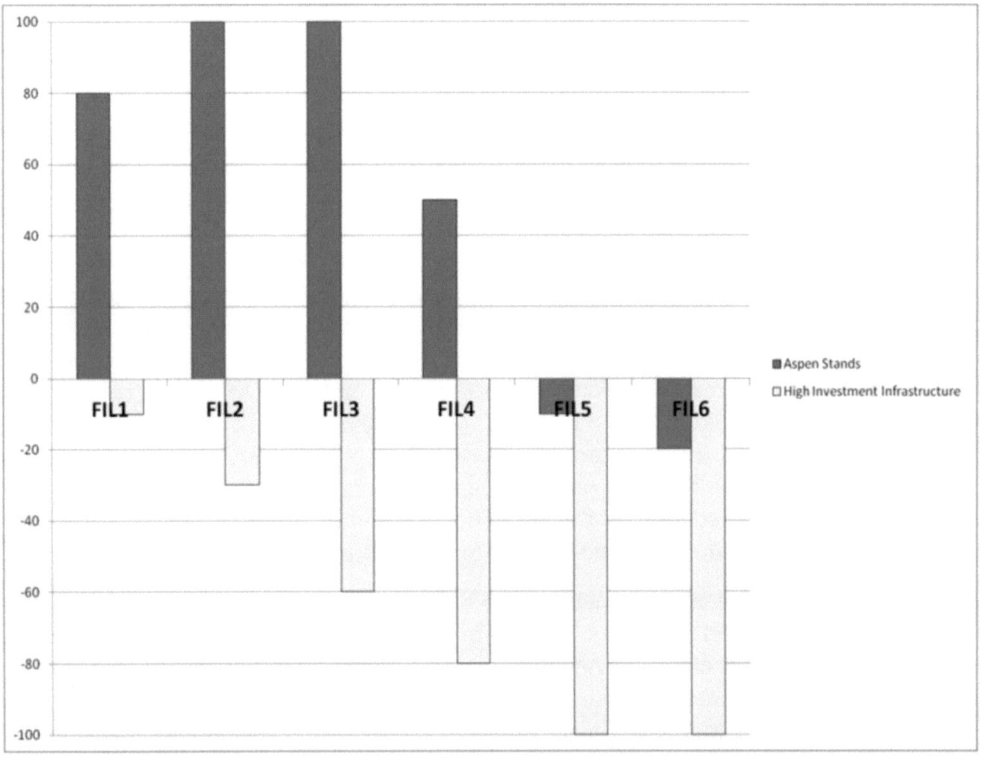

Fig. 9. Response functions plotting relative NVC (y-axis) against fire intensity level (FIL; x-axis), for stands of aspen and high investment infrastructure.

Figure 10 displays additional response functions identified as part of the wildfire risk assessment for the Lewis and Clark National Forest. This figure highlights use of an additional geospatial variable, in this case moisture conditions on the site, to further refine fire effects estimates to old growth (OG) forest stands. Dry forests typically have evolved with and tend to receive a benefit from low to moderate intensity fires. At extreme intensities, high levels of mortality and damage are expected irrespective of site moisture conditions.

5. Case study of wildfire risk

In this section we briefly review a recently published example of integrated, national-scale wildfire risk assessment (Thompson et al., 2011b). The effort leveraged tools, datasets, and expertise of the Fire Program Analysis (FPA) system, a common interagency strategic decision support tool for wildland fire planning and budgeting (http://www.fpa.nifc.gov). We aggregated results according to eight geographic areas organized largely for purposes of incident management and mobilization of suppression resources: California (CA), Eastern Area (EA), Great Basin (GB), Northern Rockies (NR), Northwest (NW), Rocky Mountain (RM), Southern Area (SA), and Southwest (SW).

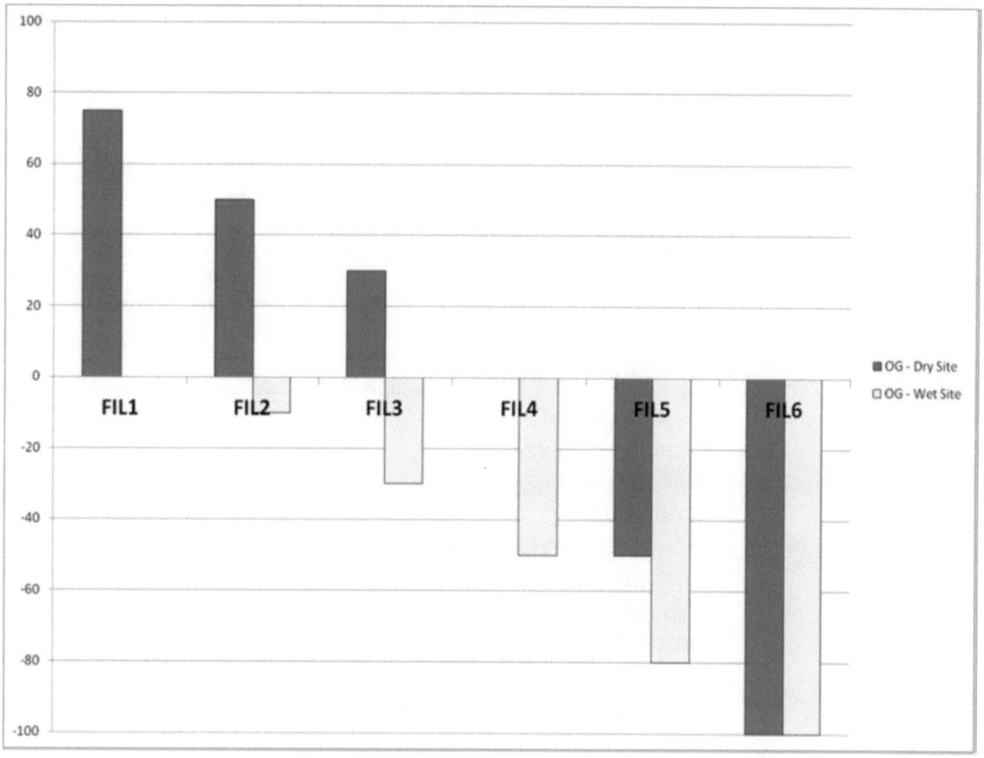

Fig. 10. Response functions plotting relative benefit/loss (y-axis) against fire intensity level (FIL; x-axis), for the Old Growth (OG) HVRA, sorted by dry/wet site.

To map wildfire hazard we used wildfire simulation outputs (burn probability and intensity) from the FSim large fire simulator (Finney et al., 2011a), mapped at on a pixel basis (270m x 270m). In cooperation with the FPA Executive Oversight Group we identified seven key HVRA themes: residential structure locations, municipal watersheds, air quality, energy and critical infrastructure, federal recreation and recreation infrastructure, fire-susceptible species, and fire-adapted ecosystems. Table 2 delineates the major HVRA themes along with identified sub-themes. We engaged ten fire and fuels program management officials from the Forest Service, National Park Service, Bureau of Land Management, Fish and Wildlife Service, and the Bureau of Indian Affairs to facilitate response function assignments.

Response functions indicated percentage NVC according to fire intensity category, as measured by flame length. As a consistent measure of NVC across HVRAs we used an area-based proxy called Total Change Equivalent (TCE). TCE effectively measures the equivalent area lost (or gained) for a particular HVRA. Since mapped pixels can support multiple HVRA layers, generation of risk estimates entailed geospatial computations for each pixel-HVRA layer combination.

HVRA Category	**HVRA Layer**	**HVRA Value Category**
Residential structure location	Low density built structures	High
	Moderate and high density built structures	Very High
Municipal watersheds	6th order Hydrologic Unit Codes	Very High
Air quality	Class I areas	Moderate
	Non-attainment areas for PM 2.5 and Ozone	Very High
Energy infrastructure	Power transmission lines Oil and gas pipelines Power plant locations Cellular tower locations	High
Recreation infrastructure	FS campgrounds FS ranger stations BLM recreation sites and campgrounds NPS visitor services and campgrounds FWS recreation assets National scenic and historic trails National alpine ski area locations	High
Fire-susceptible species	Designated critical habitat National sage-grouse key habitat	High
Fire-adapted ecosystems	Fire-adapted regimes	Moderate

Table 2. HVRA layers used in national risk assessment Modified from (Thompson et al., 2011b).

Although calculating TCE in a common area-based measure does facilitate integration of multiple HVRAs and the evaluation of alternative mitigation strategies on the basis of cost-effectiveness, TCE does not capture management priorities across HVRAs. To better integrate TCE calculations we turned to multi-criteria decision analysis techniques to assign each HVRA an importance weight. First, we adopted a categorical approach using input from the fire and fuels program management officials consulted for assistance with fire effects analysis. With guidance from the experts we assigned each HVR to one of three value categories: Moderate, High, and Very High. HVRAs assigned to the Very High category related to human health and safety, specifically concerns regarding air quality, water quality, and communities at risk. We then aggregated TCE results into a single weighted risk metric (wTCE) by assuming that the ranking of value categories maintained a simple proportional relationship. With this framework a (1, 3, 9) weight vector means that HVRAs assigned in the Very High value category are 3 times as important as resources in the High value category, which in turn are 3 times as important as resources in the Moderate value category. Clearly decision-makers can experiment with alternative value category and weight vector assignments, but our purposes were primarily to illustrate joint application of multi-criteria decision analysis and risk assessment.

Table 3 summarizes TCE values by HVRA, value category, and geographic area. In the Moderate value category the Southern Area (SA) presents the greatest risk, largely to Class I areas and concerns about air quality. Across all geographic areas fire-adapted ecosystems expect to see a benefit from fire, which on balance tend to outweigh losses to other HVRAs, leading to positive values for NVC. Within the High value category fire-susceptible species were the largest contributors to risk. The Southern Area contained the largest overall area of risk to energy infrastructure, with relatively low loss expected elsewhere. Low density built structures similarly had relatively low TCE values, with higher losses associated with the Southern Area, California, and the Southwest. Within the Very High value category non-attainment areas were by far the largest contributors to risk, and especially in California. Overall California presents the largest risk in the Very High value category, followed by the Southern Area. Lastly the bottom row presents weighted TCE (wTCE) values using the (1, 3, 9) weight vector. Consistent with results from the Very High value category, California and the Southern Area appear most susceptible to wildfire-related losses. The Great Basin ranks third, due primarily to extensively mapped sage grouse habitat. Thompson et al. (2011b) present additional results including sensitivity analysis of assigned relative importance weights, and refinements regarding the temporal effects of air quality degradation and the spatial extent of mapped habitat.

HVRA Value Category	**Geographic area**							
	CA	**EA**	**GB**	**NR**	**NW**	**RM**	**SA**	**SW**
Moderate	-0.58	0.62	3.18	1.53	5.34	0.63	-4.86	2.54
High	-9.91	-1.72	-32.99	-12.70	-19.40	-7.44	-9.54	-5.57
Very High	-55.53	-2.29	-2.54	-1.25	-1.70	-1.24	-14.44	-5.97
wTCE Totals (1, 3, 9)	**-530.11**	**-25.12**	**-118.68**	**-47.82**	**-68.12**	**-32.82**	**-163.41**	**-67.86**

Table 3. Total change equivalent (TCE) in thousands of hectares for each Geographic area and Value Category.

In summary, the case study briefly explored here demonstrates application of quantitative wildfire risk assessment. The approach is scalable, in that the same integration of burn probability maps, geospatial identification of HVRAs, and resource response functions can be applied at project-level to regional to national analyses. A number of improvements can and are being pursued, such as refining the fire simulation outputs, identifying a larger and more representative set of HVRAs, introducing more structure and engaging more experts to define response functions, and using more complex multi-criteria decision analysis methods to articulate relative importance across HVRAs.

6. Conclusion

Combining quantitative fire effects analysis with burn probability and intensity maps allows for a quantitative, actuarial representation of risk in a spatial context. Risk assessment can inform the spectrum of wildfire management activities, from real-time management of incidents to proactive fuels management to reduce losses from future fires. Comparative risk assessment enables the exploration of tradeoffs across alternative investments in prevention, fuels management, and suppression response capacity, and ideally will lead to improved efficiency in pre-suppression and suppression planning. The framework we promoted here aligns with previously established ecological risk assessment frameworks, and is increasingly being adopted for federal wildfire management in the United States. The framework can be consistently applied across planning scales, is objective, repeatable, probabilistic, and spatially-explicit. A great strength is the flexibility of the framework, in that analysts can adopt alternative approaches to characterize wildfire hazard, to characterize fire effects and response functions, and further to use alternative weighting schemes to integrate risk calculations across HVRAs. A further strength is the scalability of the framework, which can be applied from project-level planning to strategic, nation-wide analysis.

Despite the strengths of this approach there remain limitations and challenges to address. Understanding current risk is not the same as projecting future risk, which requires prediction of changes in vegetation from natural growth and from other disturbances, changes in demographics and development patterns that could expose more human communities to wildfire risk, the dynamic feedbacks of wildfires changing landscape conditions, and the influences on fire regimes of a changing climate. Characterizing risk is a necessary but not sufficient component to developing, selecting, and implementing mitigation strategies. Information about management opportunities, treatment costs, and their relation to risk factors needs to be considered, as does uncertainty related to science delivery and policy direction.

A number of promising extensions to the work presented in this chapter exist. Embedding geospatial wildfire risk analysis within optimization algorithms could inform multiple applications, for instance pre-positioning aerial firefighting resources, initial attack home base locations and dispatch strategies, fuels and vegetation management, and incident management. Fuel management in particular is a promising avenue for spatially explicit optimization approaches. Increasing use of expert systems plus appropriate fire effects models will enable improved estimates of the consequences of wildfire. Increasing use of multi-criteria decision analysis techniques will enable integrated assessments of risk across social and ecological values, and will facilitate prioritization efforts. Non-market valuation studies could further assist prioritization efforts and articulation of management tradeoffs. One very important, and highly uncertain, topic is the consideration of future wildfire risk as a function of contemporary management, land use patterns, vegetative succession and disturbance, and, importantly, climate change.

7. Acknowledgement

The authors wish to recognize and thank Julie Gilbertson-Day, Mark Cochrane, Anne Birkholz, Jon Rieck, Joe Scott, and Don Helmbrecht for various contributions to figures and tables presented in this figure. The lead author is grateful for support of the Rocky Mountain Research Station and the National Fire Decision Support Center.

8. References

Agee, J.K. & Skinner, C.N. (2005). Basic principles of forest fuel reduction treatments. *Forest Ecology and Management,* Vol. 211, No. 1-2, pp. 83-96, ISSN 0378-1127

Ager, A.A., Finney, M.A., Kerns, B.K., & Maffei. H. (2007). Modeling wildfire risk to northern spotted owl (*Strix occidentalis caurina*) habitat in Central Oregon, USA. *Forest Ecology and Management,* Vol. 246, No. 1, pp. 45-56, ISSN 0378-1127

Ager, A.A., Vaillant, N.M., & Finney, M.A. (2010). A comparison of landscape fuel treatment strategies to mitigate wildland fire risk in the urban interface and preserve old forest structure. *Forest Ecology and Management,* Vol. 259, No. 8, pp. 1556-1570, ISSN 0378-1127

Ager, A.A., Vaillant, N.M., & Finney, M.A. (2011). Integrating Fire Behavior Models and Geospatial Analysis for Wildland Fire risk Assessment and Fuel Management Planning. *Journal of Combustion.* Doi: 10.1155/2011/572452, ISSN 2090-1976

Ager, A.A., Finney, M.A., Reger, A., & Buonopane, M. (Submitted) Wildfire exposure to analysis on the national forests in the Pacific Northwest, USA. *Risk Analysis,* ISSN 0272-4332

Andrews, P., Finney, M., & Fischetti, M. (2007). Predicting Wildfires. Scientific American Vol. August, pp. 47-55, ISSN 0036-8733

Brillinger, D.R., Autrey, B.S., & Cattaneo, M.D. (2009). Probabilistic risk modeling at the wildland urban interface: the 2003 Cedar Fire. *Environmetrics* Vol. 20, pp. 607-620, ISSN 1180-4009

Bruins, R.J.F., Munns, Jr., W.R., Botti, S.J., Brink, S., Cleland, D., Kapustka, L., Lee, D., Luzadis, V., McCarthy, L.F., Rana, N., Rideout, D.B., Rollins, M., Woodbury, P., & Zupko, M. (2010). A New Process for Organizing Assessments of Social, Economic, and Environmental Outcomes: Case Study of Wildland Fire Management in the USA. *Integrated Environmental Assessment and Management* Vol. 6, No. 3, pp. 469-483, ISSN 0730-7268

Calkin, D., Ager, A.A., & Thompson, M.P. (2011). A comparative risk assessment framework for wildland fire management: the 2010 cohesive strategy science report. Gen. Tech. Rep. RMRS-GTR-262. Fort Collins, CO: U.S. Department of Agriculture, Forest Service, Rocky Mountain Research Station. 63 p..

Carmel, Y., Paz, S., Jahashan, F., & Shoshany, M. (2009). Assessing fire risk using Monte Carlo simulations of fire spread. *Forest Ecology and Management* Vol. 257, No. 1, pp. 370-377, ISSN 0378-1127

Chuvieco, E., Aguado, I., Yebra, M., Nieto, H., Salas, J., Martín, M.P., Vilar, L., Martínez, J., Martín, S., Ibarra, P., de la Riva, J., Baeza, J., Rodríguez, F., Molina, J.R., Herrera, M.A., & Zamora, R. (2010). Development of a framework for fire risk assessment using remote sensing and geographic information system technologies. *Ecological Modelling* Vol 221, No 1, pp. 46-58, ISSN 0304-3800

Cochrane, M.A., C.J. Moran, M.C. Wimberley, A.D. Baer, M.A. Finney, K.L. Beckendorf, J. Eidenshink, & Z. Zhu. (In Press) Estimation of wildfire size and risk changes due to fuel treatments. *International Journal of Wildland Fire*, ISSN 1049-8001

Collins, B.M., Stephens, S.L., Moghaddas, J.J., & Battles, J. (2010). Challenges and Approaches in Planning Fuel Treatments across Fire-Excluded Forested Landscapes. *Journal of Forestry* Vol. 108, No. 1, pp. 24-31, ISSN 0022-1201

Cruz, M.G., & Alexander, M.E. (2010). Assessing crown fire potential in coniferous forests of western North America: a critique of current approaches and recent simulation studies. *International Journal of Wildland Fire*, Vol. 19, No. 4, pp. 377-398, ISSN 1049-8001

Fairbrother, A., & Turnley, J.G. (2005). Predicting risks of uncharacteristic wildfires: Application of the risk assessment process. *Forest Ecology and Management*, Vol. 211, No. 1-2, pp. 28-35, ISSN 0378-1127

Finney, M.A. (2005). The challenge of quantitative risk analysis for wildland fire. *Forest Ecology and Management*, Vol. 211, No. 1-2, pp. 97-108, ISSN 0378-1127

Finney, M.A., Seli, R.C., McHugh, C.W., Ager, A.A.; Bahro, B., & Agee, J.K. 2007. Simulation of long-term landscape-level fuel treatment effects on large wildfires. *International Journal of Wildland Fire*, Vol. 16, No. 6, pp. 712-727, ISSN 1049-8001

Finney, M.A., McHugh, C.W., Grenfell, I.C., Riley, K.L., & Short, K.C. (2011a). A Simulation of Probabilistic Wildfire Risk Components for the Continental United States. *Stochastic Environmental Research and Assessment*, Vol. 25, No. 7, pp. 973-1000, ISSN 1436-3240

Finney, M.A., Grenfell, I.C., McHugh, C.W., Seli, R.C., Tretheway, D., Stratton, R.D., & Brittain, S. (2011b) A Method for Ensemble Wildland Fire Simulation. *Environmental Modeling and Assessment* Vol. 16, No. 2, pp. 153-167, ISSN 1420-2026

González, J.R., Kolehmainen, O., & Pukkala, T. (2007). Using expert knowledge to model forest stand vulnerability to fire. *Computers and Electronics in Agriculture*, Vol. 55, No. 2, pp. 107-114, ISSN 0168-1699

Gude, P., Rasker, R., & Van den Noort, J. (2008) Potential for Future Development on Fire-Prone Lands. *Journal of Forestry*, Vol. 106, No. 4, pp. 198-205, ISSN 0022-1201

Hessburg, P.F., Reynolds, K.M., Keane, R.E., James, K.M., & Salter, R.B. (2007). Evaluating wildland fire danger and prioritizing vegetation and fuel treatments. *Forest Ecology and Management*, Vol. 247, No. 1-3, pp. 1-17, ISSN 0378-1127

Hirsch, K.G., J.A. Podur, R.D. Jansen, R.D. McAlpine & Martell, D.L.(2004). Productivity ofOntario initial attack fire crews: results of an expert-judgment elicitation study. *Canadian Journal of Forest Research*, Vol. 34, No. 3, pp. 705-715, ISSN 0045-5067

Hudak, A.T., Rickert, I., Morgan, P., Strand, E., Lewis, S.A., Robichaud, P.R., Hoffman, C., & Holden, Z.A. (2011). Review of fuel treatment effectiveness in forests and rangelands and a case study from the 2007 megafires in central, Idaho, USA. Gen.

Tech. Rep. RMRS-GTR-252. Fort Collins, CO: U.S. Department of Agriculture, Forest Service, Rocky Mountain Research Station. 60 p..

Kaloudis, S., Tocatlidou, A., Lorentzos, N.A., Sideridis, A.B., & Karteris, M. (2005). Assessing Wildfire Destruction Danger: a Decision Support System Incorporating Uncertainty. *Ecological Modelling,* Vol. 181, No. 1, pp. 25-38, ISSN 0304-3800

Keane, R.E., Drury, S.A., Karau, E.C., Hessburg, P.F., & Reynolds, K.M. (2010). A method for mapping fire hazard and risk across multiple scales and its application in fire management. *Ecological Modelling,* Vol. 221, No. 1, pp. 2-18, ISSN 0304-3800

Keane, R.E., & Karau, E. (2010). Evaluating the ecological benefits of wildfire by integrating fire and ecosystem simulation models. *Ecological Modelling,* Vol. 221, No. 8, pp. 1162-1172, ISSN 0304-3800

Kim, Y-H., Bettinger, P., & Finney, M. (2009). Spatial optimization of the pattern of fuel management activities and subsequent effects on simulated wildfires. *European Journal of Operational Research,* Vol. 197, No. 1, pp. 253-265, ISSN 0377-2217

Knol, A.B., Slottje, P., van der Sluijs, J.P., & Lebret, E. (2010). The use of expert elicitation in environmental health impact assessment: a seven step procedure. *Environmental Health,* Vol. 9, No. 19, pp. 1-16, ISSN 1476-069X

Konoshima, M., Montgomery, C.A., Albers, H.J., & Arthur, J.L. (2008). Spatial-Endogenous Fire Risk and Efficient Fuel Management and Timber Harvest. *Land Economics,* Vol. 84, No. 3, pp. 449-468, ISSN 0023-7639

Kuhnert, P.M., Martin, T.G., & Griffiths, S.P. (2010). A guide to eliciting and using expert knowledge in Bayesian ecological models. *Ecology Letters,* Vol. 13, No. 7, pp. 900-914, ISSN 1461-023X

Littell, J.S., McKenzie, D., Peterson, D.L., & Westerling, A.L. (2009). Climate and wildfire area burned in western U.S. ecoprovinces, 1916-2003. *Ecological Applications,* Vol. 19, No. 4, pp. 1003-1021, ISSN 1051-0761

Martin, J., Runge, M.C., Nichols, J.D., Lubow, B.C., & Kendall, W.L. (2009). Structured decision making as a conceptual framework to identify thresholds for conservation management. *Ecological Applications,* Vol. 19, No. 5, pp. 1079-1090, ISSN 1051-0761

Martínez, J., Vega-Garcia, C., & Chuvieco, E. (2009). Human-caused wildfire risk rating for prevention planning in Spain. *Journal of Environmental Management,* Vol. 90, No. 2, pp. 1241-1252, ISSN 0301-4797

Noonan-Wright, E., Opperman, T.S., Finney, M.A., Zimmerman, T., Seli, R.C., Elenz, L.M., Calkin, D.E., & Fiedler, J.R. (2011). Developing the U.S. Wildland Fire Decision Support System. *Journal of Combustion.* Doi: 10.1155/2011/168473, ISSN 2090-1976

Parisien, M.A., Junor, D.A., &Kafka, V.G. (2007). Comparing landscape-based decision rules for placement of fuel treatments in the boreal mixed wood of western Canada. *International Journal of Wildland Fire,* Vol. 16, No. 6, pp. 664-672, ISSN 1049-8001

Prasad, V.K., Badarinath, K.V.S., & Eaturu, A. (2007). Biophysical and anthropogenic controls of forest fires in the Deccan Plateau, India. *Journal of Environmental Management,* Vol. 86, No. 1, pp. 1-13, ISSN 0301-4797

Prestemon, J.P., D.T. Butry, K.L. Abt, & Sutphen, R. (2010). Net benefits of wildfire prevention education efforts. *Forest Science,* Vol. 56, No. 2, pp. 181-192, ISSN 0015-749X

Preisler, H.K., & Westerling, A.L. (2007). Statistical Model for Forecasting Monthly Large Wildfire Events in Western United States. *Journal of Applied Meteorology and Climatology,* Vol. 46, No. 7, pp. 1020-1030, ISSN 1558-8424

Reinhardt, E.D., Keane, R.E., & Brown, J.K. (2001). Modeling fire effects. *International Journal of Wildland Fire,* Vol. 10, No. 4, pp. 373-380, ISSN 1049-8001

Reinhardt, E.D., & Dickinson, M.B. (2010). First-order fire effects models for land Management: Overview and issues. *Fire Ecology,* Vol. 6, No. 1, pp. 131-142, ISSN 1933-9747

Scott, J.H. (2006). An analytical framework for quantifying wildland fire risk and fuel treatment benefit. In: Andrews, P.L., Butler, B.W. (Comps), Fuels Management-How to Measure Success: Conference Proceedings, March 28–30, Portland, OR. USDA Forest Service, Rocky Mountain Research Station Proceedings RMRS-P-41, pp. 169–184.

Syphard, A.D., Keeley, J.E., & Brennan, T.J. (2011). Factors affecting fuel break effectiveness in the control of large fires on the Los Padres National Forest, California. *International Journal of Wildland Fire,* Vol. 20, No. 6, pp. 764-775, ISSN 1049-8001

Thompson, M.P., Calkin, D.E., Gilbertson-Day, J.W., & Ager, A.A. (2011a). Advancing effects analysis for integrated, large-scale wildfire risk assessment. *Environmental Monitoring and Assessment,* Vol. 179, No. 1-4, pp. 217-239, ISSN 0167-6369

Thompson, M.P., Calkin, D.E., Finney, M.A., Ager, A.A., & Gilbertson-Day, J.W. (2011b). Integrated national-scale assessment of wildfire risk to human and ecological values. *Stochastic Environmental Research and Risk Assessment,* Vol. 25, No. 6, pp. 761-780, ISSN 1436-3240

Thompson, M.P., & Calkin, D.E. (2011). Uncertainty and risk in wildland fire management: a review. *Journal of Environmental Management,* Vol. 92, No. 8, pp. 1895-1909, ISSN 0301-4797

U.S. Environmental Protection Agency. (1998). Guidelines for Ecological Risk Assessment. US Environmental Protection Agency, Washington, DC, EPA/630/R-95/002F

Vadrevu, K.P., Eaturu, A., & Badarinath, K.V.S. (2010). Fire risk evaluation using multicriteria analysis - a case study. *Environmental Monitoring and Assessment,* Vol. 166, No. 1-4, pp. 223-239, ISSN 0167-6369

Venn, T.J., & Calkin, D.E. (2011) Accommodating non-market values in evaluation of wildfire management in the United States: challenges and opportunities. *International Journal of Wildland Fire,* Vol. 20, No. 3, pp. 327-339, ISSN 1049-8001

7

Theories and Methods for the Emergency Rescue System

Jianfeng Li[1,2], Wenmao Liu[2] and Bin Zhang[1]
[1]*School of Environment, Tsinghua University, Beijing,*
[2]*Institutes of Education, Tsinghua University, Beijing,*
[3]*Beijing Municipal Institute of Labor Protections,*
Beijing Academy of Science and Technology, Beijing,
P.R. China

1. Introduction

According to the "China State Plan for Rapid Response to Public Emergencies" (hereinafter referred to as "Plan"), which was published by the Central Government of the People's Republic of China, "public emergencies" refer to those emergencies that happened suddenly, and would (or might) cause heavy casualties and property loss, damage ecological environment, bring severe harms to our society and threat public safety. In the "Plan", public emergencies were divided into four categories: natural disasters, accidental disasters, public emergencies and social security events.

Since long time ago, the progress of human society has been achieved at the cost of deteriorating our living environment. Consequently, the number of natural or manmade disasters has been increasing. Earthquakes, floods, hurricanes, nuclear leakages, sudden outbreak of infectious diseases, fires and explosions attacked the human-beings one after another. For example, the Great Hanshin Earthquake in Japan in 1995, the "September 21" Earthquake in Taiwan in 1998, the "September 11" Terrorist Attack in US in 2001, the "August 14" Power Failure in US and Canada in 2003, and the disastrous Indian Ocean Tsunami in 2005, have brought severe losses to local economy, peoples' life and property.

As we all know, public emergencies, particularly natural disasters, are unavoidable. But we could reduce the loss of disasters to a minimum, or even eliminate the negative impact of disasters, by designing an appropriate emergency rescue system. For example, in 2005, the southern United States was attacked by Hurricane Katrina. The local government failed to allocate emergency resources in a timely manner. Consequently, the local people didn't have enough emergency supplies, such as food, drinking water, the necessities of life and medicine. Due to the severe shortage of emergency supplies, many disaster-stricken people resorted to violence. Riots occurred in many places, making the situation even worse.

Another example is the 7.6-magnitude earthquake happened on South Asian Subcontinent in October 2005. The disaster-stricken areas were faced with several problems: 1) Water supply was interrupted. The local residents didn't have food to eat. 2) Hospitals were shut down. The residents were in urgent need of medical care. 3) The traffic conditions were poor in disaster-stricken areas. The disaster-stricken people didn't have enough emergency resources to make their living. Consequently, they ransacked shops for food and medicine and severely undermined the local social order.

These two examples have fully revealed the importance of designing a sophisticated emergency rescue system. The loss of public emergencies would be greatly reduced by understanding the distribution of disaster-stricken people and providing appropriate emergency resources to them. Otherwise, the public emergencies would be uncontrollable. To make things worse, the situation of disasters might be more serious, and even lead to the breakout of secondary disasters.

2. Problem statement

When we design an emergency rescue system, we need to coordinate the manpower with the financial, material resources. It is a complicated process to optimally allocate various elements within a system. It involves a wide range of contents. Repeated researches should be made on several theories and methods. Designing of an emergency rescue system covers the following four aspects (See Figure 1):

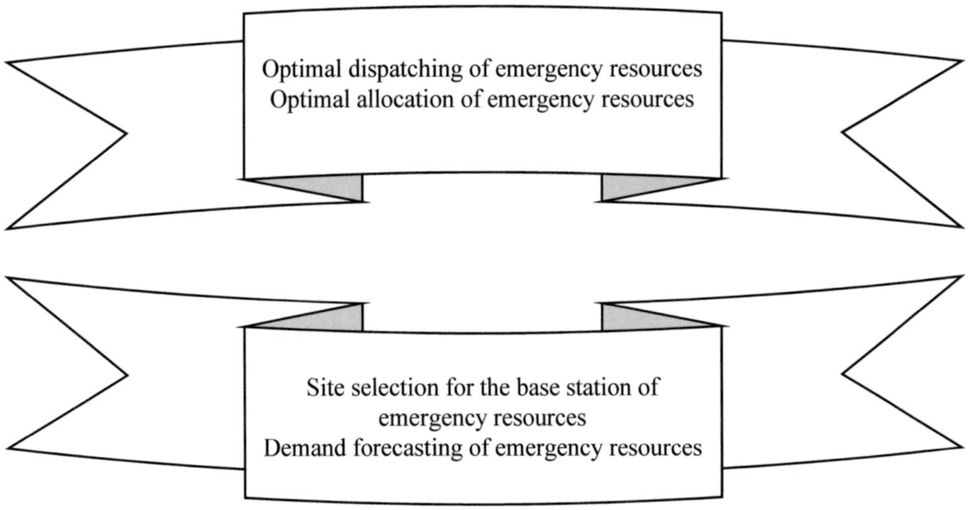

Fig. 1. Research Course for the Emergency Rescue System.

These four aspects have been cross-linked each other essentially.

1. Demand Forecasting of Emergency Resources

In recent years, unconventional emergencies frequently broke out, severely endangered people's life and property. How to timely predict people's demand on resources after the disasters? This issue has become an important problem for us.

Here, a precise predictive method has been designed by combining the Fuzzy Set Theory with the Learning Rules of Hebb Neural Network, Multiple Linear Regression and Case

Reasoning. By applying this method, we have solved the problems of information insufficiency and inaccuracy when we predict the resource demand after unconventional emergencies, and could correctly predict people's demand on resources.

2. Optimal Site Selection for the Base Station of Emergency Resources

If the resource demand has been determined, sufficient emergency resources need to be transported to the emergency base stations (Emergency rescue station). To achieve this goal, how many base stations for emergency resources should be established, and where should we establish these stations, these issues will be worth considering. In other words, we should optimally select the sites of base stations and find appropriate locations within a certain region as the base stations of emergency resources. The number of location should also be suitable. When disasters break out, we could allocate resources from these base stations to deal with the emergencies. By optimally selecting the sites of base stations, we could not only reduce costs, but could also ensure the timeliness of emergency resources, making these resources arrive at the emergency scenes quickly, safely and timely.

Here, a summarization has been made on relevant site selection knowledge, and the Operations Research theories have been applied based on the existing site selection methods. A multistage model of site selection has been designed to make an optimal planning on the number and location of base stations. Example analysis has also been made to verify the results of calculation. It has been proved that this model is simple, convenient for use, and could get results quickly. This model would be suitable for the site selection and planning of base stations of emergency resources.

3. Appropriate Allocation of Emergency Resources

The emergency resources deployment is a hardcore of emergency management. After the happening of the public emergency, it is important to study how to deliver the emergency resources to base stations quickly. When we've determined the location of base stations, we should optimally allocate emergency resources. More to this point, it should predetermine the number, type and quality of resources for each base station. Otherwise, there's an important constraint condition for us to consider: the costs.

This chapter proposed the dynamic optimal process of emergency resources deployment planning, making use of Markov decision processes, and discovered the optimal deployment planning to guarantee the timelines.

4. Optimal Dispatching of Emergency Resources

Aiming at solving the resource allocation problems in case of emergency events, this chapter presented an optimum mathematical simulation model based on the dynamic programming.

In accordance with the number of emergency base stations, the given model tries to divide the resource allocation procedure into the some stages. The stated variable stands for the amount of the emergency resource available for allocation can be used at the beginning of each stage. As is depicted in the dynamic programming theory, the remaining resource of the previous stage may have a strong influence on the succeeding stage. During each stage, three factors may restrict the object function, that is, the remaining resource, the decision, and the demand. The total function is the sum of the object function of each stage. In addition, the concrete case can be used to confirm the model's validity and practicability. The results of our repetitive experimental application of the model show that it works perfectly for its duty in improving the efficiency of emergency management and overcoming the problem of wasting emergency resource as well as low efficiency in emergency rescue.

3. Methods

3.1 Emergency resources demand prediction using case-based reasoning

3.1.1 Research background

Public emergencies usually bring great negative impacts on economy and society, cause damage on casualties and property, bring destructions on ecological environment and human living environments, have adverse impacts on social order and public safety, and even arise social and political instability. Moreover, due to the change and influence of multiple factors, the type, occurrence probability and influence degree of public emergencies are increasing.

The demand on emergency resources refers to the minimum guarantee requirements for effective response to public emergencies. The so-called effective response refers to that the response on public emergencies should be efficient, and it also refers to that the emergency resources should be used with high efficiency. While the minimum guarantee requirements refer to that the smallest demands are needed when public emergencies are successfully solved. Obviously, an optimized idea is involved in the determination of emergency resource demand, meaning that under some given parameters such as type, intensity and influencing range of Emergency response, the smallest resource demand required for the successful response to public emergencies.

Currently, there are few researches on this aspect, in most cases the emergency decision maker subjectively decides whether the quantity, quality and type of emergency resources are rational and can meet the requirements of emergency. Besides, due to the particularity of emergency process, the effect of cost is smaller than that of time effect, so that in many cases, no efforts are spared to conduct the emergency rescue. But the method is easily to cause the irrational demand of emergency resources, so that it is unscientific and will cause groundless waste of numerous resources, meaning that a scientific prediction method is sorely needed to achieve the prediction on the demand of emergency resources.

Case-based reasoning (CBR) is a relatively new problem solving technique that is attracting increasing attention. For a long time, expert systems or knowledge-based systems (KBS) are one of the success choices in Artificial Intelligence (AI) research. The *first generation* KBS, and today's systems, are based upon an explicit model of the knowledge required to solve a problem. When it comes to so called *second generation* systems, a deep causal model was adopted to enable a system to reason using first principles [1]. But whether the knowledge is shallow or deep an explicit model of the domain must still be elicited and implemented often in the form of rules or perhaps more recently as object models. However, knowledge elicitation is a difficult process, often being referred to as the knowledge elicitation bottleneck; implementing KBS is a difficult and slow process requiring special skills; and once implemented they are difficult to maintain [2-6].

Over the last few years an alternative reasoning paradigm and computational problem solving method has increasingly attracted more and more attention. Case-based reasoning (CBR) solves new problems by adapting previously successful solutions to similar problems. CBR is attracting attention because it seems to directly address the problems outlined above. CBR does not require an explicit domain model and so elicitation becomes a task of gathering case histories, implementation is reduced to identifying significant features that describe a case, an easier task than creating an explicit model, by applying database techniques largely volumes of information can be managed, and CBR systems can learn by acquiring new knowledge as cases thus making maintenance easier.

The work Schank and Abelson in 1977 is widely held to be the origins of CBR [7]. They proposed that our general knowledge about situations is recorded as scripts that allow us to set up expectations and perform inferences. Whilst the philosophical roots of CBR could perhaps be claimed by many what is not in doubt is that it was the work of Roger Schank's group at Yale University in the early eighties that produced both a cognitive model for CBR and the first CBR applications based upon this model [8]. Janet Kolodner developed the first CBR system called CYRUS [9-11]. An alternative approach came from Bruce Porter's work, at The University of Texas in Austin, into heuristic classification and machine learning resulting in the PROTOS system [12-13].

In the U.S., Edwina Rissland's group at the University of Massachusetts in Amherst developed HYPO [14]. This system was later combined with rule-based reasoning to produce CABARET [15].

In Europe, the first one is that of Derek Sleeman's group from Aberdeen in Scotland. They studied the uses of cases for knowledge acquisition, developing the REFINER system [16]. Mike Keane, from Trinity College Dublin, undertook cognitive science research into analogical reasoning [17]. Michael Richter and Klaus Althoff in the University of Kaiserslautern applied CBR to complex diagnosis [18]. This has given rise to the PATDEX system [19] and subsequently to the CBR tool S3-Case. In the University of Trondheim, Agnar Aamodt has investigated the learning facet of CBR and the combination of cases and general domain knowledge resulting in CREEK [20-21].

In the UK, CBR seemed to be particularly applied to civil engineering. A group at the University of Salford was applying CBR techniques to fault diagnosis, repair and refurbishment of buildings [22]. Yang & Robertson [23] in Edinburgh developed a CBR system for interpreting building regulations, a domain reliant upon the concept of precedence. Another group in Wales applied CBR to the design of motorway bridges [24].

Further, there are active CBR groups in Israel [25-26], India [27] and Japan [28].

3.1.2 Methods for emergency resource demand prediction

According to the characteristics of emergency resource demand prediction process, both the risk analysis and case-based reasoning method are introduced into the process, accordingly a case-based reasoning method for emergency resource demand prediction based on risk analysis is obtained, which improves the scientificity of emergency resource demand. The case-based reasoning flow for emergency resource demand prediction based on risk analysis is shown in Fig.2.

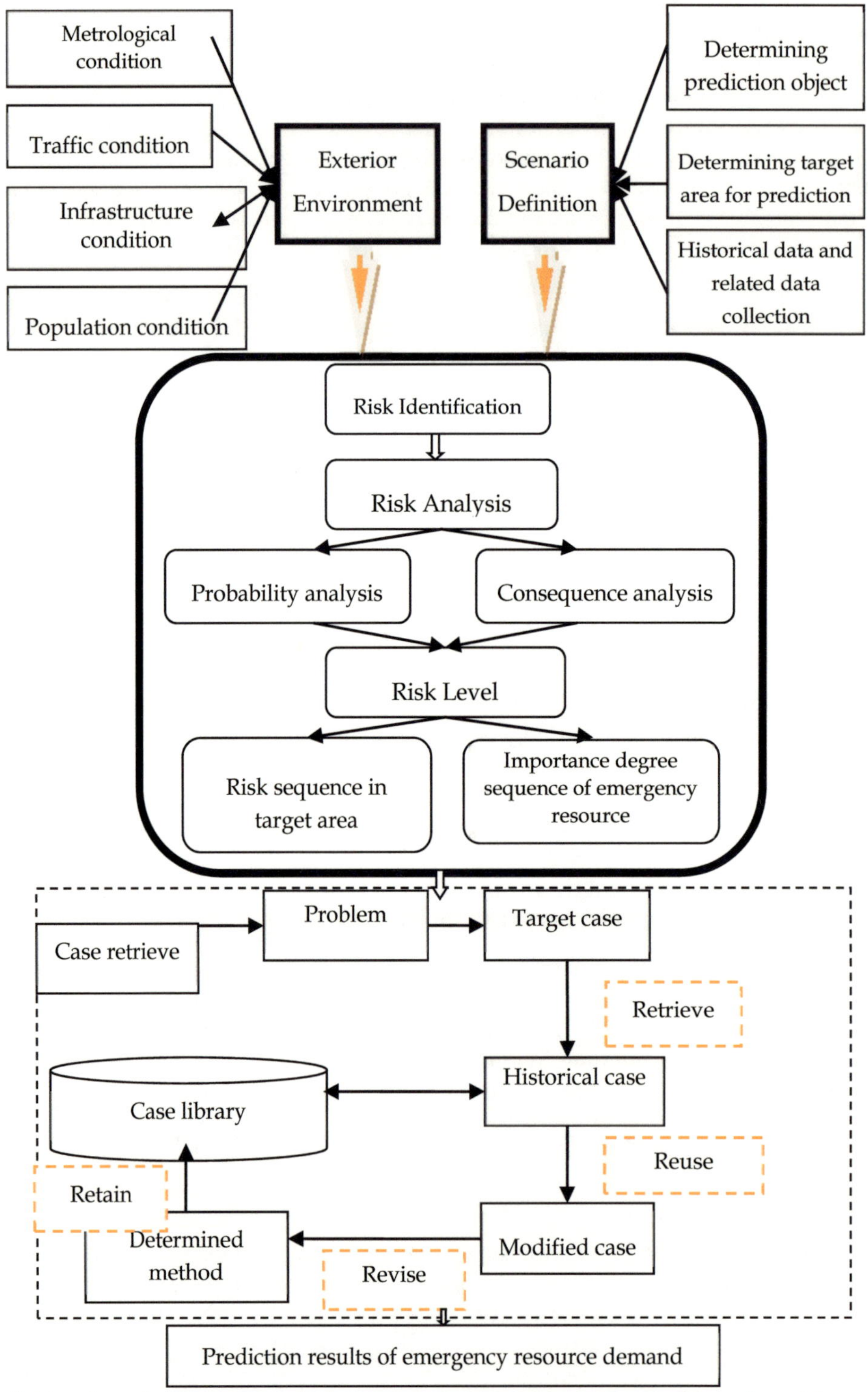

Fig. 2. Case-based reasoning prediction on emergency resource demand based on risk analysis.

The case-based reasoning is a comprehensive form of three types of human thoughts including imaginal thinking, logical thinking and creative thinking. From the view of reasoning method, the case-based reasoning is an analogy reasoning from one case (old case) to another case (new problem), while from the view of knowledge, the case-based reasoning is a method based on memory in which old experiences are used to guide the problems. The CBR is generally composed of four main processes, including retrieve, reuse, revise and retain [29-30], so that CBR is also called 4R.

Therefore, in this paper, the case-based reasoning prediction method associated with risk analysis process is used to conduct demand prediction on the quantity, quality and type of emergency resources. After conducting risk analysis on target area, characteristic values of risk in these are can be obtained, including possible incident type, incident results, occurrence probability of incident, etc., accordingly the case-based reasoning process can be used for emergency resource demand prediction.

1. Expression of case

The case generally includes two parts, including case attribute description and case solution, of which the former one is the index structure of case and the latter one is the answer of case. While the emergency resource demand prediction is composed of three parts, including characteristic description of Emergency response, characteristic description of emergency rescue plan and description of emergency resource demand, all of which can be determined based on the results of risk analysis, namely risk probability and risk results.

- Characteristic description of Emergency response: it includes some characteristic information of Emergency response, including type, intensity, natural environment surround the occurrence site, population density, losses, duration time of hazard, etc., all of which depict and describe the characteristic attributes of Emergency response.
- Characteristic description of emergency rescue plan: it includes the characteristic attributes of emergency object, emergency rescue method, emergency procedure, etc. If there is a difference in the emergency object, way, technique and process of Emergency response of the same type, the material demands will be different too.
- Description of emergency resource demand: it includes the quantity, quality and type of emergency resources.

On the whole, in order to obtain complete data, the case should be described in detail as can as possible under the specific condition. Generally, one case can be composed of several attributes, all of which can be further divided, while the whole case library is composed of associated cases at different attribute levels. Therefore, in the emergency resource demand prediction, the case can be modeled as follows:

$$case\ (F, P, D)$$

In the formula: $F=(f_1, f_2, \ldots, f_n)$, f_n is a characteristic attribute of Emergency response, which can be obtained according to the results of risk analysis; $P=(p_1, p_2, \ldots, p_n)$, p_n is a characteristic attribute of emergency rescue plan; D is the demand attribute of emergency resource.

2. Case-based reasoning process of emergency resource demand
a. Characterization of emergency resource demand case

Given that there are n cases in the case library, the case i is expressed as C_i (i=1,2,…,n). Its characteristic factor set $B=\{b_1, b_2,\ldots,b_m\}$. Therefore, the membership function of case C_i to the characteristic factor b_j (j=1,2,…,m) is expressed as $n_{C_i}(b_j)$, and the characteristic vector corresponding to the case C_i in the case library is as follows:

$$V_{C_i} = \{n_{C_i}(b_1), n_{C_i}(b_2), \ldots, n_{C_i}(b_m)\} = \{n_{C_i}(b_j) \mid j = 1, 2, \ldots, m\} \tag{1}$$

Given that the characteristic vector set of prediction plan is T, which can be expressed as the formula below:

$$V_T = \{n_T(b_1), n_T(b_2), \ldots, n_T(b_m)\} = \{n_T(b_j) \mid j = 1, 2, \ldots, m\} \tag{2}$$

b. Emergency resource demand case retrieve-similarity calculation

According to the organization form of case, the nearest neighbour method is used. The nearest neighbour method is a method in which the cumulative sum of characteristic weights of the input case that is matched with the existing case in the case library is used to retrieve the case, namely that:

$$\frac{\sum_{i=1}^{n} w_i sim(b_i^I, b_i^R)}{\sum_{i=1}^{n} w_i} \tag{3}$$

In the formula above: w is the important weight value of characteristic factor, *sim* is the similarity function, b^I and b^R is the input case value and retrieve case value of characteristic factor *i*.

In the similarity matching of cases using the characteristics of case, the effect of each characteristic is different, so that in the similarity calculation, it is necessary to assign different weights to each characteristic factor.

Given that the influencing weight set of the characteristic factor set $B=\{b_1, b_2,\ldots,b_m\}$ is $\{w_1, w_2,\ldots,w_m\}$, and the following condition is satisfied:

$$\sum_{j=1}^{m} w_j = 1, \quad j = 1, 2, \ldots, m \tag{4}$$

Consequently, the similarity can be calculated by the formula below:

$$sim(A, B) = \frac{\sum_{j=1}^{m} w_j (n_A(x_j) \wedge n_B(x_j))}{\sum_{j=1}^{m} w_j (n_A(x_j) \vee n_B(x_j))} \tag{5}$$

In the formula (5), $\wedge$ is the maximum lower limit, and $\vee$ is the minimum upper limit.

c. Weight calculation of characteristic factor

Generally, under different decision-making environments, the same characteristic factor has different effects on the decision output. Given that $n(b)$ represents the value of case when the characteristic factor is b. If there is a large difference in distribution of $n(b)$ in the case library C (C ={c_1, c_2,…,c_n}), which indicates that the factor has great effect on classification identification, and it should be assigned a larger weight value. On the contrary, if there is a small difference in distribution of $n(b)$ in the case library C (C ={c_1, c_2,…,c_n}), which indicates that the factor has little effect on classification identification, and it should be assigned a smaller weight value.

Therefore, each case in the case library can be classified into one type. Given the case C_i takes the value of $n(b_j)$ when the characteristic factor is b_j, the membership function of the case to the characteristic factor b_j can be expressed as $n_{C_i}(b_j)$, and the formula below can be obtained:

$$\bar{n}(b_j) = \frac{1}{n}\sum_{j=1}^{n} n_{C_i}(b_j) \tag{6}$$

Thus, the mean square deviation is expressed as the formula below:

$$\delta(b_j) = \left[\frac{\sum_{i=1}^{n}(n_{C_i}(b_j) - \bar{n}(b_j))^2}{n}\right]^{\frac{1}{2}} \tag{7}$$

The weight w_j of each characteristic factor can be obtained by the formula below:

$$w_j = \frac{\delta(b_j)}{\sum_{j=1}^{m}\delta(b_j)}, \quad j = 1, 2, \ldots, \ m \tag{8}$$

After the weight of characteristic factor is obtained, the similarity function is combined with the weight of characteristic factor, and the formula below can be obtained:

$$sim(T, C_i) = \frac{\sum_{j=1}^{m} w_j(n_T(x_j) \wedge n_{C_i}(x_j))}{\sum_{j=1}^{m} w_j(n_T(x_j) \vee n_{C_i}(x_j))} \geq \eta, \quad i = 1, 2, \ldots, n \tag{9}$$

In the formula above, the similarity between target case T and C_i can be expressed as $sim(T,C_i)$ ($\in[0,1]$), η is a threshold value, and the learning strategies of case are divided into following types:

a. $\forall$ sim (T, C_i) =0, $i\in[0, 1]$, the new case doesn't match with all cases in the case library, and it can be added into the case library;
b. $\exists$ sim (T, C_i) =1, $i\in[0, 1]$, the new case is completely similar to certain case, and it can not added into the case library;
c. $\forall$ sim $(T, C_i) <\eta$, $i\in[0, 1]$, the new case can be added into the case library;
d. $\forall$ sim $(T, C_i) >\eta$, $i\in[0, 1]$, the solution for the case with maximum similarity (max[sim (T, C_i)]) is converted into the solution for new case.

All cases in accordance with the similarity calculation formula are the similar cases, among which the one with the maximum sim (T, C_i) is the most similar case. Accordingly, the material demands of the most similar case are taken as the prediction results of material demand when the Emergency response occurs.

3.2 Application of multi-stage location planning model in optimizing location of emergency resource base stations

3.2.1 Research background

Urban Planning refers to the specific method or process of predicting urban development and managing various resources to adapt to its development, to guide the design and development of built environments. While modern urban planning is trying to study the impact which a variety of economic, social and environmental factors have on the change of land using patterns, and develop planning reflecting the continuous interaction. Currently, in the process of making urban planning, parties have paid more and more attention to urban safety planning, in which the optimization of location planning of emergency logistics base station is one of the very important contents[31-33].

Emergency Logistics refers to the special logistics activities through which the necessary emergency supplies are provided to minimize the loss caused by unforeseen accidents and disasters in the shortest time, while an important role of emergency logistics base station (also known as emergency resource base station) is to provide adequate and timely emergency response resources to potential unforeseen accident or disaster sites[34-37].

The optimization of location planning of emergency resource base station includes determining reasonable position and scale of emergency resource base station, and since the special construction of emergency logistics base station costs a lot due to its specialization, so the optimization of the location planning of emergency resource base station should mainly start with two aspects, one is to determine the appropriate scale or the reasonable amount of emergency resource base stations, the other is to solve the problem of spatial distribution, namely, location optimization[38-40].

City T, as largest coastal open city in Northern China, currently has jurisdiction over 15districts including Heping, Hedong, Nankai, Hexi, Hebei, Hongqiao, Tanggu, Hangu, Dagang, Dongli, Xiqing, Jinnan, Beichen, Wuqing, Baodi, and 3 counties such as Jinghai,

Ninghe and Jixian. It has a total area of 11,919.7 square kilometers, and a resident population of 1023.67 million. Table 1 shows the statistics of accidents this year in City T.

Time	Accident	Death	Injured	Population/ 10 thousand	Accident rate 0/000	Mortality rate 0/000	Injured rate 0/000
2004	10	32	27	1001	0.999	3.197	2.697
2005	9	37	25	1001	0.899	3.696	2.498
2006	1	9	3	1001	0.010	0.899	0.300
2007	9	21	5	1024	0.879	2.051	0.488

* Data derives from 2004~2007 Statistics of accident distribution in cities and provinces China in *Journal of Security and Environment*

Table 1. Statistics of accidents in recent years in City T*.

As can be seen from the table, City T is a metropolis with a good security situation, but in order to take preventive measures, it is very important and very necessary to carry out the optimization of location planning of emergency logistics base stations in City T. Therefore, in this paper, taking City T as the object, the author makes use of multi-stage location planning optimization model to study the location planning of emergency resource base stations in this city and seek a reasonable location planning program to provide a decision making basis for the future construction and development of City T.

3.2.2 Multi-stage location planning optimization model of emergency resource base station

An important role of emergency logistics base station is to provide adequate and timely emergency response resources to potential unforeseen accident or disaster sites. And the optimal planning of emergency logistics base station helps to make rational use and allocation of spaces and emergency resources, to reduce the risks the city, as well as conduce to the efficient, orderly and sustained operation of urban economy, social activities and construction activities.

The optimization of location planning of emergency resource base station includes determining reasonable position and scale of emergency resource base station. Therefore, the optimization of the location planning of emergency resource base station should mainly start with two aspects, one is to determine the appropriate scale of emergency resource base stations, and the other is to solve the problem of spatial distribution.

The first step is to use set covering model to determine the minimum number of emergency resource base stations which can meet the needs of all demand sites; and the second step is to use maximum coverage model to determine the optimal sites of the minimum number of emergency resource base stations among the options, to meet the needs of all demand sites to the maximum.

a. Scale optimization - set covering model

Coverage model is one of the most basic models of optimal planning of emergency resource base stations. The meaning of coverage refers to that the services scope of emergency resource base stations set up should be able to cover all sites requiring service. And it is one of the common goals of optimal planning of emergency logistics base station to cover all demand sites with the minimum number of emergency logistics base stations.

Set covering model is simple, but highly practical. It can be used to determine the most efficient number the emergency resource base stations covering all demand sites. Since the investment in emergency resource base stations can be quite expensive, so decision-makers need to keep a minimum number of base stations at the same time of taking providing services of necessary level to each demand site into account, therefore, they need to determine the reasonable number of emergency logistics base station under the limitation of covering distance or covering time.

The binary decision variable x_j is set as follows: When the candidate site j is selected, $x_j = 1$; otherwise $x_j= 0$. if the set of candidate sites which can cover all the demand sites i is $N_i = \{j \mid d_{ij} \le S\}$ (or $N_i = \{j \mid t_{ij} \le R\}$), the minimum number of the necessary facilities which can cover all the demand sites may be decided by set covering model:

$$\begin{aligned} \min \quad & z = \sum_{j \in J} x_j \\ s.t. \quad & \sum_{j \in N_i} x_j \ge 1 \quad \forall i \in I \\ & x_j \in (0,1) \quad \forall j \in J \end{aligned} \tag{10}$$

In which the objective function can minimize the number of base stations, the constraint 1 can ensure that each demand site is covered by at least one emergency resource base station, and it is one of the basic objectives of optimal planning of emergency resource base station. The constraint 2 limits the decision variables x_j as integer variables between (0, 1).

Set covering model is of integer linear programming model, mathematically, it is a typical NP-hard model. Generally, its solution can be obtained through relaxing integer limiting requirement against x_j, using the procedure of general linear programming, and in most cases, integer solution of general problems can be directly obtained.

b. Space layout optimization - maximum coverage model

After the first phase of finding model solution, the number of emergency resources base stations providing services to all the demand sites in the whole region can be determined. Being clear about the number of base stations, the goal of the second stage is to optimize the spatial layout of these emergency resource base stations, to enable them to meet emergency requirements.

Maximum coverage model is an extension and expansion of set covering model. This model can be used to consider the maximization of demand site value coverage (population or other indicators). Maximum coverage model was originally presented by Church and ReVelle, where d_i refers to the demand of node i, p refers to the available emergency resources. Binary variables y_i can be used to present whether the demand site is overwritten or not, when the demand site is overwritten, $y_i = 1$, otherwise $y_i = 0$.

$$
\begin{aligned}
& \textbf{\textit{Maximize}} \ \sum_{i \in V} d_i y_i \\
& \textbf{\textit{subject to}} \\
& \sum_{j \in W_i} x_j \geq y_i \quad (i \in V) \\
& \sum_{j \in W} x_j = p \\
& x_j \in \{0,1\} \quad (j \in W) \\
& y_i \in \{0,1\} \quad (i \in V)
\end{aligned}
\tag{11}
$$

Maximum coverage model can be used to seek the best possible using method of available resources, but does not guarantee to cover all demand sites. Therefore, maximum coverage model can be used to determine the optimal solution of maximum coverage on the base of the optimal solution of set coverage.

3.3 Appropriate allocation of emergency resources

3.3.1 Research background

At present, our country is in an important opportunity period for economic and social development, which is also the crucial period to implement the third-step strategic deployment of the modernization construction; therefore, the important task of our country is to maintain the long-term harmonious and stable social environment and stable and united situation. As the most important link in emergency handling, the contingency plan strengthens the research in respect of optimized resource allocation, which has very important significance for promoting the technological level of dealing with unexpected accidents and emergency management capacity of our country, guaranteeing public safety in our country and establishing reasonable and efficient contingency plans for national public safety.

The emergency resource management process of unexpected public events is in fact a set of decisions and decision implementation processes under a series of goal constraint conditions. These series of decisions and decision implementation processes mean "when and which resources at which place to allocate, and what to do". It is necessary to invest in enhancing urban comprehensive emergency capability, which proposes the problems of optimized allocation and dispatch of limited emergency resources, and the problem solving relates to whether the limited resources can exert the greatest effect, whether the emergency rescue system can achieve the desired goal and so on. At present, the researches related to emergency resource allocation practically aim at single resource optimization, such as the emergency service vehicle dispatch or vehicle relocation problems. When the emergency service vehicle system receives the service demand, it dispatches its emergency response unit (such as police car, fire engine, ambulance and so on) to the service demand zone. After an emergency service vehicle is dispatched for service, it is necessary to study the vehicle relocation problems to guarantee the defence demand in remaining zones. Studies on the emergency resource allocation beyond service vehicles are still at the initial stage. In China, Liu Chunlin et al (1999) studied the minimized transportation time problems of emergency resource allocation in continuous consumption system and one-off consumption system [41]; Liu Chunlin et al (2000) studied the emergency problems when the required time from the depot to emergency zone is a fuzzy number. In overseas countries, Fiorucci et al (2005) studied the emergency resource allocation and scheduling problems before and after the fire through building a dynamic model; Fiedrich et al (2000) studied the problems of simultaneous resource allocation for different disaster relief tasks and so on through building a dynamic programming model [42]. At the present stage, there are fewer studies on optimized allocation of many resources under unexpected accident disasters, moreover the majority of studies only take shortest emergency time as the optimized objective of the system, and the optimized method is too simple, and lacks consideration of the complexity of the emergency process; in addition, static models are more widely used in the studies, which lack emergency resource allocation parameters reflecting the accident disaster development status.

The present research aims at enhancing the urban emergency management capacity, establishes the emergency resource allocation model in view of many accident disasters, so as to effectively integrate various emergency resources, and reduce the investment cost of emergency resource management.

3.3.2 Optimized resource allocation model

3.3.2.1 Decision model based on dynamic programming

The emergency resource allocation process is divided into *N* corresponding stages in view of the accident disaster emergency management characteristics, using the dynamic programming method, and according to the number of emergency zones (the number is supposed as *N*), based on which, a mathematical model is built, so as to optimize the emergency resource allocation. In the emergency process, a certain amount of resources are allocated to meet the emergency demand, and various parameter variables are expressed as follows.

k is the emergency stage $(k=1,2,\cdots,N)$; x_k is the state variable in the dynamic programming model, representing the gross amount of allocated emergency resources at the k_{th} stage; u_k is the decision variable in the model, representing the alternative decision scheme; w_k represents the emergency resource demand at the k_{th} stage with given probability distribution. D_k is the set of all decision variables from 1st stage to the k_{th} stage.

Suppose that $w_1,w_2,\cdots,w_N$ are independent random variables depending on the disaster situation at emergency zones. The relationship between allocatable resource x_k, emergency resource demand w_k and emergency decision variables u_k in the emergency process is shown in Figure 3.

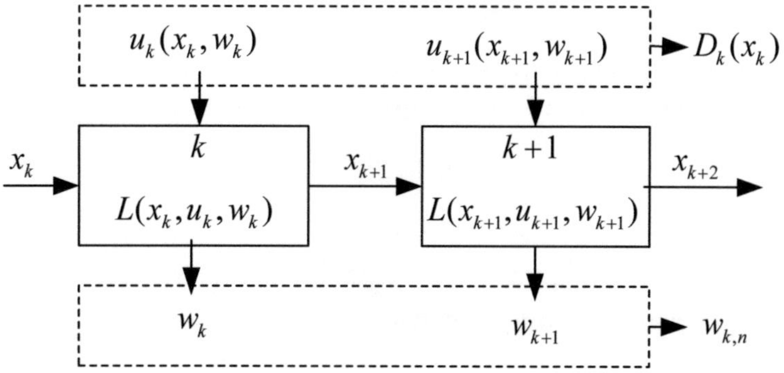

Fig. 3. The relation of proximate emergency stages based on dynamic programming.

For the $(k+1)_{th}$ stage, the dynamic system has

$$x_{k+1} = f(x_k, u_k, w_k)(k = 0,1,2,\cdots,N-1) \tag{12}$$

Where, *N* is the number of emergency zones in the emergency process.

The decision function sequence composed of the decision $u_k(x_k)(k = 1,2,\cdots,N)$ at each stage is called the whole process strategy, strategy for short. Strategy refers to the set of all emergency resource allocation decisions established at any emergency stage, which is only related to the stage and state in that stage process, and is expressed as $p(x_k)$

$$p(x_k) = (u_1(x_1), u_2(x_2), \cdots, u_N(x_N)) \tag{13}$$

For the objective function *J* with given state and decisions, it can be expressed as

$$J = J(x(0), x(1), \cdots, x(N); u(0), u(1), \cdots, u(N)) \tag{14}$$

An appropriate *u* value (namely, appropriate decision sequence) is selected, so as to minimize *J* (or use other evaluation standards of *J*, such as the maximization of *J*), and optimize the objective function. Function *J* is called the criterion function.

3.3.2.2 Establishment of an optimized resource allocation model

The decision-making process at the N_{th} stage is determined using the following factors.

$$(x(0), x(1), \cdots, x(N); u(0), u(1), \cdots, u(N)) \tag{15}$$

$x_{k+1} = f(x_k, u_k, w_k)(k = 0,1,2,\cdots,N-1)$, and $x(0)$ is the known initial state, suppose that all $x(k)(k = 1,2,\cdots,N)$ can be expressed as $x(0)$ and $u(k)(k = 0,1,\cdots,N-1)$, so the criterion function

$$J = J(x(0), u(0), u(1), \cdots, u(N)) \tag{16}$$

x(0) is the given initial state, decision $u(k)(k = 0,1,\cdots,N)$ is free variables, so the simultaneous nonlinear equation

$$\frac{\partial J}{\partial u(k)} = 0,(k = 0,1,\cdots,N) \tag{17}$$

To solve practical problems, it is necessary to analyze and calculate the form of limit criterion function *J* in multistage decision process. In the present model

$$u(k) = u(x(k), w(k), k) \tag{18}$$

i.e., the existing decision is only the function of existing state and stochastic disturbance.

The criterion function of emergency resource allocation has Markov properties, i.e., the objective function has the following attributes

$$J = \sum_{k=0}^{N} L(x(k), u(k), w(k)) \tag{19}$$

Where, $L(x(k), u(k), w(k))$ is the objective function at each stage of the emergency process. In this model, *L* is a nonnegative function depending on the state and sum of decision items at a single stage. *J* is the objective function of the whole emergency process, equivalent to the sum of objective functions at all stages.

In general, it is known that a group of states $x(k) \in X$, and *X* is available emergency resources, then a new group of states $x(k+1)$ can be obtained according to $x(k)$ with the computing formula as

$$x(k+1) = f(x(k), u(k), w(k)) \tag{20}$$

At the same time, $J(x(k+1), k+1)$ can also be calculated

$$J(x(k+1), k+1) = L(x(k+1), u(k+1), w(k+1) + J(x(k), k) \tag{21}$$

So the total cost function of the emergency process of any systematic sample can be expressed as

$$\begin{cases} J_{k+1}(x_{k+1}) = \min \sum_{k=1}^{N-1} L_k = \min(L_{k+1}(x_{k+1}, u_{k+1}, w_{k+1}) + J_k(x_k)) \\ \text{s.t. } u_k \in p_k(x_k) \\ L_0(x_1) = 0 \\ \dfrac{\partial J}{\partial u(k)} = 0 \end{cases} \tag{22}$$

3.4 Optimal dispatching of emergency resources

3.4.1 Research background

It is necessary for the emergency command department to make the emergency resource scheduling decisions after the occurrence of sudden public events. It is necessary for emergency decision-makers to determine future resource scheduling according to the emergency resource demand situation at the present stage, and multi-stage emergency resource scheduling with the event development and changes and according to the emergency effect at the last stage and present situation. Therefore, the emergency resource scheduling is a dynamic process. Under the situation that the emergency resource site layout and allocation is known, emergency managers are concerned about the problems of how to formulate optimized scheduling scheme, guarantee the timeliness of emergency resource scheduling, and minimize the resource arrival time [43-44]. As a result, it is necessary to formulate beforehand the optimized scheme of emergency resource scheduling in the light of the specific scene of sudden public events, so as to start the emergency resource scheduling scheme as early as possible and guarantee the timeliness of emergency rescue action.

The Markov decision process can select an action from the available action set to make a decision according to the observed state at each moment. Meanwhile, the decision makers can make another new decision according to the newly observed state, and repeat such process [45]. Therefore, this section plans to study the dynamic optimization of emergency resource scheduling of sudden public events using the Markov decision process, so as to provide a basis for optimized emergency resource scheduling under sudden public events

3.4.2 Dynamic Markov decision of emergency resource scheduling

Due to a series of characteristics of sudden public events, such as nonrepeatedness, uniqueness, gradual evolution and so on, the decision-making problems for emergency resource scheduling have three main characteristics: sudden public event is dynamically changing; information about the event development is from fuzziness to clearness and from incompleteness to completeness, namely the future state is uncertain; the scheme formulated under incomplete information can be easily adjusted in time under complete information.

The optimized emergency resource scheduling can be more scientifically and reasonably realized by referring to the Markov decision analysis method, but sudden public events are not evolved and developed according to the pre-established direction. Therefore, emergency measures can be only taken according to previous experience, emergency plan and real-time information at the scene of accident (usually incomplete), and be adjusted according to unceasing improvement of the information in the emergency process.

The application of Markov decision analysis method in the optimized emergency resource scheduling process of sudden public events is shown in Figure 4. The whole decision-making process is how to select a scheme to cope with the uncertainty development state of the sudden public event, until the sudden public event is completely under control.

Basic thought of the Markov process is to infer the future state distribution according to the probability distribution of current state, and make judgment and decisions accordingly.

$X(t)$ is used to express the system state, the state sequence $\{X(t);t \in T\}$ is a stochastic process, $U_{(i)}^{m}$ is the decision set of the state i at the *n*th stage. Suppose that P_{ij} is the one-step state transition probability, $f_n(i,\pi_n)$ represents the expected total reward when the system state shifts from $X(n)=i$ at the *n*th stage to the process end; r_{ij} represents the corresponding reward when the state shifts from $X(n)=i$ to the next state $X(n+1)=j$, then there is

$$f_n(i,\pi_n)=q(i)+\sum_{j=1}^{n}p_{ij}f_{n+1}(i,\pi_{n+1}), i=1,2,\cdots,m; n=1,2,\cdots \quad (23)$$

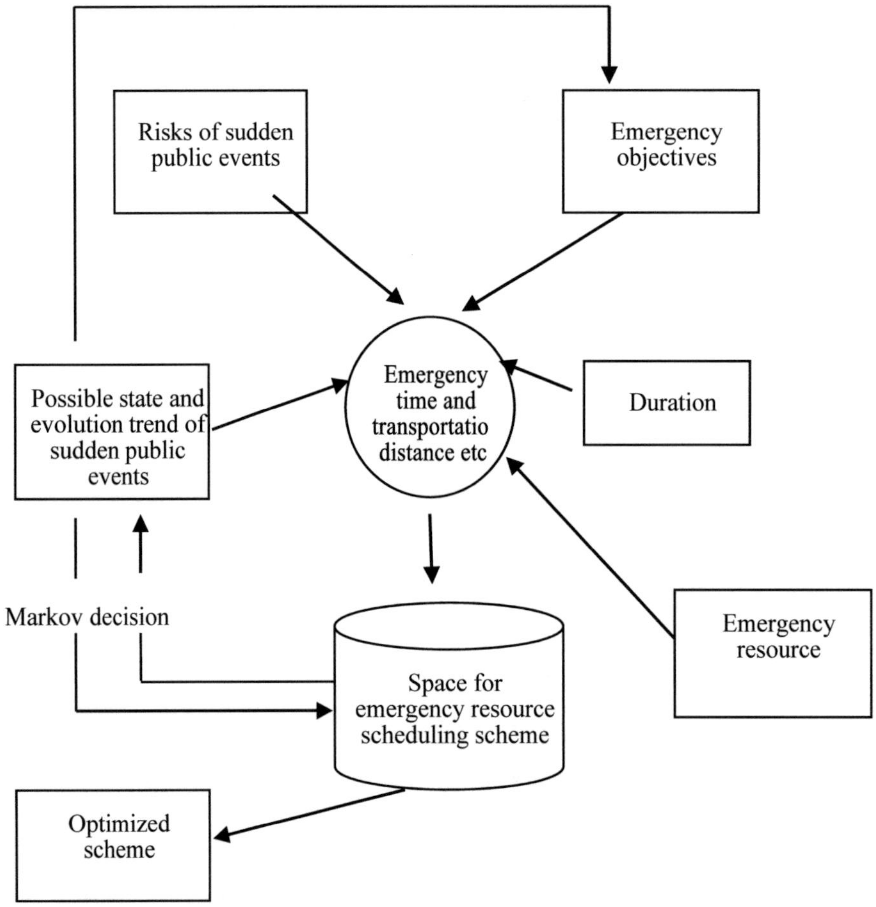

Fig. 4. Diagram of Markov decision processes in the emergency resources scheduling of the sudden public events.

π_n represents the sequence $\{\delta_n,\delta_{n+1},\cdots\}$, $\pi_n=(\delta_n,\pi_{n+1})$ of the decision regulation δ from the *n*th stage to the process end, where δ_n is the decision regulation at the *n*th stage.

If $q(i)+\sum_{j=1}^{n}p_{ij}r_{ij}, i=1,2,\cdots,m$, $q(i)$ represents the expected reward when the state i shifts once, namely the real-time expected reward of the state, then the above formula can be rewritten as

$$f_n(i,\pi_n) = \sum_{j=1}^{n} p_{ij} r_{ij} + \sum_{j=1}^{n} p_{ij} f_{n+1}(i,\pi_{n+1}), i = 1,2,\cdots,m; n = 1,2,\cdots \quad (24)$$

This formula is the basic equation for Markov decision problems.

To research the transient state behavior the ergodic Markov chain, it is necessary to obtain its basic equations set using z transform analysis method. z transform can transform the difference equation to corresponding generalized equation. There is one-to-one correspondence between the function and its z transform, and meanwhile the primary function can be mutually converted with its z transform. Therefore, the following formulae can be obtained through z transform

$$\begin{aligned} nv + f_i &= q_i + \sum_{j=1}^{m} p_{ij}[(n-1)v + f_j] i = 1,2,\cdots,m \\ nv + f_i &= q_i + \sum_{j=1}^{n} p_{ij} f_i \quad i = 1,2,\cdots,m \end{aligned} \quad (25)$$

This is the basic equations set for Markov decision problems, which can be obtained through the following algorithms.

a. An initial strategy π_n is selected, a decision regulation δ_n is selected for each state $i(i = 1,2,\cdots,m)$, so that its decision $u_{(i)}^{k} = \delta_n(i)$, and let $n = 0$;
b. For the known strategy π_n, let $f_m^{(n)} = 0$, the corresponding strategy profit $v^{(n)}$ and corresponding value $f^{(n)}(i = 1,2,\cdots,m; n = 0,1,2,\cdots)$ can be obtained through solving the basic equations set for Markov decision problems;
c. A new strategy regulation δ_{n+1} is sought using the $f_m^{(n)}$ obtained from the last strategy, so that for each state i,

$$q_i^{\delta_{n+1}(i)} + \sum_{j=1}^{n} p_{ij}^{\delta_{n+1}(i)} f_j^{(n)} - f_i \quad (26)$$

achieves its maximal value, and a new strategy π_{n+1} is obtained accordingly;

d. If the obtained strategy Π_{n+1} is completely equal to the strategy Π_n obtained through the last iteration, namely $\Pi_{n+1}=\Pi_n$, then the iteration is stopped, and the optimized strategy is obtained. Otherwise, return to step 2 and let n = n+1.

4. Case study

4.1 Case study – Emergency resources demand prediction using case-based reasoning

The prediction process above can be applied not only in the prediction on emergency resource demand for the public emergencies that have not yet happen, but also in the prediction on emergency resource demand for occurred public emergencies.

Given that the city T plans to conduct a prediction on the demand of emergency resource when the earthquake occurs, and there are four cases for this type of Emergency response in the case library, expressing as $C=(C_1, C_2, C_3, C_4)$, and each case includes the demand information of quantity, quality and type of corresponding emergency resource, as shown in Table 2.

Case	Emergency resource type	Emergency resource quantity	Emergency resource quality
C_1	Tent	100,000	Excellent
	Clean water	150,000kg	Excellent
	Blood plasma	2000 ml	Excellent
C_2	Instant noodle	140,000	Excellent
	Quilt	100,000	Excellent
	Cotton dress	100,000	Excellent
C_3	Stretcher	5000	Excellent
	Tent	150,000	Excellent
	Food	200,000 kg	Excellent
C_4	Cloth	120,000	Excellent
	Drinking water	200,000 kg	Excellent
	Fresh vegetables	150,000 kg	Excellent

Table 2. Simplified Instance of Case Information Library.

Given that the emergency rescue plans for this type of Emergency response are the same. Through the risk analysis, five characteristic factors reflecting the characteristics of Emergency response are selected, meaning that the characteristic factor set B is composed of hazard intensity, disaster-affected population, direct economic losses, stricken area and duration time of disaster, and the membership function of four cases to five characteristic factors is as follows respectively:

$$
\begin{aligned}
n_{C_1}(b) &= \frac{0.8}{b_1}+\frac{0.7}{b_2}+\frac{0.4}{b_3}+\frac{0.3}{b_4}+\frac{0.6}{b_5} \\
n_{C_2}(b) &= \frac{0.6}{b_1}+\frac{0.6}{b_2}+\frac{0.8}{b_3}+\frac{0.9}{b_4}+\frac{0.7}{b_5} \\
n_{C_3}(b) &= \frac{0.4}{b_1}+\frac{0.6}{b_2}+\frac{0.8}{b_3}+\frac{0.8}{b_4}+\frac{0.6}{b_5} \\
n_{C_4}(b) &= \frac{0.9}{b_1}+\frac{0.8}{b_2}+\frac{0.8}{b_3}+\frac{0.8}{b_4}+\frac{0.7}{b_5}
\end{aligned} \tag{27}
$$

Given that an Emergency response occurs now, and it needs to conduct a prediction on its emergency resource demands. Given that the emergency resource demand prediction plan for this Emergency response expressed as T, and its membership function can expressed as the formula below:

$$n_T(b) = \frac{0.5}{b_1} + \frac{0.7}{b_2} + \frac{0.7}{b_3} + \frac{0.8}{b_4} + \frac{0.6}{b_5} \qquad (28)$$

It can be calculated that $\overline{n}(b_1)$=0.68, $\overline{n}(b_2)$=0.68, $\overline{n}(b_3)$=0.7, $\overline{n}(b_4)$=0.65 and $\overline{n}(b_5)$=0.65, further it can be obtained that $\delta(b_1)$=0.192, $\delta(b_2)$=0.083, $\delta(b_3)$=0.173, $\delta(b_4)$=0.05 and $\delta(b_5)$=0.05.

Therefore, the weight value of each characteristic factor can be calculated as follows: w_1=0.35, w_2=0.15, w_3=0.32, w_4=0.09 and w_5=0.09. According to the similarity calculation method, the similarity of each case is calculated as follows: $sim(T, C_1)$=0.36, $sim(T, C_2)$=0.66, $sim(T, C_3)$=0.86 and $sim(T, C_4)$=0.66. All above are shown in Table 2, and the similarity of each case is ordered as follows:

$$sim(T, C_1) < sim(T, C_2) = sim(T, C_4) < sim(T, C_3) \qquad (29)$$

It can be seen from the calculations above that this Emergency response is similar to the case C_3 in the case library, so that the emergency resource demand prediction results of this Emergency response are similar to that of the case C_3. Consequently, the conclusions of prediction on this emergency resource demand can be drawn by correcting and adjusting the emergency resource demand analysis results of the case C_3.

Attribute No.	Weight of each attribute	Case 1	Case 2	Case 3	Case 4
Disaster intensity	0.35	Extra large	Large	Large	Large
Disaster-affected population (10,000 peoples)	0.15	0.8	0.5	0.4	0.6
Direct economic losses (10,000 yuan)	0.32	100	11	10.6	9
Stricken area (km^2)	0.09	5	3	2	2
Duration time of disaster (day)	0.09	10	6	6	5
Similarity with current case		0.36	0.66	0.86	0.66

Table 3. Characteristic Factor Information of each Case.

4.2 Case study – Application of multi-stage location planning model in optimizing location of emergency resource base stations

In this section, the author makes an optimization of location planning of emergency resource base stations in City T according to the specific circumstances of City T. City T has jurisdiction over 15districts including Heping, Hedong, Nankai, Hexi, Hebei, Hongqiao, Tanggu, Hangu, Dagang, Dongli, Xiqing, Jinnan, Beichen, Wuqing, Baodi, and 3 counties such as Jinghai, Ninghe and Jixian.. The distribution of major districts and counties (the six districts including Heping, Hedong, Nankai, Hexi, Hebei, Hongqiao are called as a unified urban) is shown in Figure 5.

Fig. 5. Administrative Map of City T.

For the sake of computational convenience, and the visual presentation method, in this paper, distances between these administrative districts are presented with straight-line distances, and the central areas of these administrative districts with high concentration of population are determined as the ends of distance calculations, Table 4 shows the straight-line distances between these districts and the population distribution of various administrative districts.

Distance (km)	Urban district	Tanggu	Hangu	Dagang	Dongli	Xiqing	Jinnan	Beichen	Wuqing	Baodi	Jixian	Ninghe	Jinghai
Urban district	0	39.3	53	36.3	9.7	17.3	20.3	13.3	32.3	67	104.3	57	31.7
Tanggu	39.3	0	27.7	26	30	56.7	23.3	49.3	65.3	81.7	115.3	36	63.3
Hangu	53	27.7	0	60.7	50.7	69.3	45.7	58	67.3	61	95	8.7	83.3
Dagang	36.3	26	60.7	0	28.7	49.3	17.3	49.3	68.3	96.7	132.7	61.3	46.7
Dongli	9.7	30	46	28.7	0	26.7	12	21.7	40	70	106.7	50.7	37.3
Xiqing	17.3	56.7	69.3	49.3	26.7	0	36	13.3	26.7	68.7	106	72	24
Jinnan	20.3	23.3	45.7	17.3	12	36	0	32.7	51.7	80	116.7	52	40
Beichen	13.3	49.3	58	49.3	21.7	13.3	32.7	0	19.3	54	94.7	60	36
Wuqing	32.3	65.3	67.3	68.3	40	26.7	51.7	19.3	0	44	80	66.7	50
Baodi	67	81.7	67.3	96.7	70	68.7	80	54	44	0	37.3	61	92.7
Jixian	104.3	115.3	95	132.7	106.7	106	116.7	94.7	80	37.3	0	87.3	130
Ninghe	57	36	8.7	61.3	50.7	72	52	60	66.7	61	87.3	0	88
Jinghai	31.7	63.3	83.3	46.7	37.3	24	40	36	50	92.7	130	88	0
Population (10000)	389	45	17	32	30.4	31	42	34	82	65	78	36	50

Table 4. The straight-line distances between centers of various administrative districts.

The 15 administrative districts are respectively defined in Table 5. To divide the distances by the speed of 80km / h, the travel times are listed in Table 6.

Urban district	Tanggu	Hangu	Dagang	Dongli	Xiqing	Jinnan	Beichen	Wuqing	Baodi	Jixian	Ninghe	Jinghai
A1	A2	A3	A4	A5	A6	A7	A8	A9	A10	A11	A12	A13

Table 5. Signs of various administrative districts.

Time (min)	Urban	Tanggu	Hangu	Dagang	Dongli	Xiqing	Jinnan	Beichen	Wuqing	Baodi	Jixian	Ninghe	Jinghai
Urban	0	29.48	39.75	27.23	7.28	12.98	15.23	9.98	24.23	50.25	78.23	42.75	23.78
Tanggu	29.48	0	20.78	19.50	22.50	42.53	17.48	36.98	48.98	61.28	86.48	27.00	47.48
Hangu	39.75	20.78	0	45.53	38.03	51.98	34.28	43.50	50.48	45.75	71.25	6.53	62.48
Dagang	27.23	19.50	45.53	0	21.53	36.98	12.98	36.98	51.23	72.53	99.53	45.98	35.03
Dongli	7.28	22.50	34.50	21.53	0	20.03	9.00	16.28	30.00	52.50	80.03	38.03	27.98
Xiqing	12.98	42.53	51.98	36.98	20.03	0	27.00	9.98	20.03	51.53	79.50	54.00	18.00
Jinnan	15.23	17.48	34.28	12.98	9.00	27.00	0	24.53	38.78	60.00	87.53	39.00	30.00
Beichen	9.98	36.98	43.50	36.98	16.28	9.98	24.53	0	14.48	40.50	71.03	45.00	27.00
Wuqing	24.23	48.98	50.48	51.23	30.00	20.03	38.78	14.48	0	33.00	60.00	50.03	37.50
Baodi	50.25	61.28	50.48	72.53	52.50	51.53	60.00	40.50	33.00	0	27.98	45.75	69.53
Jixian	78.23	86.48	71.25	99.53	80.03	79.50	87.53	71.03	60.00	27.98	0	65.48	97.50
Ninghe	42.75	27.00	6.53	45.98	38.03	54.00	39.00	45.00	50.03	45.75	65.48	0	66.00
Jinghai	23.78	47.48	62.48	35.03	27.98	18.00	30.00	27.00	37.50	69.53	97.50	66.00	0

Table 6. The travel times between centers of various administrative districts.

Table 7-9 show the minimum numbers of emergency resource base stations covering all the administrative districts and specific sites, in the case of that emergency response time standards are 30min, 45min and 60min respectively.

1. When the emergency response time standard is set as 30min

District	Districts within a 30-min drive
A1	A2,A4,A5,A6,A7,A8,A9,A13
A2	A1,A3,A4,A5,A7,A12
A3	A2,A12
A4	A1,A2,A5,A7
A5	A1,A2,A4,A6,A7,A8,A9,A13
A6	A1,A5,A7,A8,A9,A13
A7	A1,A2,A4,A5,A6,A8,A13
A8	A1,A5,A6,A7,A9,A13
A9	A1,A5,A6,A8,
A10	A11
A11	A10
A12	A2,A3

Table 7. The coverage for different administrative districts in T=30min.

Using WinQSB to find the solution of set covering model, the optimal solution of z = 4 is obtained, that is, to cover all demand sites, four emergency resource base stations are necessary. Then using MCLP model, and having the p-value increase continuously from 1 to 4, using WinQSB, solution of the model can be obtained as follows:

When p = 1, the solution is: A5 = 1, Y1 = Y2 = Y4 = Y6 = Y7 = Y8 = Y9 = Y13 = 1, z = 705, it means that if an emergency resource base station is built in the Dongli District, it can cover the urban district, Tanggu, Dagang, Xiqing, Jinnan, Beichen, Wuqing and Jinghai in 30 minutes, the total number of the population in these districts is 7.05 million.

When p = 2, the solution is: A1 = A2 = 1, Y1 = Y2 = Y3 = Y4 = Y5 = Y6 = Y7 = Y8 = Y9 = Y12 = Y13 = 1, z = 788.4, it means that if two emergency resource base stations are built in the urban district and Tanggu respectively, they can cover the urban district, Tanggu, Hangu, Dagang,Dongli, Xiqing, Jinnan, Beichen, Wuqing, Ninghe and Jinghai in 30 minutes, the total number of the population in these districts is 7.884 million.

When p = 3, the solution is A1 = A2 = A10 = 1, Y1 = Y2 = Y3 = Y4 = Y5 = Y6 = Y7 = Y8 = Y9 = Y11 = Y12 = Y13 = 1, z = 866.4, it means that if two emergency resource base stations are built in the urban district, Tanggu and Baidi respectively, they can cover the urban district, Tanggu, Hangu, Dagang, Dongli, Xiqing, Jinnan, Beichen, Wuqing, Jixian County, Ninghe and Jinghai in 30 minutes, the total number of the population in these districts is 8.664 million.

When p = 4, the solution is A2 = A5 = A10 = A11 = 1, Y1 = Y2 = Y3 = Y4 = Y5 = Y6 = Y7 = Y8 = Y9 = Y10 = Y11 = Y12 = Y13 = 1, z = 931.4, it means that if four emergency resource base stations are built in Tanggu,Dongli, Baidi and Jixian County respectively, they can cover all administrative districts in 30min, the total number of the population in these districts is 9.314 million.

2. When the emergency response time standard is set as 45min

District	Districts within a 45-min drive
A1	A2,A3,A4,A5,A6,A7,A8,A9,A12,A13
A2	A1,A3,A4,A5,A6,A7,A8,A12
A3	A1,A2,A5,A7,A8,A12
A4	A1,A2,A5,A6,A7,A8,A13
A5	A1,A2,A3,A4,A6,A7,A8,A9,A12,A13
A6	A1,A2,A4,A5,A7,A8,A9,A13
A7	A1,A2,A3,A4,A5,A6,A8,A9,A12,A13
A8	A1,A2,A3,A4,A5,A6,A7,A9,A10,A12,A13
A9	A1,A5,A6,A7,A8,A10,A13
A10	A8,A9,A11,A13
A11	A10
A12	A1,A2,A3,A5,A7,A8

Table 8. The coverage for different administrative districts in T=45min.

Through the same calculation process as above, the optimal solution of z = 2 is obtained, that is, to cover all demand sites, two emergency resource base stations are necessary. Then using MCLP model and having the p-value increase continuously from 1 to 2, using WinQSB, solution of the model can be obtained as follows:

When p = 1, the solution is A8 = 1, Y1 = Y2 = Y3 = Y4 = Y5 = Y6 = Y7 = Y9 = Y10 = Y12 = Y13 = 1, z = 819.4, it means that if a emergency resource base station is build in Beichen, it can cover the urban district, Tanggu, Hangu and Dagang, Dongli, Xiqing, Jinnan, Wuqing, Baodi, Ninghe and Jinghai within 45min, the total number of the population in these districts is 8.194 million.

When p = 2, the solution is A8 = A10 = 1, Y1 = Y2 = Y3 = Y4 = Y5 = Y6 = Y7 = Y8 = Y9 = Y10 = Y11 = Y12 = Y13 = 1, z = 931.4, it means that if two emergency resource base stations are built in Beichen and Baodi respectively, they can cover all the administrative regions in 45-min, the total number of the population in these districts is 9.314 million.

3. When the emergency response time standard is set as 60 min

District	Districts within a 60-min drive
A1	A2,A3,A4,A5,A6,A7,A8,A9,A10,A12,A13
A2	A1,A3,A4,A5,A6,A7,A8,A9,A12,A13
A3	A1,A2,A4,A5,A6,A7,A8,A9,A10,A12
A4	A1,A2,A3,A5,A6,A7,A8,A9,A12,A13
A5	A1,A2,A3,A4,A6,A7,A8,A9,A10,A12,A13
A6	A1,A2,A3,A4,A5,A7,A8,A9,A10,A12,A13
A7	A1,A2,A3,A4,A5,A6,A8,A9,A10,A12,A13
A8	A1,A2,A3,A4,A5,A6,A7,A9,A10,A12,A13
A9	A1,A2,A3,A4,A5,A6,A7,A8,A10,A11,A12,A13
A10	A1,A3,A5,A6,A7,A8,A9,A11,A12
A11	A9,A10
A12	A1,A2,A3,A4,A5,A6,A7,A8,A9,A10

Table 9. The coverage for different administrative districts in T=60min.

Through the same calculation process as above, the optimal solution of z = 2 is obtained, that is, to cover all demand sites, two emergency resource base stations are necessary. Then using MCLP model and having the p-value increase continuously from 1 to 2, using WinQSB, solution of the model can be obtained as follows:

When p = 1, the solution is A9 = 1, Y1 = Y2 = Y3 = Y4 = Y5 = Y6 = Y7 = Y8 = Y10 = Y11 = Y12 = Y13 = 1, z = 849.4, it means that if a emergency resource base station is build in Wuqing, it can cover the urban district, Tanggu, Hangu, Dagang, Dongli, Xiqing, Jinnan, Beichen, Baodi, Jixian County, Ninghe and Jinghai within 60min, the total number of the population in these districts is 8.494 million.

When p = 2, the solution is A2 = A9 = 1, Y1 = Y2 = Y3 = Y4 = Y5 = Y6 = Y7 = Y8 = Y9 = Y10 = Y11 = Y12 = Y13 = 1, z = 931.4, it means that if two emergency resource base stations are built in Tanggu and Wuqing respectively, they can cover all the administrative regions in 60 min, the total number of the population in these districts is 9.314 million.

4. Analysis of the result

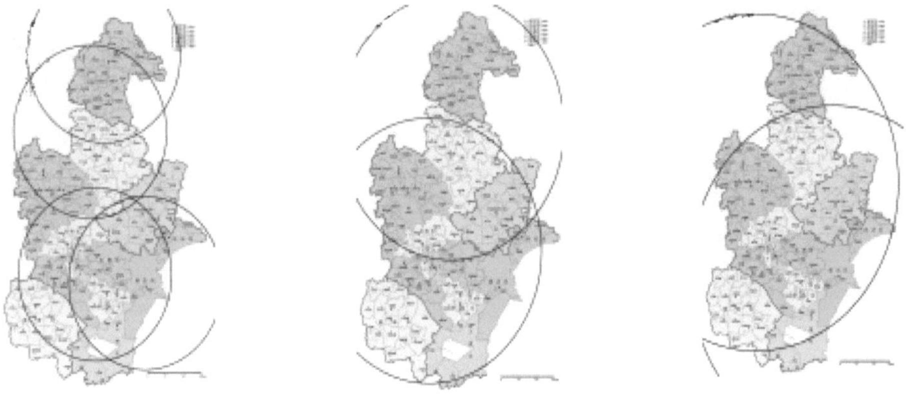

30min 45min 60min

Fig. 6. Results Map of City T.

Comparing the above results under the three emergency response time standards, we can find that:

1. Though the solutions of set covering model and maximum coverage model, it is found that within the area of a 30-min drive (response time standard), four emergency resource base stations are the most reasonable and can meet the demand, and when the number of base stations is less than 4, the emergency demand can not be met, and when the number is more than 4, unnecessary waste and redundant coverage will be made, and through MCLP model, the construction site of the four base stations can be determined as Tanggu, Dongli, Baodi and Jixian.
2. When the emergency response time standards are set as 45min and 60min, the result shows a maximum of two emergency resources is enough to cover all administrative regions of City T, but for different time standards, the base stations should be built in different districts, when the time standard is 45min, they should be built in Beichen and Baodi, and when the time standard is 60min, they should be built in Tanggu and Wuqing.
3. Because this result is obtained through simplifying the actual problems appropriately, can provide a reference for the actual decision-making, but there may be some errors, so the research of planning method more precise and close to the actual is required.

4.3 Case study – Appropriate allocation of emergency resources

It is supposed that there are 4 dangerous emergency zones in the emergency process, namely $N = 4$; total available amount of emergency resources is 12, namely $X = 12$, and X represents the total amount of available resources. Risk value when each scene is allocated with different amounts of emergency resources is listed in Table 10.

Emergency resources are dispatched according to the emergency resource allocation model, so as to achieve the optimization objective of minimizing the sum of risk values at various emergency zones in the following processing steps:

1. Suppose that the emergency response process can be divided into 4 stages, namely 4 emergency zones, i.e. $k=1,2,3,4$;
2. The emergency resource allocation objective at each stage is L_k, and the total emergency objective is expressed as formula (7) when the emergency stage reaches $k+1$;
3. There is only one state variable $x(k)$ in the example, and $\sum_{k=1}^{N} x(k) = 12$;

Emergency resource amount	Risk value of zones			
	A	B	C	D
2	20	31	27	38
3	16	26	25	35
4	11	21	23	33

Table 10. Emergency resource amount and the risk value of zones.

4. Suppose that in the available allocation scheme, the emergency resource amount provided for each emergency zone changes from 2 to 4, namely $2 \le u_k \le 4$;
5. In that case, formula $x(k+1) = f(x(k), u(k), w(k))$ can be written as $x_{k+1} = u_k + x_k$;
6. Parameters are substituted into formula (8) to obtain

$$\begin{cases} J_{k+1}(x_{k+1}) = \min\limits_{2 \le u_k \le 4} \sum\limits_{k=1}^{N-1} L_k = \\ \qquad \min\limits_{2 \le u_k \le 4} (L_{k+1}(x_{k+1}, u_{k+1}, w_{k+1}) + J_k(x_k)) \\ J_0(x_1) = 0 \end{cases}$$

The optimized emergency resource allocation is calculated under the supposed scene of accidents with the mathematical mode of optimized emergency resource allocation, so as to obtain the optimized allocation result.

For the risk zone A, alternative resource allocation decision is 2-4, corresponding total risk value is from 20 to 11, so risks can be minimized to 11 with the decision 4 in the risk zone A (Table 11). The optimized decision scheme of zones B, C and D can be obtained with the same method (Table 12-14). Under 4 dangerous scenes, optimized allocation of limited resources is 4 in zone A, 4 in zone B, 2 in zone C and 2 in zone D, thus the total risk value is reduced to 97.

x_1	Corresponding risk value of zone A at the first stage of the resource allocation strategy u_1			$\min\sum L_1$
	2	3	4	
2	20+0	-	-	20
3	20+0	16+0	-	16
4	20+0	16+0	11+0	11

Note: x_1 is the corresponding amount of resource demand under each resource allocation decision at the first stage; $\min\sum L_1$ is the minimum risk value corresponding to different decisions at the first stage.

Table 11. Resource allocation strategies and corresponding risk value of zone A (the first stage).

x_2	Corresponding risk value of zone B at the second stage of the resource allocation strategy u_2			$\min\sum L_2$
	2	3	4	
4	31+20	-	-	51
5	31+16	26 + 20	-	46
6	31+11	26 + 16	21+20	42
7	31+11	26 + 11	21+16	37
8	31+11	26+ 11	21+11	32

Note: x_2 is the corresponding amount of resource demand under each resource allocation decision at the second stage; $\min\sum L_2$ is the minimum risk value corresponding to different decisions at the second stage.

Table 12. Resource allocation strategies and corresponding risk value of zone B (the second stage).

x_3	Corresponding risk value of zone C at the third stage of the resource allocation strategy u_3			$\min\sum L_3$
	2	3	4	
8	27 + 42	25 + 46	23 + 51	69
9	27 + 37	25 + 42	23+46	64
10	27 + 32	25 + 37	23+42	59

Note: x_3 is the corresponding amount of resource demand under each resource allocation decision at the third stage; $\min\sum L_3$ is the minimum risk value corresponding to different decisions at the third stage.

Table 13. Resource allocation strategies and corresponding risk value of zone C (the third stage).

x_4	Corresponding risk value of zone D at the fourth stage of the resource allocation strategy u_4			$\min\sum L_4$
	2	3	4	
12	38 + 59	35 + 64	33 + 69	97

Note: x_4 is the corresponding amount of resource demand under each resource allocation decision at the fourth stage; $\min\sum L_4$ is the minimum risk value corresponding to different decisions at the fourth stage.

Table 14. Resource allocation strategies and corresponding risk value of zone D (the fourth stage).

4.4 Case study – Optimal dispatching of emergency resources

In order to validate the dynamic optimization process of emergency resource scheduling of sudden public events with the Markov decision process, here the rationality and practicability of dynamic optimization method of emergency resource scheduling based on the Markov decision process is proved through analysis and explanation by examples.

Now it is supposed that an earthquake disaster takes place in a city, which is likely to cause two secondary disasters S_1 and S_2, namely the state space of this earthquake disaster is $S=\{S_1,S_2\}$. where, S_1 and S_2 represent the initial event S, namely secondary accidents are likely to be obtained from evolution of the earthquake disasters.

Then it is supposed that only one emergency resource R is required in emergency of this sudden public event, and the emergency time standard T is 2 time units. So long as enough amounts of emergency resources is transported to sudden public event sites within the standard time under the state of an event, then the sudden public event can be under control. If the amount of resource R transported to the scene of accident is insufficient, then the sudden public event can only be partially controlled (expressed as the availability of the emergency resources a), and the range of values of a is $0\%\sim100\%$. It is also supposed that the relationship between the demand and the availability of the emergency resource R is shown in Table 15.

State S	Demand of the emergency resource R	Availability $a/\%$
S_1	120	100
	80	80
S_2	200	100
	120	80
	80	40

Table 15. Relationship between the demand and the availability of the resources under different states of S.

It is supposed that 4 emergency resource sites around the sudden public events can cover this event site in 2 time units, as shown in Figure 7. As can be seen from the Figure 7, only the site A is in 1 time unit, while the site B is in the space of 1.5 time units, and both the site C and site D are in 2 time units. The amount of stored emergency resources at each site is also shown in the Figure 7.

Now it is stipulated that the emergency satisfaction is defined as the emergency success. Under the circumstances, the emergency resource site A is closest to the scene of accident X, therefore it is necessary to choose the site A, so as to satisfy the shortest emergency time. When the accident is under the state of S_1, 80 units of emergency resource R are transported from the emergency resource site A to the event site X, so the availability of the resources is only 80% under the state of S_1, which is unable to completely satisfy the emergency demand. Under the circumstances, it is necessary to be supported by the sites B, C and D. When the sudden public event is under the state of S_1, 40 units of emergency resources can be transported from the emergency resource sites B, C and D to the site X, so that the total accumulated amount achieves 120 units, and the amount of resources achieves the availability of 100%; but it is necessary to transport 120 units of resources from the sites B, C and D to the site *X* under the state of S_2, so that the total accumulated amount achieves 200 units, and the availability achieves 100%.

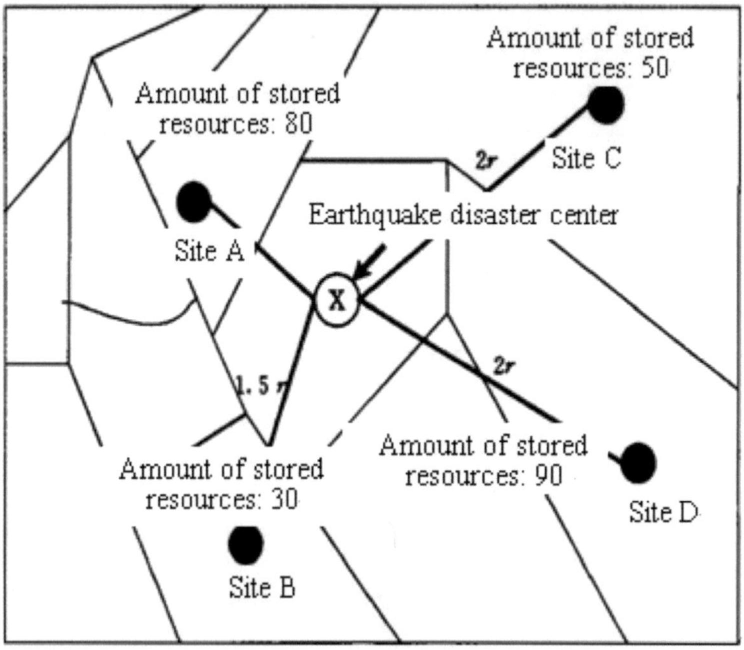

Fig. 7. Distribution map of the emergency resources sites.

Therefore, the state space in this example is $S = \{S_1, S_2\}$, the decision space can be expressed as the scheduling scheme, and the reward can be expressed as the cost or scheduling duration, as shown in Table 16.

			Site A	Site B	Site C	Site D	Total	Availability
Amount of stored emergency resources at each site			80	30	50	90	250	
State S_1	Scheme I	Transportation quantity	80	30	0	10	120	100%
		Cost	10	30	0	30	70	
	Scheme II	Transportation quantity	80	30	10	0	120	100%
		Cost	10	30	20	0	60	
State S_2	Scheme III	Transportation quantity	80	30	20	70	200	100%
		Cost	10	30	40	150	230	
	Scheme IV	Transportation quantity	80	30	50	40	200	100%
		Cost	10	30	80	100	220	

Table 16. Statistical table of the transportation quantity and cost.

The state transition probability of sudden public events can be obtained using the Domino effect analysis method, as shown in Table 17. This process is complex, so it is unnecessary to go into details in this section.

State i	Decision	Transition probability		Reward		Expected timely reward
	$V_{(i)}^{K}=\delta(i)$	$p_{i1}^{\delta(i)}$	$p_{i2}^{\delta(i)}$	$r_{i1}^{\delta(i)}$	$r_{i2}^{\delta(i)}$	$q_{i}^{\delta(i)}$
S_1	*Scheme I*	0.3	0.7	-10	-3	-5.1
	Scheme II	0.8	0.2	-12	-11	-11.8
S_2	*Scheme III*	0.6	0.4	-9	-8	-8.6
	Scheme IV	0.5	0.5	-4	-5	-4.5

Note: The reward uses emergency consumption time, and is negative as a consequence.

Table 17. The transition probability and reward.

The solution process is as follows:

There are two states in this case, and two decisions under each state, namely the scheduling schemes. $u_{(1)}^{1}$ represents selecting the scheduling scheme I when the event is under the state of S_1; $u_{(1)}^{2}$ represents selecting the scheduling scheme II when the event is under the state of S_2; $u_{(2)}^{1}$ represents selecting the scheduling scheme III when the event is under the state of S_2; $u_{(2)}^{2}$ represents selecting the scheduling scheme IV when the event is under the state of S_2.

Expected timely reward: $q_1^1=\sum_{j=1}^{m}p_{ij}r_{ij}=0.3\times(-10)+0.7\times(-3)=-5.1$, likewise $q_1^2=-11.8$, $q_2^1=-8.6$, $q_2^2=-4.5$.

In the first step, select the initial strategy π_0; let $\delta_0(1)=u_{(1)}^{1}$, $\delta_0(2)=u_{(2)}^{1}$, that is, select the scheduling scheme I under the state of S_1, and select the scheduling scheme II under the state of S_2, then there is $P=\begin{bmatrix}0.3 & 0.7\\ 0.6 & 0.4\end{bmatrix}$, $Q=\begin{bmatrix}-5.1\\ -8.6\end{bmatrix}$.

In the second step, calculate the fixed value, and estimate the initial strategy

$$\begin{cases} v+f_1=-5.1+0.3f_1+0.7f_2 \\ v+f_2=-8.6+0.6f_1+0.4f_2 \end{cases},$$

let $f_2=0$, $v^{(0)}=-6.99$, $f_1^{(0)}=2.69$, $f_2^{(0)}=0$ is obtained through solving the equations set.

The third step is the strategy improvement program, in which the improvement strategy π_1 is obtained.

For the state S_1, select a strategy $u_1^{(k)}$, so as to maximize $q_1^k + p_{11}^k f_1^{(0)} + P_{12}^k f_1^{(0)}$, that is

$$\begin{cases} -5.1+0.3\times2.69+0.7\times0-2.69=-6.99 \\ -11.8+0.8\times2.69+0.2\times0-2.69=-12.388 \end{cases}$$

Select the strategy $u_{(1)}^1$. Scheme I is used for emergency resource scheduling under the state of S_1.

For the state S_2, select a strategy $u_2^{(k)}$, so as to maximize $q_2^k + p_{21}^k f_1^{(0)} + p_{22}^k f_2^{(0)}$, that is

$$\begin{cases} -8.6+0.6\times2.69+0.4\times0-0=-6.99 \\ -4.5+0.5\times2.69+0.5\times-0=-3.15 \end{cases}$$

Select the strategy $u_{(2)}^2$. Scheme IV is used for emergency resource scheduling under the state of S_2.

The improvement strategy is obtained as $\delta_1(1) = u_{(1)}^1$, $\delta_1(2) = u_{(2)}^2$ from the above computing results. The strategy π_1 is different from π_0, so no optimized strategy is obtained and it is necessary to go on iteration.

The fourth step is fixed value operation for the purpose of obtaining $v^{(1)}$, $f_1^{(1)}$, $f_2^{(1)}$

$$\begin{cases} v^{(1)}+f_1^{(1)}=-5.1+0.3f_1^{(1)}+0.7f_2^{(1)} \\ v^{(1)}+f_2^{(1)}=-4.5+0.5f_1^{(1)}+0.5f_2^{(1)} \end{cases}$$

Let $f_2^{(1)} = 0$, then $v^{(1)} = -4.75$, $f_1^{(1)} = -0.5$, $f_2^{(1)} = 0$ is obtained through solving the equations set.

In the fifth step, seek the improvement strategy π_2.

For the state S_1, there is

$$\begin{cases} -5.1+0.3\times(-0.5)+0.7\times0+0.5=-4.75 \\ -11.8+0.8\times(-0.5)+0.2\times0+0.5=-11.7 \end{cases}$$

So the strategy $u_{(1)}^1$ is still taken.

For the state S_2, there is

$$\begin{cases} -8.6+0.6\times(-0.5)+0.4\times0-0=-8.9 \\ -4.5+0.5\times(-0.5)+0.5\times0-0=-4.75 \end{cases}$$

So the strategy $u_{(2)}^2$ is still taken.

As a result, $\delta_1(1) = u_{(1)}^1$, $\delta_1(2) = u_{(2)}^2$ is obtained, which is exactly the same as the previous iteration results, so the optimized strategy is obtained as π_1. That is, take the scheduling scheme I when the sudden event is under the state of S_1, and take the scheduling scheme IV when the sudden event is under the state of S_2.

5. Conclusion and future researches

When we design an emergency rescue system, we need to coordinate the manpower with the financial, material resources. It is a complicated process to optimally allocate various elements within a system. It involves a wide range of contents. Repeated researches should be made on several theories and methods. Designing of an emergency rescue system covers the following four aspects which have been cross-linked each other essentially, that are 1) Demand Forecasting of Emergency Resources;2) Optimal Site Selection for the Base Station of Emergency Resources;3) Appropriate Allocation of Emergency Resources;4) Optimal Dispatching of Emergency Resources. Here, it proposed the overall and detailed methods to fulfill these four aspects.

In the future it is necessary to develop a computer system, so that these methods can adapt to the dynamic optimization process of emergency resource scheduling scheme under complex conditions such as many times of derivation and many kinds of resources etc., and it can more greatly satisfy the actual need.

6. Acknowledgments

The authors appreciate the support of No.47 China's Postdoc Foundation(NO:20100470305) and the national science foundation for the youth (NO:41105099).

7. References

[1] Clancey, W.J., (1985). Heuristic Classification. *Artificial Intelligence,* 27: pp289-350.

[2] Bachant, J., & McDermott, J., (1984). R1 Revisited: Four years in the Trenches. *The AI Magazine,* 5(iii).

[3] Coenen, F. & Bench-Capon, T.J.M. (1992). Maintenance and Maintainability in Regulation Based Systems. *ICL Technical Journal,* May 1992, pp.76-84.

[4] Watson, I.D., Basden, A. & Brandon, P.S. (1992a). The Client Centered Approach: Expert System Development. *Expert Systems* 9(iv): pp.181-88.

[5] Watson, I.D., Basden, A. & Brandon, P.S. (1992b). The Client Centered Approach: Expert System Maintenance. *Expert Systems* 9(iv): pp189-96..

[6] Watson, I.D., & Abdullah, S. (1994). Developing Case-Based Reasoning Systems: A Case Study in Diagnosing Building Defects. In, Proc. *IEE Colloquium on Case-Based Reasoning: Prospects for Applications*, Digest No: 1994/057, pp.1/1-1/3.

[7] Schank, R.C. & Abelson, R.P. (1977). *Scripts, Plans, Goals and Understanding*. Erlbaum, Hillsdale, New Jersey, US.

[8] Schank, R. (1982). *Dynamic memory: a theory of reminding and learning in computers and people*. Cambridge University Press, Cambridge, UK.

[9] Kolodner, J. L. (1983a). Maintaining Organization in a Dynamic Long-Term Memory. *Cognitive Science,* 7(iv): pp.243-80.

[10] Kolodner, J. L. (1983b). Reconstructive Memory: A Computer Model. *Cognitive Science,* 7(iv): pp.281-28.

[11] Kolodner, J. L. (1993). *Case-Based Reasoning*. Morgan Kaufmann.

[12] Porter, B.W. & Bareiss, E.R. (1986). PROTOS: An experiment in knowledge acquisition for heuristic classification tasks. *In Proceedings of the First International Meeting on Advances in Learning (IMAL)*, Les Arcs, France, pp.159-74.

[13] Bareiss, E. R.,(1988). *PROTOS: A Unified Approach to Concept Representation, Classification, and learning*. Ph.D. thesis, Department. of Computer Science, University of Texas.

[14] Ashley, K.D. (1988). Arguing by Analogy in Law: A Case-Based Model. In D.H. Helman (Ed.), *Analogical Reasoning: Perspectives of Artificial Intelligence, Cognitive Science, and Philosophy*. D. Reidel.

[15] Rissland, E.L., & Skala , D.B. (1989). Combining case-based and rule-based reasoning: A heurestic approach. In, *Eleventh International Joint Conference on Artificial Intelligence*, IJCAI-89: pp. 524-30, Detroit, Michigan.

[16] Sharma, S. & Sleeman, D. (1988). REFINER: A Case-Based Differential Diagnosis Aide for Knowledge Acquisition and Knowledge Refinement. In, *EWSL 88; Proc. European Working Session on Learning*: pp201-10.

[17] Keane, M., (1988). Where's the Beef? The absence of pragmatic factors in theories of analogy. In, *ECAI-88*: pp.327-32.

[18] Althoff, K.D. (1989). Knowledge acquisition in the domain of CBC machine centres: the MOLTKE approach. In, *EKAW-89, Third European Workshop on Knowledge-Based Systems*, Boos, J., Gaines, B. & Ganascia, J.G. (eds.), pp.180-95. Paris, July 1989.

[19] Richter, A.M. & Weiss, S. (1991). Similarity, uncertainty and case-based reasoning in PATDEX. In, *Automated reasoning, essays in honor of Woody Bledsoe. Kluwer* R.S. Boyer (ed.): pp249-265.

[20] Aamodt, A., (1989). Towards robust expert systems that learn from experience - an architectural framework. In, EKAW-89: Third European Knowledge Acquisition for Knowledge-Based Systems Workshop, Boose, J., Gaines, B. & Ganascia J.-G. (eds.), pp.311-326. Paris, July 1989.

[21] Aamodt, A. (1991). A Knowledge intensive approach to problem solving and sustained learning, PhD. dissertation, University of Trondheim, Norwegian Institute of Technology, May 1991. University Microfilms PUB 92-08460.

[22] Watson, I.D., & Abdullah, S. (1994). Developing Case-Based Reasoning Systems: A Case Study in Diagnosing Building Defects. In, Proc. *IEE Colloquium on Case-Based Reasoning: Prospects for Applications*, Digest No: 1994/057, pp.1/1-1/3.

[23] Yang, S., & Robertson, D. (1994). A case-based reasoning system for regulatory information. In, Proc. *IEE Colloquium on Case-Based Reasoning: Prospects for Applications*, Digest No: 1994/057, pp.3/1-3/3.

[24] Moore, C.J., Lehane, M.S. & Proce, C.J. (1994). Case-Based Reasoning for Decision Support in Engineering Design. In, Proc. *IEE Colloquium on Case-Based Reasoning: Prospects for Applications*, Digest No: 1994/057, pp.4/1-4/4.

[25] Oxman, R.E., (1993a). PRECEDENTS: Memory structure in design case libraries. In CAAD Futures 93, Elsevier Science Publishers.

[26] Oxman, R.E., (1993b). Case-based design support: Supporting architectural composition through precedent libraries. Journal of Architectural Planning Research.

[27] Venkatamaran, S., Krishnan, R. & Rao, K.K. (1993). A rule-case based system for image analysis. In, Proc. *1st. European Workshop on Case-Based Reasoning, Posters & Presentations*, 2: pp.410-15.

[28] Kitano, H. (1993). Challenges for massive parallelism. In, Proc. *13th. Int. Conference on Artificial Intelligence*, IJCAI-93: pp813-34.

[29] Roger. C. S，David. B. L．*Creativity and Learning in a Case-based Explainer*．AI，1989，40：353~385

[30] Sun Mingxi, Wu Junqing, Ai Guoqiang, et.,al. *Practical Prediction Method and Instance Analysis*. Beijing: Science and Technology Document Press, 1993.

[31] Liu Mao, Wu Zongzhi, *Introduction to Emergency Rescue - Emergency Rescue System and Programs*. BeiJixianng: Chemical Industry Press, 2004

[32] Xiong wei, *Operational Research*. BeiJixianng: Mechanical Industry Press, 2005

[33] Wayne L. Winston, *Operations Research Application and Algorithms*. BeiJixianng: Tsinghua University Press, 2006

[34] Donald Waters. *Logistics: An Introduction to Supply Chain Management*. BeiJixianng: Electronic Industry Press, 2006

[35] He Jixiananmin, Liu Chunlin, Cao Jixianedeng et al., *Emergency Management and Emergency Response System - Location, Scheduling and Algorithms*. BeiJixianng: Science and Technology Press, 2007

[36] Zhu Tan, Liu Mao and Zhao Guomin. *Study of points of urban public security planning and developing*. China's Development 2003 (4) pp.10-12

[37] Liu Mao, Zhu Tan and Zhao Guomin. *Study of urban public security emergency systems*. China's development 2003 (4) pp.13-16

[38] Chen Zhizong and You Jixiananxin. *Modeling of hierarchy location for urban disaster prevention and mitigation facilities*. Journal of Natural Disasters, 2005 Volume 14 (2) pp.131-135

[39] Sun Ying, Chi Hong, Jixiana Chuanliang. *Non-linear mixed integer planning model of multipath emergency resource scheduling*. in Operations Research and Management, 2007, pp.16.

[40] Zhou Xiaomeng, Jixianang Lizhen, Zhang Yunlong. *Study of emergency resource optimal allocation quantitative model for unforeseen accidents*. Journal of Safety and Environment, 2007, pp.6.

[41] Larson R E, Carstea J L. *Dynamic programming principle* [M]. Chen Weiji, Wang Yongxian, Yang Jiaben. Trans, Beijing:Tsinghua University Press, 1984.

[42] Bertsekas D P. *Dynamic programming: deterministic and stochastic models*[M].Trans. Xi'an: Xi'an Jiaotong University Press, 1990.

[43] Feinberg E A, Shwartzz A. *Markov decision models with weighted discounted criteria* [J]. Math O R, 1994, 19:152-168.

[44] Bouakiz M, Kebir Y. *Target-level criterion in Markov decision processes*[J]. J Optim Theory Appl, 1995, 86:1-15.

[45] Chen M, Filar J A, Liu K. *Semi-infinite Markov decision processes*[J]. Mathematical Methods of Operations Research, 2000, 51:115-117.

8

Absorption and Accumulation of Heavy Metal Pollutants in Roadside Soil-Plant Systems – A Case Study for Western Inner Mongolia

Lu Zhanyuan[1], Zhi Yingbiao[2,3*], Wang Zai-lan[4], Hua Yupeng[2], Hong Ge[2], Emmy Camada[5] and Yao Yiping[1]

[1]*Inner Mongolia Academy of Agricultural Science, Hohhot,*
[2]*Ordos College, Inner Mongolia University, Ordos,*
[3]*College of Life Science, Nanjing University, Nanjing,*
[4]*School of Environment and Natural Resources, Renmin University of China, Beijing,*
[5]*Chinese Culture Center of San Francisco, CA,*
[1,2,3,4]*China*
[5]*USA*

1. Introduction

Soil - plant system is the biosphere and pedosphere whose the basic structural unit of soil is the main target. Soil - plant systems enable human productivity but suffer from pollution damage caused by humans. Currently, the annual loadings of harmful metals in soil are (10^4 t / a): Hg 0.83, Cd 2.2, Cr 89.6, Pb 79.6, Ni 32.5, Cu 95.4, Zn 137.1, As 8.1, Se 4.1 [1,2]. Contaminated soil will directly, or by amplifying through the food chain, affect the normal function and growth of plant and even human health. At the same time, the ecosystem, through a series of physical, chemical and biological processes in the environment, provides a purification of pollutants. Beyond the loading capacity of the environmental pollution load capacity and super-threshold, its biological production will be affected, resulting in severe loss of productivity and may even directly or indirectly endanger human life and health. Phytoremediation is considered a green technology for the removal of heavy metal pollution in the ascendant [3]. With the rapid development of the national economy and the subsequent traffic pollution, negative environmental effects are becoming increasingly apparent, especially for roadside soil - plant systems. The the evidence is apparent in Shanxi Province where 5,000 km of roadside farmland was polluted by coal dust, reducing food productivity by 2 800 × 10^4kg [4]. In recent years, car exhaustion and road dust caused heavy metal pollution on the soil-plant systems on both sides of the roads, and consequently, the heavy metal content has brought stress on the structure and function of ecosystem, which increasingly exposed agricultural issues. Currently, research on the domestic and international distribution of heavy metals in soil

* Corresponding Author

on both sides of the road have been reported [5-10]. At the same time, studies on heavy metal pollution on soil-plant systems on both sides of the road have been carried out [11-17], but only limited studies are conducted on the heavy metal absorption and accumulation on the highway [18]. With the rapid development of the regional economy, the western region of Inner Mongolia produces more coal, building materials, and chemical products, and the rapid increase in traffic and road transport vehicles increased significantly [19]. In 1992, the average traffic in the Inner Mongolia Autonomous Region were 731 vehicles per day. By the end of 2004, the average number of traffic had increased to 5 171 per day, of which 2 533 are State Roads with G109 vehicles / day, G210 are State Roads with 4946 vehicles / day, and G110 are National Roads with 6 739 vehicles / day. With heavy road traffic and associated problems of automobile exhaust emissions becoming more evident, it is necessary to have studies on the accumulation of heavy metals and the potential ecological risk assessment. Therefore, in this study western Inner Mongolia Transport heavy G110 National Road, G210 National Highway and the G109 National Road are selected as study objects, along with highway green vegetable pine (*Pinus tabulaeformis* Carr.) and lobular Yang (*Populus simonii* Carr.) Heavy metal absorption and accumulation and heavy metals in rhizosphere soil and the relationship between distribution and morphology to the road along the soil - plant system bioremediation of heavy metals are investigated to provide theoretical reference.

2. Material and methods

2.1 Material

2.1.1 Characteristics of the study area

According to the study area (106 ° 35 '~ 111 ° 36'E, 38 ° 56' ~ 42 ° 28'N) and vegetation characteristics, soil type and parent material conditions, the different sampling points are divided into three types (Figure 1). Specifically: Hohhot - Baotou - Wuyuan - Linhe - Dengkou located Tumochuan and Hetao plains, there it is high underground water level and high salinity, vegetation intrazonal Salt meadow-based, low land vegetation,and zonal soil type is meadow soil, and now most of the land has been opened Cultivated as farmland: Wuhai - Uda - Etuog County - Hangjin County there the vegetation dominated by grassland desertification, desert region in western China is a special area, zonal soil type is brown soil; Dongsheng - Zhunger, Yijinhuoluo - Dongsheng - Dalate located Ordos Plateau in the east, the vegetation's Stipa (*Stipa bungeana*) constructive species of warm temperate typical steppe, but the long-term cultivation, overgrazing, native natural vegetation has been destroyed, therefore, there is only a broken hilly slope, hills, zonal soil type is chestnut soil [20].

2.1.2 Sample collection and processing

The 192 copies of simonii green vegetables' roots, stems, leaves from three loose sides of the road and samples of 384 (2 × 192) copies from the area around the root rhizosphere soil [12] were collected in the western region of Inner Mongolia National Highway G210, G110 and G109 within 20 m on both sides of the plant and soil samples by plum-shaped distribution (Figure 1).

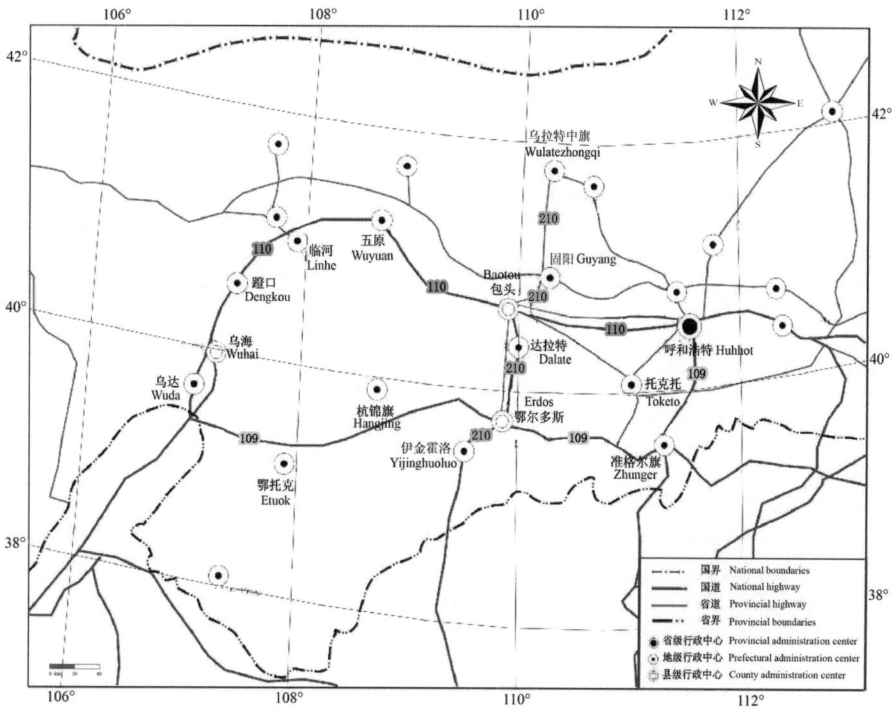

Fig. 1. The map of the sampling station.

And the simultaneous acquisition of the plants as control (CK) is from relatively clean area around the off-road and industrial pollution. Samples of plants was washed with distilled water. Plant roots, stems, leaves and rhizosphere soil samples were naturally air-dried, ground, sieved, bagged, spared, and cold stored [21].

2.1.3 Determination

Plant (roots, stems and leaves mixed sample, and overall) and soil heavy metal content was assessed by using atomic absorption spectrophotometer Spectr AA DUO, AF 610A, atomic fluorescence spectrometer; traces of heavy metals Cd, Pb, Cu, Zn, Ni, Cr were determined by dry ash method, and perform GB/T5009 in Cd, Pb, Cu, Zn, Ni, Cr atomic absorption spectrometry determination in a muffle oven 500°C dry ashing, with 1:1 nitric acid dissolved in 2 ml. The way of metal determination is as follows: the plant samples adopt nitric acid - perchloric acid digestion, atomic fluorescence spectrometry; Hg uses aqua regia digestion, atomic fluorescence spectrometry; Se uses nitric acid - perchloric acid digestion, atomic fluorescence spectrometry; rhizosphere soil of heavy metals Cu, Zn, Pb, Ni and Cr speciation experiment uses synchronized Tessier sequential extraction method [22] for the determination of the specific processes shown in Figure 2. Soil pH measured with pH meter method, the soil and water ratio of 1:1 [21].

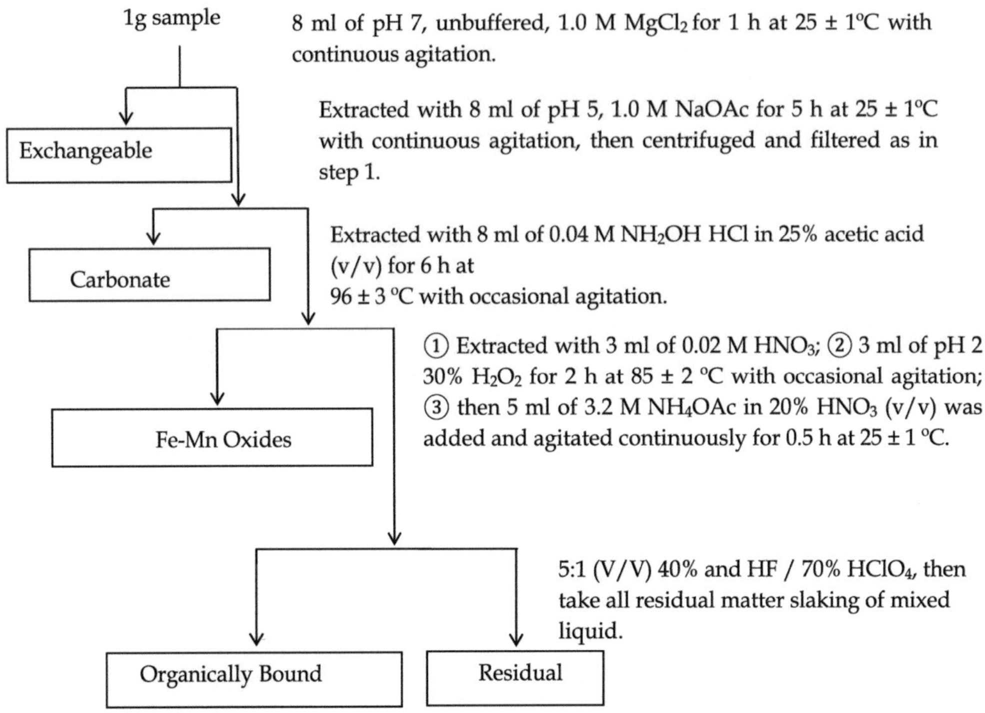

Fig. 2. Sequential extraction procedure for the speciation of heavy metal in sediment.

3. Results and analysis

By comparison among the main green vegetable's mixed samples of roots, stems, leaves in pine and poplar along the highway from the western Inner Mongolia with heavy metals in the background medium value of plants' rhizosphere soil [23,24], the heavy metals of Cr, Ni, Cu, As, Pb levels in pine and poplar is below the world soil median chemical composition. Meanwhile, the results display the character, that is, the determination value is less than Chinese soil (A level) element of the background value and is less than Inner Mongolia's background value of soil elements and also less than different soil types in study area (meadow soil, brown soil, and chestnut soil). But Zn content (110.82 mg·kg^{-1}) of poplar is significantly higher than the world median chemical composition of soil (9.0 14 mg·kg^{-1}) and China Soil (A level) element of the background value (74.2 mg·kg^{-1}), as well as Inner Mongolia soil element background values (59.1mg·kg^{-1}). Simultaneity, Zn content of poplar is significantly higher than the study area in different soil types (such as meadow soil (59.10 mg·kg^{-1}), brown soil (56.2 mg·kg^{-1}) and chestnut soil (66.9 mg·kg^{-1}) ; While Zn content of the pine (77.48 mg·kg^{-1}) is close to the Chinese soil (A level) element of the background value, but it is higher than different types of soil meadow in the study area.

Average concentration of heavy metals Zn and Hg in green plants and rhizosphere soil is similar, while other elements shown in the rhizosphere soil are higher than average levels in plants, for example, Cu in the soil content of plant is 1.6 times; The content of Pb and As in the soil is 7 times the plant; Cd, Cr, Ni in the soil are 10 times as in plants. The soil is alkaline in study area where the soil pH is 8.1. Heavy metals in different plants are relevant to soil pH and to the plant's selective absorption of different heavy metals and to the certain speciation of heavy metals in the soil [25]. Delorme *et al* [26] state that the excess accumulation of blue food is a plant containing the rhizosphere of *Thlaspi caerulescens* than *Trifolium pratense* rhizosphere microorganisms, including heavy metals Cd and Zn-resistant bacteria more than is due to lower pH of the rhizosphere.

There are differences in the content of heavy metals in different organs of plants, and this is also reflected in differences between different plants. For lobular, Hg distribution in stems and leaves and roots is more consistent. Cd, Cu, Zn, As and Se content is higher in the leaf stems and leaves than the roots; For pine, Hg content is evenly distributed among the stems, leaves and roots; Pb, Cr, Ni, Cu in the stems and leaves were higher than in the root; Cd, Zn and As were higher than in the roots stems and leaves. This difference displays the direct relationship between the elements in the soil and speciation of metals and the plant's selective absorption of various elements.Generally, contents of heavy metals in different organs at different levels of plants follow the heirarchy of Zn> Cu> Ni, Cr, As, Pb> Cd> Hg.

The same plant in different organs were also exhibit different degrees of heavy metal enrichment in its different organs. the content ratio of the three kinds of heavy metals Cr, Ni and Pb in simonii is greater than 1 but the content ratio of Cu, Zn, As and Cd in the of stems and leaves and roots is less than 1.(roots). while the content of Cr, Ni and Pb in the above-ground parts of simonii (stems and leaves is stronger than that in the underground parts (roots).Meanwhile, the aerial parts (stems and leaves) of Cu, Zn, As and Cd enrichment is weaker than the underground part (root). As previous studies showed, Cu, Zn, Pb, Cd accumulated in the aerial parts of woody plants is greater than the amount seen in underground parts, and the cumulative amount of Mn element is greater than the underground part of the aerial parts [27]. The primary organ to observe heavy metals in plants is the root, and when the presence of heavy metals in air pollution is high, the plant's leaves can also absorb a significant amount of metal [28].

There is a large difference in heavy metals distributed over different species, and absorption and utilization behavior varies over different environments and different plant cell types [29-34]. Exchangeable, carbonate bound, Fe-Mn oxides and organic matter bound heavy metals in plants can be directly or in a certain redox conditions, absorbed and utilized [35-37].

4. Discussion

Along the highway of western Inner Mongolia, studies on the main green vegetable, leaf pine, rhizosphere Yang, heavy metal content, distribution, morphology and soil heavy metal bioavailability draw the following conclusions:

1. The green plants in roadside of heavy metals content for Cr, Ni, Cu, As, Pb is lower than the background value. However, Zn content was significantly higher than background values. Rhizosphere soil of heavy metals adsorption of Cd was the highest.
2. There are some differences in heavy metals perceived in the plant content of different organs of the body, but this is also reflected in the difference exhibited between different plants. With the increasing atomic number (Cr → Pb), two kinds of heavy metals in the roots and stems of poplar and pine have shown "N" shaped changing

tendency. and In addition, heavy metals in different plants and in the different organs have the tendency of Zn> Cu> Ni, Cr, As, Pb> Cd> Hg.

3. The different organs of the same plant have various degrees of enrichments in the different heavy metals. Five kinds of heavy metals' availability phase in the plants' rhizosphere soil are different in the total amount of the percentage in sequence and heavy metals in different organs of the different plants.
4. The study reveals that the plants' absorption and utilization of heavy metal in rhizosphere soil is relevant to the percentage of availability phase, that is to say, the greater the percentage of availability phase is, the bigger enrichment of the heavy metal in the plants appears under a certain environmental condition where plants' transpiration pull and the transportation of water and nutrient shows the equal characteristics along the highway of western Inner Mongolia.

5. Acknowledgments

This work was funded by the Inner Mongolia Natural Science Foundation of China (No. 2011MS0501),by the Inner Mongolia Grassland Elite Foundation of China (201140) ,by the Inner Mongolia Talent Development Foundation of China (2011RC1228) The Postdoctoral Science Foundation of China (20060400551) and .We are very grateful to Emmy Camada of Chinese Culture Center of San Francisco in USA for the comment on draft and grammar about the manuscript. Meanwhile, we thank Cui Yan and Liu Jianping for their support in determination of heavy metal at Inner Mongolia Academy of Agricultural Science, and colleagues Mao Jianshuai, Wang zala for their assistance in the fieldwork.

6. References

[1] Dai S G, Yuan G C, Wang X R. 1997. Environmental Chemistry, Beijing: Heigher Education Press,.217 ~ 130.

[2] He S Y, Jing Y Y. 2000.Environmental Chemistry (3rd edition), Shanghai: East China University of Science and Technology Press, 127 ~ 130.

[3] Wei Z Y, Chen T B. 2001, Hyperaccumulators and phytoremediation of heavy metal contaminated soil: a review of studies in China and abroad.Acta Ecologica Sinica, 21 (7):1196 ~ 1203.

[4] Lu Y S, Luan S J, Zhu Y. 2001. Aanalyse of Environmental influence, Beijing:Higher education press, 145 ~ 180.

[5] Münch D. 1992. Soil contamination beneath asphalt roads by polynuclear aromatic hydrocarbons, zinc, lead and cadmium. Science of the Total Environment, 126(1/2):49 ~ 60.

[6] Teichman J. 1993. A survey of lead contamination in soil along interstate 880, Alamed County, California. American Industrial Hygiene Association Journal, 54(9):557 ~ 559.

[7] Härkönen J, Valkonen E, Kukkonen J, Rantakrans E, *et al.* 1995. An operational dispersion model for prediction pollution form a road.International Journal of Environment and Pollution, 5(4/6):602 ~ 610.

[8] Volker R. 2002. Influence of heavy metals on the microbial degradation of diesel fuel. Chemosphere, 49(6):559 ~ 568.

[9] Unal A. 2004. Quantification of highway vehicle emissions hot spots based upon on-board measurements.Journal of the Air & Waste Management Association, 54 (2):130 ~ 140.

[10] Jamriska M. 2004. Diesel bus emissions measured in a tunnel study. Environmental Science & Technology, 38:(24):6701 ~ 6709.

[11] Pulford I D. 2003. Phytoremediation of heavy metal-contaminated land by trees.Environment International, 29(4):529 ~ 540.

[12] Chen H M, Chen N Y, Chen Y X, *et al.* 1996. Heavy Metal Pollution of Soil-plant System.Beijing:Science Press, 210 ~ 247, 314 ~ 317.

[13] Macleod M. 2004. Application of contaminant fate and bioaccumulation models in assessing ecological risks of chemicals: a case study for gasoline hydrocarbons.Environmental Science & Technology, 38(23):6225 ~ 6233.

[14] Yin Y L, Song J, Luo Y M. 2005. Assessment on variation of heavy metals in the greenbelt soils alongside urban-rural roadway in Nanjing city. Acta Pedologica Sinica, 42(2):206 ~ 209.

[15] Zhu J J, Cui B S, Yang Z F, *et al.* 2006. Spatial distribution and variability of heavy metals contents in the topsoil along roadside in the Longitudinal Range-Gorge Region in Yunnan Province.Acta Ecologica Sinica, 26 (1):146 ~ 153.

[16] Cheng X, Yu F, Peng S S, *et al.* 2006. Study on vertical change of highway side slope plllution with heavy metal in Karst area Carsologica Sinica, 25(3):242 ~ 245

[17] Wang Z L, He J, Zhi Y B, *et al.* 2006b,The distribution characteristics of heavy metals pollutants in soil-plant system along highway, Journal of Nanjing Forestry University (Natural Science Edition), 30(4):15 ~ 20.

[18] Cai Z Q, Ruan H H, Ye J Z. 2001, A preliminary study on the absorption and accumulation in Oak (*Quercus variabilis*) stand in suburb of Nanjing. Journal of Nanjing Forestry University (Natural Science Edition), 25(1):18 ~ 22.

[19] Zhang Q M, Zhao M. 2003. The analyzing and appraising if the traffic actuality in Inner Mongolia.Economic Geography, 23(1):102 ~ 105,112.

[20] Liu Z L. eds. 1985, Inner Mongolia Vegetation, Beijing: Science Press, 332 ~ 367.

[21] Bao S D. 2005. Soil and Agricultural Chemistry Analysis (3rd edition).Beijing:Chinese Agricultural Press. 14 ~ 22,178 ~ 199.

[22] Tessier A, Campbell P G C, Bisson M. 1979. Sequential extraction procedure for the speciation of particulace trace metals. Analytical Chemistry, 51:844 ~ 851.

[23] Jia Z B, Huo W Y, Zhao Z J,Tao S. 2000. Secondary phase enrichment factor for evaluation of heavy metal pollution of sediment in the Chai river. Acta Scientiarum Naturalium Universitatis Pekinensis, 36(6):808 ~ 812.

[24] Wei F S. 1990. The element background values of Chinese soil. Beijing: Chinese Environmental Science Press. 20 ~ 86,330 ~ 380.

[25] Li X Y. 2001. Soil Chemistry, Beijing: Higher Education Press, 387 ~ 398.

[26] Delorme T A, Gagliardi J V, Angle J S, *et al.* 2001. Influence of the zinc hyperaccumulator *Thlaspi caerulescens* J. & C. Presl and the nonmetal accumulator *Trifolium pratense* L. on soil microbial populations. Canadian Journal of Microbiology, 47(8):773 ~ 776

[27] Sun F, Zhong Z C. 1998. Accumulation and biological cycle of heavy metal elements in *Gordonia acuminata* evergreen broad-leaved forest of mountain Jinyun in Chongqing.China Environmental Science, 18(2):111 ~ 116

[28] Wang D Y, Mou S S, Qing C L. 1998. The effect of atmospheric mercury on the accumulation of mercury in soil-plant system. China Environmental Science, 18 (2):194 ~ 198
[29] Zhang H, Ma D S. 1998. An approach to the characteristics of heavy metal phase as well as the capacity of desorption and adsorption in soil about heavy metal pollution formed by highway, Environmental Chemistry, 17(6):564 ~ 568.
[30] Dong Y T, Cui Y S, Wang Q R. 2003. Uptake of Cd, Zn and Pb by two susceptible plants under monoand multiple-contamination conditions.2003,Acta Ecologica Sinica, 23 (5):1018~1024.
[31] Chen H M. 2002. Behavior of Chemicals in Soils and Its Relation to Environmental Quality, Beijing: Science Press, 1 ~ 22, 46 ~ 78.
[32] Zhang H, Ma D S. 1997. Fractionation of heavy metals in soils from Nanjing.Acta Scientiae Circumstantiae, 17 (3):346 ~ 351.
[33] Wu C N, Wang Y, Zhao L. 2004. The speciation characteristics of heavy metals pollutants along highway in Shanghai-Nanjing (Zhenjiang segment), Environmental Chemistry, 23(4):465 ~ 466.
[34] Chen H M. 2005. Environmental Agrology, Beijing: Science Press, 219 ~ 236.
[35] Ernst W H O. 1996. Bilavailability of heavy melals and deconrarniation of soils by plants. Applied Geochemistry, 11:163 ~ 167
[36] Zhi Yingbiao, Wang Zailan, Wang Zhongsheng, *et al.* 2007. The absorption and accumulation of heavy metals pollutants in plant along highway in western Inner Mongolia. Acta Ecologica Sinica, 27 (5):1863 - 1872.
[37] Zhi Yingbiao, Wang Zailan, Ma Zhong, *et al.* 2007. The speciation and bioavailability of heavy metals pollutants in soil along highway in Erdos. Acta Ecologica Sinica, 27 (5): 2030 - 2039.

Section 2

Risk Assessment in Human Health

9

Non-Invasive Matrices Use in Pollution Evaluation at Nanoscale Levels – A Way Forward in Ecotoxicological Studies

Melinda Haydee Kovacs[1], Dumitru Ristoiu[2] and Cezara Voica[1]
[1]*National Institute for Research and Development of Isotopic and Molecular Technologies Cluj-Napoca,*
[2]*Environmental Science and Engineering Faculty, Babes-Bolyai University Cluj-Napoca,*
Romania

1. Introduction

„For the first time in the history of the world, every human being is now subjected to contact with dangerous chemicals, from the moment of conception until death" - R. Carson, Silent Spring.

This is a quote that summarizes the reality at this moment regarding our safety in our surrounding environment, wherever we are. Unfortunately, pollutants imminence is more increased in developing countries, given that they register higher level of pollutants in all environmental compartments owing to their poor pollution control and pollutants monitoring.

In most cases, environment contamination is a result of humanity's lifestyle resulted from industrial-, agricultural activities and extended urbanization. For most of us it is more and more difficult to imagine our lifestyle from every day without using and profit of products stocked by chemical industry such as pharmaceuticals, petrochemicals, agrochemicals and many other consumer chemicals (Bhandari et al, 2009). Unfortunately together with the rise of chemical manufacture and its use has also come increasing public awareness and concern regarding presence of these chemicals in our surrounding environment. Concernments are mainly caused by the physicochemical properties of these chemical compounds and their possible negative consequence linked to human health and biota.

Owing to awareness's that was attributed by scientific communities and mass media to environmental pollution and living things exposure to such chemicals, has made a clutter regarding the terms like contamination and pollution, terms that tend to be use as synonyms (Hansen, 1993). On our days has been made an agreement between scientific experts, decision-makers and inspectors from different authorities at worldwide level, in that the term contamination should be used where a chemical is present in a given sample with no concrete evidence of harm while the term pollution could be used in cases where was demonstrated that the presence of the chemical compounds caused harm for humans or to other living things (Neuzil et al, 1996).

Doubtlessness any chemical compounds can become a pollutant in any kind of environmental media (water, soil, air, etc.) causing negative effects if it is at a high enough concentration. Despite the fact that any chemical compounds can be a pollutant, certain chemicals have been identified in regulations or by international agreements as being priority chemicals for control (Harrison, 2001). These chemicals have been selected based on their: frequency found in surrounding environment at global scale; persistence in different media; toxic effects at low concentration (damage of selected organs, mutagenic and teratogenic effects) and carcinogenesis; and not in the last case based on their increased bioaccumulation capacity (Pierce et al, 1998).

Living organisms including humans are exposed to chemical contaminants, called also as environmental toxicants, via environmental media. Exposure might be occurred by breathing, ingestion by drinking liquids or eating nourishments that contain chemical contaminants, or through skin contact - see figure 1.

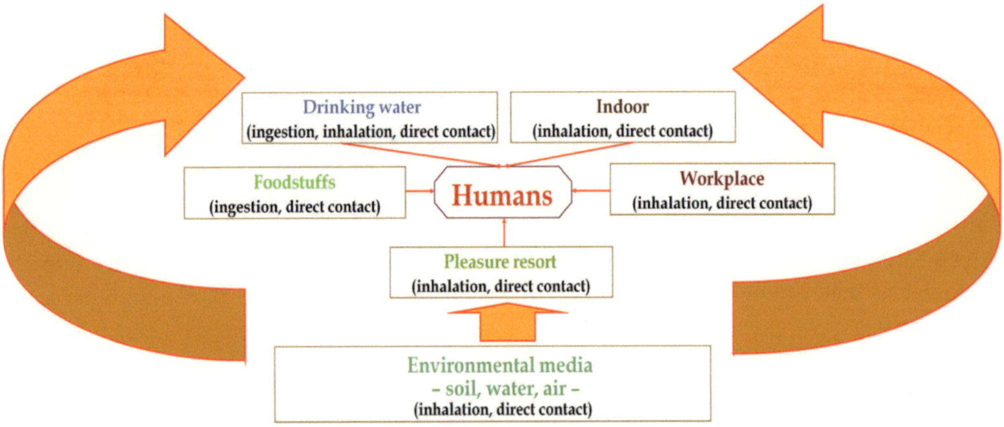

Fig. 1. Humans exposure pathways to organic chemical compounds contaminants.

When living organisms are exposed to environmental contaminants, a cascade of chemical, biological and biochemical events take place in them. These effects intensity depends strongly on exposure conditions - thus when we talk about exposure we must to take into account the dose of chemical contaminants (how much by the chemical substances are ingested, inhaled or imbibed through skin contact), the time period of exposure and the way through that living beings get contact with them. Also is necessary to look at the other chemical contaminants hereat the people are exposed (mixture effects) as well their age, gender, diet, lifestyle, family trails and state of health. Even in our day is difficult to give concrete answers to questions regarding organic chemical contaminants movement and distribution in the living environment - within individual living beings, communities and ecosystems - where multiple factors and events come in to play resulting usually in unwanted events and effects (Walker, 2009). In the last decades many studies underlined that contaminants conspecific to organic chemical classes result in different types of harmful effects on humans or other existences (Byres, 2006).

As regards living beings, concernment regarding our surrounding environment contamination is amplified by issues as movement of toxic contaminants from contaminated media in food products whatever we refers to vegetables or animals grown for consumption purposes. Therefore, humans are not exposed to chemical contaminants just through environmental media or workplaces (particular cases) but also along food chains (Walker, 2009).

Considering the lack of knowledge regarding the additive harm effects of such pollutants on biota (even if we talk about human subjects or animal subjects), becomes more necessary to pay a special attention to identifying, both qualitatively as well quantitatively, the presence of organic chemical contaminants in living beings. Therefore, is important to develop new, accurate and fine analytical methods for analysis of chemical contaminants at very low levels (micro- and nano- orders) from complex biological and environmental matrixes.

Unfortunately even at the present moment, when we want to evaluate the presence of such chemicals in living organisms toward formation of their ecotoxicological profiles, almost all analysis methods use invasive sampling protocols which in a large number of cases, especially in studies on animals, result in causing harm or death of the studied subjects. Introducing of analytical methods which use non-invasive sampled matrixes as saliva, hair, nail, milk, etc., will blot out problems regarding: causing harm (to human or animal subjects) or loss of studied subjects as animals; available number of analysis (referring to the number of human and animal subjects monitoring); and sample collection process.

Development of analytical methods for non-invasively collected biological samples will help in future to provide a better image regarding the contamination with complex chemicals of biota as well will help to get a better understand about their eco-toxicological hall-mark on any living organisms (whether we refer to human or animal subjects).

2. Current obstacles in biota monitoring

Starting from the early of XX century, scientific communities at worldwide scale started to put questions regarding the possibility of humans' exposure to unwanted chemical contaminants as well on the facts that could resulted after exposure events to such chemical contaminants. Thus in 1973, was initiated through World Health Organization the Environmental Health Criteria Programme which presented at that time these emerging objectives:

- to identify new or potential pollutants;
- to assess the biological/zoological specimens which are the most suitable for contaminants monitoring thus to get information on the relationship between exposure to environmental chemical contaminants and human health, in order that in future to could provide guidelines for setting useful exposure limits;
- to develop guidelines for sampling, sample preparation, analytical requirements and storage relative to biomonitoring;
- to draw up recommendation for further research and development;
- to promote the harmonization of toxicological and epidemiological methods in order to have internationally comparable results (WHO, 2006).

Nowadays, in medicine and ecotoxicology fields, it is widely accepted that contaminants effects testing must involve animals because in most cases researches performed on animals helped scientists to discover and developed lifesaving treatments and medications, and to observe which chemical compounds might be dangerous to humans if they get contact with it through ingestion, inhalation and/or dermal contact. But even if we consider all benefits that were obtained in humans health science through animal testing, not everyone consider animal testing as the best method for toxicology. There are two major reasons for that they are not totally accepted, once due to the fact that toxicological study performed on animal subjects induce terrible pain on them, most of the time resulting in the animal subjects death, and secondly due to the fact that not always the approximation between the selected testing animal species physiology and their anatomical profile is similar enough with that of humans (Watson, 2009).

Leaving aside whether if the chosen animal subjects in the study correspond entirely in term of physiological and anatomical structure to humans, when we want to have certainty that a chemical compounds is able to leave some imprint on a living things, study on any kind of animal subject becomes priority. Based on the presence of a chemical compounds on a animal subject, scientist can made analogy about the possibility of this chemical compound presence in human body.

Taking into account all that were presented before as well the controversies that have arisen in the recent years regarding testing on animal subjects, today more and more researchers are trying to develop novel testing methods that are not invasive to the subjects in question.

Development of such of kind methods is beneficial even for human biomonitoring, permitting a larger number of studies on a wider range of subjects as age, gender and geographical living position.

2.1 "Old sampling methods" in monitoring of living things exposure to organic contaminants

Human biomonitoring used for determination of chemical substances in body fluids was for the first time introduced in the early of 1930's by occupational medicine for health protection of exposed workers (Angerer et al, 2007). The first biological sample matrixes that were used was blood and urine, (blood being the first biological sample collected through invasive mode while urine was the first biological sample matrix that was sampled through non-invasive ways). The benefits of using biological samples in order to get information regarding human exposure made that this mode of humans health evaluation in terms of exposure to chemical contaminants to expand, so today it is used in areas like public health politics, environmental medicine, toxicology and other science that relate to the impact of chemical compounds on human health.

Usually, testing the qualitative as well quantitative presence of any kind of chemical compounds in living organisms, involve the use of sample matrixes as blood, tissues or bones. Although in terms of pollutants biomonitoring from living things it is considered that blood samples offer the most accurate information regarding the quantitative presence of a pollutants or its metabolites (due to the fact that it's get contact with the whole organism), this test specimen is done by venipuncture (piercing artery or vein) which cause pain and discomfort to subjects (Beebee, 2008). As regards bone or tissue sampling methods, these are more painful causing in most cases severe injuries.

Moreover, in terms of ecotoxicological studies performed on animal subjects in most situations the performed studies were completed by their sacrifice. Considering these, in the recent years at world wide it has brought up repeatedly in discussions the ethical aspects throughout these studies are conducted (Watson, 2009). But considering the importance of such kind of study (mainly as regards environmental contaminants biomonitoring or their toxicological evaluation on living things) by which obtained information make possible to take measures (pollutants presence regulation through law, contaminated environment remediation) or get treatment measures (in case of diseases) made them until now acceptable and necessary even in conditions of sacrifice of studied animal subjects.

Today a large number of countries introduce regulation regarding the way through that are performed these studies, imposing clear conditions as limiting the number of animal subjects subjected for study as well regarding the mode through that they are treated during the experiment. Unfortunately, considering these requirements in most cases the necessity of a large number of data (which obviously involve uses of a large number of subjects) is increasingly difficult to be fulfilled in researcher studies.

2.2 Current trends in living things exposure monitoring

Taking into account the requirements imposed by legislation but also by the need of a large number of subjects submitted for study so that to could draw realistic and convincing conclusions induced the need to seek new ways for ecotoxicological evaluation of environmental contaminants.

In an attempt to meet these ideas has emerged and formed a new path in ecotoxicological assessment of chemical pollutants, namely qualitative and quantitative evaluation of chemical contaminants from biological matrices from living beings using noninvasive sampling ways. This new direction in ecotoxicological biomonitoring of environmental contaminants not only allows the monitoring of animal subjects exposed but make also possible to assess the impact of those chemical contaminants on humans that leave in contaminated areas. Also this way of sampling of biological matrix from living beings allows an extension of studied subjects in terms of both as species, age, gender and living environment.

3. Choice of sample matrix type collected non-invasively

As in any case of chemical pollutants monitoring a sample must to render an accurate picture about chemical compounds composition present in the environmental media from which the sample was taken at the sampling moment. In order of eliminating the problems of invasive sampling method of biological matrix from living beings, now sample matrixes were chosen on the following principles:

- primarily to not cause any kind of discomfort, pain or sacrifice to subjects submitted for study
- to be stable in time without altering the sample and the chemical structure, physicochemical properties and quantities of the monitored chemical contaminants
- to be a representative "mirror of the footprint" leaved by the chemical contaminants on the studied organisms

- to allow to be sampled:
  - in sufficient quantities to could perform the chemical analysis in a reliable, accurate and sensitive mode
  - to allow sampling on a more extensive field as number of samples, subject species, geographical position, duration, etc.

3.1 Noninvasively collected biological sample matrices from inhabitants

In case of human biomonitoring as regards their exposure levels to environmental contaminants as matrices sampled in a non-invasive way may be taken into accounts these types of samples: hair, nail, saliva, urine, milk, etc.

From these, the most often used sample in toxicological and environmental medicine studies was *urine* - these type of samples being used for different analytical assessment (medical, toxicological, ecotoxicological) with years ago, especially when water-soluble chemicals and environmental contaminants were the targets of study. Choosing of this type of matrices from very old times is because urine provides good information about the chemical content uptake by the human body and not in the last because it was easy to be collected and the available amount was sufficient to perform the analysis in good conditions. Also it is available at large scale as subject's geographical position, age and gender.

Hair, nowadays is another well used biological matrix presenting as major advantage its stability as matrix (comparing with other biological matrix), and easy collection, transport and storage. Hair as biomonitoring matrix is able to provide information both about short or long time exposure to contaminants of the studied subject.

Human breast milk is also o biological matrix that is collected through noninvasive ways. This sample type is used extensively when it is required information from both the mother as well her child. This sample is extremely suitable specially in monitoring of lipophilic chemicals considering its increased content in fat.

Nails and *saliva* are also noninvasively collected matrices that could be used in biomonitoring considering its low cost and the ease way through that they could be collected permitting the surveillance of a large number of subjects.

Once with the extension and evolution of this way of ecotoxicological and environmental epidemiology studies new biological sample matrixes collected also without cause any discomfort or damage on the studied subjects started to be introduced. Such of matrixes are sweat, faeces, semen, placenta and breath.

3.2 Noninvasively collected biological sample matrices from animal subjects

In case of animals monitoring previous studies has showed that uses of hair and feathers are useful in environmental contaminants impact evaluation on them. Also milk analysis could be used in term to evaluate the contaminants transfer from mammals to their offspring from the first days of their early life. Such kind of analysis helps to conclude the exposure pathways and its impact even on newborn subjects, giving important information for the future, which will help to understand subjects' development and pathological profile from their maturity period.

4. Sample preparation and analytical methods selection for analyzing organic contaminants in biological samples collected non-invasively

Through these studies two sample matrixes were chosen in order to evaluate the surrounding environmental contaminants on inhabitants, namely hair - in order to get a general overview about organic contaminants impact on humans exposed to them, and milk - throughout to could form an image regarding the infants' exposure from early age life. As regards animal exposure biomonitoring hair was chosen for mammalian animal subjects and feather in case of poultries biomonitoring studies. Therefore in our biomonitoring study we were able to collect high enough samples on an extended range of subjects (as species, age and geographical position) without causing any harm even when we talk about human biomonitoring or animal biomonitoring studies.

4.1 Sample collection and its preparation for analysis

In case of human biomonitoring as was suggested two sample matrixes were used, once - hair samples which made possible to be applied for any kind of subjects as age, gender, etc. and secondly - breast milk samples which as it matrix nature shows it was possible to be applied just for a special class of subjects - namely nursing mothers, way throughout we were able to found information about mother exposure as well about the infant exposure. Both methods don't involve any kind of physical or physiological trauma on the involved subjects.

Subjects number and key information regarding them	Gender		Age	Working space exposure	Living place exposure	Smoker or diet habit
	Female	*Male*				
Cluj district						
Rural sites	16	13	5-62	2	-	17
Urban sites	15	18	9-71	1	10	20
Industrial sites	9	10	22-49	6	19	13
Salaj district						
Rural sites	6	10	10-66	1	-	9
Urban sites	22	26	15-73	2	-	35
Industrial sites	5	4	33-43	1	9	5
Bistrita Nasaud district						
Rural sites	15	12	3-60	-	-	13
Urban sites	10	16	15-45	-	-	18
Industrial sites	-	1	39	-	1	1

Table 1. General information about the studied inhabitants from the three selected regions.

As regards animal subjects that were implied in our biomonitoring programs, hair was used in case of mammalians and feather in case of poultry subjects subjected for study.

Totally 208 number of human subjects were involved in these biomonitoring studies. Details regarding the human biomonitors are given in table 1. Shortly, 98 females were involved in this study from which 18 were child and 26 were nursing mother from which milk samples were collected also. As regards male, 110 subjects were monitored from which 20 persons were child. From this 208 subjects 131 persons declared smoking and diet pattern from which 68 are smokers and 24 are involved in diet programs due to health problem and 39 persons are involved also in diet programs owing to their weight (in this case 78 % being females between 14 - 35 years).

Animal subjects were also chosen from both rural and urban sites as well from industrial sites. Just animals and poultries grown for consumption purposes in farms of the monitored inhabitants were considered in this study. As animal species were considered pig, cow and chicken, species that are the most popular in these three studied district. They are basically the most important source of meat for any humans from this part of Romania; they didn't missing from any rural farms and any plates of inhabitants from these regions.

Collected hair samples - 5 mg amount/subject - including the entire length of the hair were cut with uncontaminated scissors. These collected samples were placed in plastic bags (without adding of any stabilizing agent) and transported to laboratory. There washing process was done in order to avoid external contamination of the matrixes. After washing processes the samples were cute in small species (1-2 mm) from that 2.5 mg were placed in a 10 mL glass vials. This procedure was applied for both human and animal subjects' hair samples. Similarly was done in case of feather samples. In case of human scalp hair was used as sampling zone of human body while in case of animal subjects dorsal hair were used for biomonitoring.

As regards human breast milk analysis, 40 mL of samples were collected from every nursing mom in sterile glass vials to which prior was added $K_2Cr_2O_7$ as stabilizing reagent after that the vials were closed with Teflon lined screw cap and put in a freezer at 4 °C until analysis. No milk samples were stored more than three days from the sampling moment.

4.2 Organic chemical contaminants extraction

Considering the previous research studies performed in these areas three main pollutants category were the targets in these biomonitoring studies, namely organochlorine compounds, mono- and polycyclic aromatic hydrocarbons and organometallic compounds.

Mono- and polycyclic aromatic hydrocarbon compounds from hair samples were extracted using headspace-solid phase microextraction technique (HS-SPME). To the 2.5 mg of hair samples that was placed in the 10 mL glass vials were added 15 μL of HCl and 100 μL of aqueous sodium dodecylsulphate and 5 mL of hexane. This aliquot it was subjected to ultrasonication at 50 °C for 20 minutes after that the vials was subjected to centrifugation at 7 500 rpm for 10 minutes after that the hexane was separated and the remained aliquot was subjected again to centrifugation with hexane in the same condition as before. The two phases were put together finally and subjected to HS-SPME extraction.

Extraction method performance was increased through optimization of several extraction parameters as: selection of suitable fiber coating type, using or not of a derivatizing agent, desorption time and temperature, extraction time, applied agitation mode and salt effects. In this case using of a 100 μm thickness polydimethyl siloxane (PDMS) fiber increased the extraction method capacity comparing with extraction performance parameters obtained through using other types of fiber or different film thickness of the fibers.

As regards desorption time and gas chromatograph injector temperature optimization in order to increase the sensitivity of extraction procedure five different desorption times were studied, namely 30 seconds and one, two, three and five minutes, respectively. In case of all monoaromatic hydrocarbon compounds it was observed that short time of desorption as 30 seconds or one minutes is not enough to reach a complete desorption therefore the time that is required for a good desorption of these compounds was established at two minutes. In case of polycyclic aromatic hydrocarbons the optimal desorption time was established at 5 minutes. As gas chromatograph injector temperature significant results were obtained when it was set at 230 °C.

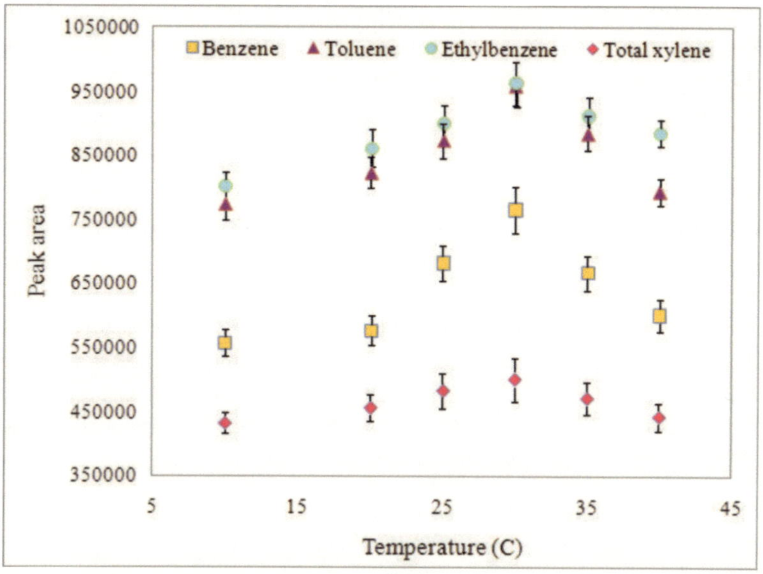

Fig. 2. Monoaromatic hydrocarbon analytes response after HS-SPME extraction method performance under different extraction temperature.

Considering the extraction temperature, a lots of study demonstrated that extraction temperature has multiple impact on the HS-SPME procedures. Liu et al, 2005, asserted that the high temperature increase the diffusion coefficients of the analytes in any kind of sample matrix and from it to the headspace, which reduce significantly the extraction time but in the same time have a negative influence on the partition coefficients of analytes in the fiber coating, reducing it, thus the efficiency of the extraction being reduced well. In this case the experiment was conducted for six different temperatures: 10, 20, 25, 30, 35 and 40 °C - see figure 2. In this figure it is observed that a good response for all mono aromatic hydrocarbon compounds is obtained at 30 °C, decreasing for higher or lower temperature.

Extraction time represents another factor that could chasten or impair the extraction method sensitivity, so we take it in consideration also. In order to establish the most suitable extraction time for the mono- and polycyclic aromatic hydrocarbon compounds ten different time were examined as follows: 3, 5, 10, 15, 20, 25, 30, 35, 40, 50 and 60 minutes, respectively. Compounds response increased up to 15 minutes extraction in case of monoaromatic hydrocarbon compounds after that was registered a continue decreasing. In case of polycyclic aromatic hydrocarbon compounds the maximum response was achieved at 20 minutes extraction, after that presented the same pattern of decreasing as the monoaromatic hydrocarbon compounds Also the equilibrium was obtained at different time in case of all aromatic hydrocarbon compounds, therefore this equilibrium was approached more rapidly in case of monoaromatic hydrocarbon compounds (10 - 15 minutes) and after longer periods in case of polycyclic aromatic hydrocarbon compounds. According with these results 25 minutes was established as the most suitable extraction time for these compounds and as equilibration time 20 minutes was established as being satisfying.

The extraction method performance was evaluated based on the determination of recovery factor which was between 82-109 %b in case of monoaromatic hydrocarbon compounds and between 78-119 % in case of polycyclic aromatic hydrocarbon compounds.

Organochlorine compounds presence from *hair samples* were extracted as follows: to the 2.5 mg of hair samples that was placed in the 10 mL glass vials were added 15 µL of HCl and 100 µL of aqueous sodium dodecylsulphate and 5 mL of isooctane. This aliquot it was subjected to ultrasonication at 40 °C for 10 minutes after that the vials was subjected to centrifugation at 3 500 rpm for 10 minutes after that the isooctane was separated and the remained aliquot was subjected again to centrifugation with isooctane in the same condition as before. The two phases were put together finally and subjected to rotary evaporation at in a water bath at 40 °C until 1 mL aliquot it was obtained. This obtained aliquot then was subjected to HS-SPME extraction procedure followed finally by instrumental analysis. Shortly the optimal HS-SPME parameters were 2 and 5 minutes as desorption time respectively as extraction time. The optimal extraction temperature was found at 30 °C.

The extraction method performance in case of organochlorine compounds was evaluated based on the determination of recovery factor which was between 92-121 %, considering all studied compounds.

In case of *organometallic compounds* also HS-SPME extraction method was used with almost similar condition as in case of mono- and polycyclic aromatic hydrocarbon compounds, but in this case a dervatization step using sodium tetraethyl borate as derivatization agent was introduced before instrumental analysis. After derivatization process the following parameters were considered as optimum for HS-SMPE extraction process where desorption time and the extraction time were set at 7 and 12 minutes, respectively. The extraction temperature was set at 75 °C.

Target chemical compounds extractions from breast milk analysis were performed through HS-SPME extraction procedure using a PDMS fiber coating with 100 µm film thickness. The lipids from milk were separated according to Johansen et al, 2004 and Behrooz et al, 2009 without any significant modification in methods. Extraction method used for organochlorine compounds as well for mono- and polycyclic aromatic hydrocarbon compounds was the same as that was described by Skrbic et al in 2010.

4.3 Instrumental analysis of organic chemical contaminants from samples collected through non-invasive procedures

All sample extracts were analyzed through gas chromatographic method. In case of organochlorine compounds analysis electron capture detector was used while in case of organometallic compounds quadrupole mass spectrometer was used as detector. In case of mono- and polycyclic aromatic hydrocarbons flame ionization detector was used in order to perform the analysis. Used working parameters are presented shortly in the follows.

In case of *mono- and polycyclic aromatic hydrocarbon analysis* these were performed on Trace GC Ultra gas chromatographic apparatus (Thermo Electron Corporation) equipped with a flame ionization detector (FID) and a split/splitless injector. After desorption of target analytes the SPME fiber was injected in the injector with a constant 230 °C temperature. As column was used TR 5 % phenyl methylsiloxane column, provided by Thermo Electron Corporation, having as characteristics 0.53 mm ID x 0.50 µm film thickness x 30 m length.

The oven temperature program has set as follows: 40 C for 5 minutes, followed by increases with 5 $°C·min^{-1}$ at 150 °C and held at this temperature for 3 minutes, subsequent by an increases of temperature with 10 $°C·min^{-1}$ at 220 °C and kept at this temperature for 5 minutes. The FID detector temperature was set at 300 °C.

As regards *organochlorine compounds* previous researches has demonstrated that electron capture detector in gas chromatographic analysis has an increased sensibility to halogenated organic compounds detection (ca. 10^{-13} $g·mL^{-1}$), this detector being considered probably the most sensitive gas chromatographic detector that is available (Quan et al, 2002; Chen et al, 2004; Khajeh et al, 2006). Considering these the analysis of extracts was performed in the followings through Trace GC Ultra gas chromatography (Thermo Electron Corporation) equipped with a ^{63}Ni electron capture detection system (Thermo Electron Corporation).

Good response were obtained for all target chlorinated organic compounds in case of TR-V1 Trace GC capillary column with cyanopropylphenyl polysiloxane phase type and with 0.53 mm I.D. x 3.0 µm film thickness x 30 m column length using the following working conditions: nitrogen was used as carrier gas with 30 $mL·min^{-1}$ flow. Split ratio was set at 1:2 while the injector temperature and detector temperature were set at 220 and 250 °C, respectively. Oven temperature programe through out were done the analysis was as follows: 40 °C (with 3 minutes hold time) increased with 7 $°C·min^{-1}$ at 100 °C (hold for 3 minutes time period) after that increased with 10 $°C·min^{-1}$ at 220 °C, where this temperature was mantained constant for 7 minutes.

Organometallic compounds were detected also through gas chromatographyc technique but this time a quadrupole mass spectrometer was used as detector. Therefore a Focus GC engaged to a DSQ II quadrupole mass spectrometer provided by Thermo Electron Corporation was used as analytical instrument for organometallic compounds separation and quantification. A split/splitless injector in the splitless mode was used. Its temperature was set at 260 °C. The target analytes were separated using a TR 5-MS 5 % phenyl polysilphenylene-siloxane capillary column with the following characteristics: 0.25 mm ID x 0.25 µm film thickness x 30 m length. This column was inserted into the mass spectrometer and the interface temperature between GC and MS was set at 270 °C. For the GC oven the next temperature program was found to be optimum in case of organometallic compounds analysis: 35 °C (hold for 2 minutes), raised with 5 $°C·min^{-1}$ at 90 °C and kept at this temperature for one minute after that continue the temperature increasing with 12 $°C·min^{-1}$ until 220 °C, temperature that once achieved was maintained constant for five minutes. As carrier gas high purity helium was used with 0.8 $mL·min^{-1}$ flow rate. The analysis through mass spectrometer detector was done in electron impact mode at 70 eV and through full scan monitoring mode.

5. Case studies

It is well known that a large scale of chemical pollutants as well their incaution management and damaging land uses make in menace the quality of our surrounding environment at world wide scale. Usually the environmental contaminants (having as main source anthropogenic activities) have great effects on the entire environmental system as draw down loss of habitats, diminish of biodiversity or adverse effects on humans that are exposed to these contaminant through food web chain, health care and/or recreational/working activities.

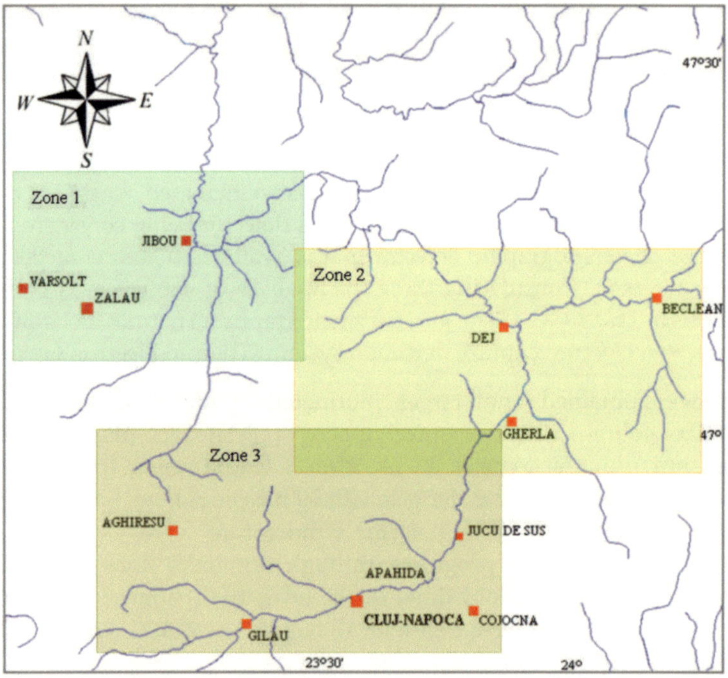

Fig. 3. Map of the studied regions during period of 2008 - 2010.

The damage of environment quality and living things health status has as result the increasing of necessity to monitor these environmental contaminants. More than that becomes a necessity to monitor their qualitative and quantitative presence even in organisms of living subjects whether it is humans or animals in order to could get regulatory, remedial and treatment decisions.

These studies were conducted in north western part of Transylvania including rural, urban and industrial zones - see map presented in figure 3. In figure 3 zone 1 represent a part of Salaj district which include both rural as well urban sites, zone 2 represent the north part of Cluj district and the western part of Bistrita-Nasaud district, this region including industrial, rural and urban sites while zone 3 represent the south part of Cluj district including urban as well rural areas. In idea to get a clear and realistic image regarding environmental contaminants presence and the degree of exposure of living beings to those contaminants environmental (soil, water) and biota samples (vegetables, biological samples from animal and human origins) were collected monthly in period of 2008-2010.

5.1 Living things exposure to organochlorine compounds – Case study of paper mills

During previous studies it was observed an increased presence of organochlorine compounds in regions near Dej city (zone 2 from figure 3). Main organochlorine compounds that were detected were chlorophenols (dichlorophenol, trichlorophenol and pentachlorophenol) and chlorinated solvents (trichloroethylene, tetrachloroethylene and carbon tetrachloride).

Presence of these compounds in environmental samples as soil and water were between range of 0.3 - 32 $\mu g \cdot kg^{-1}$ and 0.5 - 25.1 $\mu g \cdot L^{-1}$, respectively, usually with higher levels in case of soil samples.

Major amounts being detected in case of dichlorophenol and trichloroethylene compounds, usually with higher amounts in case of soil samples - see table 2 and figure 4.a. and 4.b.

Period	Chlorophenols average values					
	Monochlorophenols		Dichlorophenols		Trichlorophenols	
	Soil [$\mu g \cdot kg^{-1}$]	Water [$\mu g \cdot L^{-1}$]	Soil [$\mu g \cdot kg^{-1}$]	Water [$\mu g \cdot L^{-1}$]	Soil [$\mu g \cdot kg^{-1}$]	Water [$\mu g \cdot L^{-1}$]
2008	0.5 - 12.5	0.3 - 8.9	0.8 - 19.2	0.5 - 14.5	0.9 - 31.9	0.4 - 17.5
2009	0.3 - 18.5	0.5 - 17.2	1.8 - 22.5	0.4 - 16.8	0.5 - 21.9	0.3 - 19.5
2010	0.4 -16.2	0.4 - 10.2	0.5 - 12.5	0.6 - 8.59	0.3 - 9.2	0.7 - 3.2

Table 2. Chlorophenols average amounts in soil and water samples.

Increased presence of these compounds was attributed to the presence of paper mills that discharge its waste water in Somes River (one of the main river that cross northern part of Transylvania which is used as water source both for drinking water purposes and for irrigation networks of agricultural lands). This was demonstrated by the results of environmental monitoring before and after the closing of factory in 2009, when the factory was closed for a period of two months for rehabilitation - see figure 4a and 4b, when it was observed a decreasing tendency of organochlorine compounds presence in environmental samples (as soil and surface water) from the moment when the factory was closed.

In order to evaluate these pollutants impact on living beings it was observed that are quite strong correlation between the amount of these chlorinated compounds in surrounding environment and the amount of organochlorine compounds detected in animals and humans hair samples collected from subjects that live in the corresponding environment - see figure 5. But such dependence was obtained just in this special case, for other regions such correlation (in case or organochlorine compounds) wasn't applicable.

Consulting the obtained results it was observed that the amount of these compounds were lower in case of poultries feathers and higher in the case of cow hair. Also in case of human hair subjects their levels were higher even 30 - 50 percent than in other living beings. This could be explained by the cumulative effects of these pollutants, therefore once with age increases are possible to increase also the levels of environmental contaminants from animals and humans body.

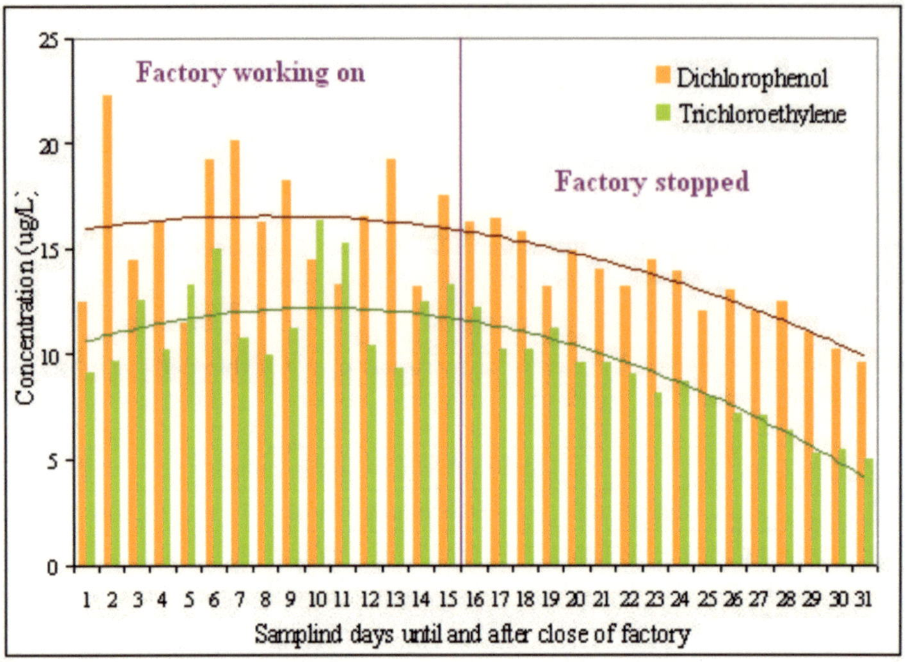

Fig. 4a. Organochlorine compounds amount in surface water while paper mills operating on and off.

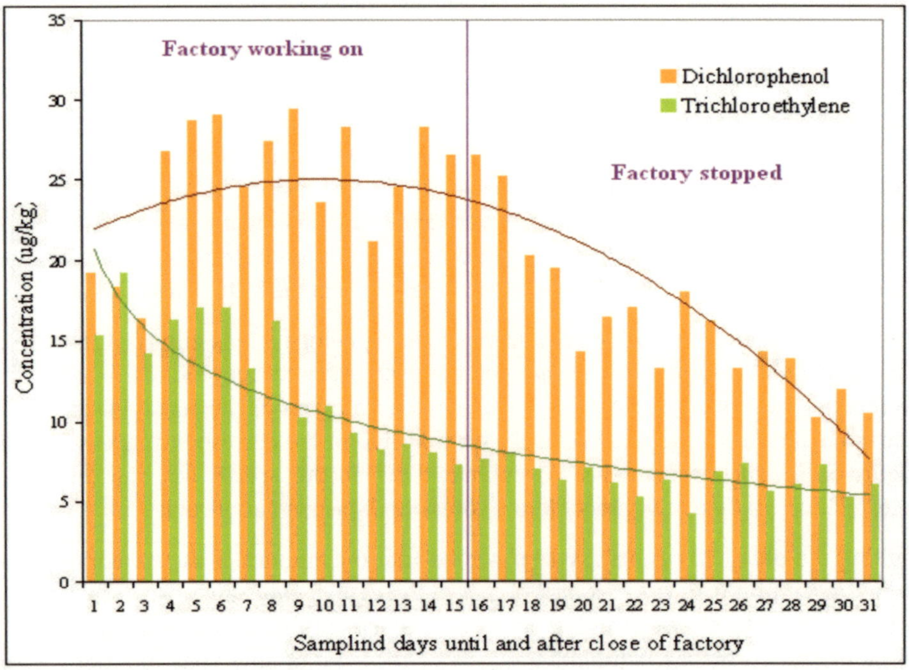

Fig. 4b. Organochlorine compounds amount in soil samples while paper mills operating on and off.

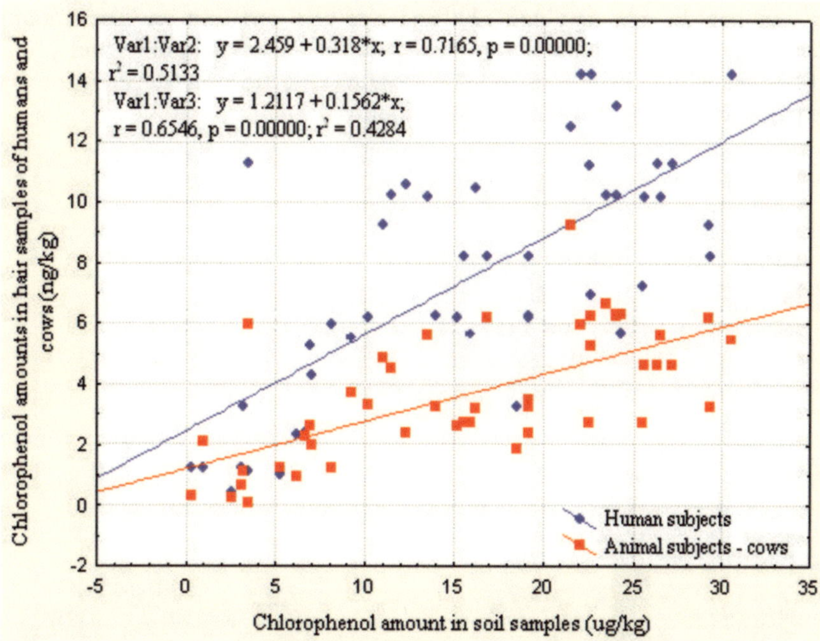

Fig. 5. Organochlorine amount dependence from humans and animal body by the surrounding environment.

As regards the amounts detected in human breast milk samples, the amount of chlorinated compounds were higher than in case of hair samples of the same subjects with almost 10 - 15 %. This fact could be attributed to the increased lipid content of breast milk, much more than exist in hair samples, therefore chlorophenol compounds being more easily bounded to the fat from the organisms.

5.2 Inhabitants and home grown animals' exposure to mono- and polycyclic aromatic hydrocarbon compounds

As regards monoaromatic hydrocarbon compounds the most detected compounds in the most increased amounts were benzene and toluene, if we take in consideration the fact that their uses was restricted in the last decades. Average values of mono- and polycyclic aromatic hydrocarbons detected in soil and water samples collected from this studied regions are presented in table 3. Higher amounts in case of anthracene and benzo[a]pyrene were detected usually.

Environmental contaminant	Environmental compartment	Average value during time period	
		2009	2010
Monoaromatic hydrocarbon	Soil	0.3 - 19.2 $\mu g \cdot kg^{-1}$	0.2 - 22.3 $\mu g \cdot kg^{-1}$
	Water	0.4 - 16.7 $\mu g \cdot L^{-1}$	0.8 - 15.9 $\mu g \cdot L^{-1}$
Polycyclic aromatic hydrocarbon	Soil	0.2 - 22.9 $\mu g \cdot kg^{-1}$	0.4 - 16.9 $\mu g \cdot kg^{-1}$
	Water	0.4 - 15.2 $\mu g \cdot kg^{-1}$	0.5 - 13.8 $\mu g \cdot kg^{-1}$

Table 3. Mono- and polycyclic aromatic hydrocarbons average amounts in soil and water samples.

In order to compare the results obtained after human hair analysis and the amount of these compounds detected in vegetables samples it was observed a quite good correlation as regards the humans' uptake values to hydrocarbons and the amounts of these compounds which are present in vegetables samples – see figure 6.

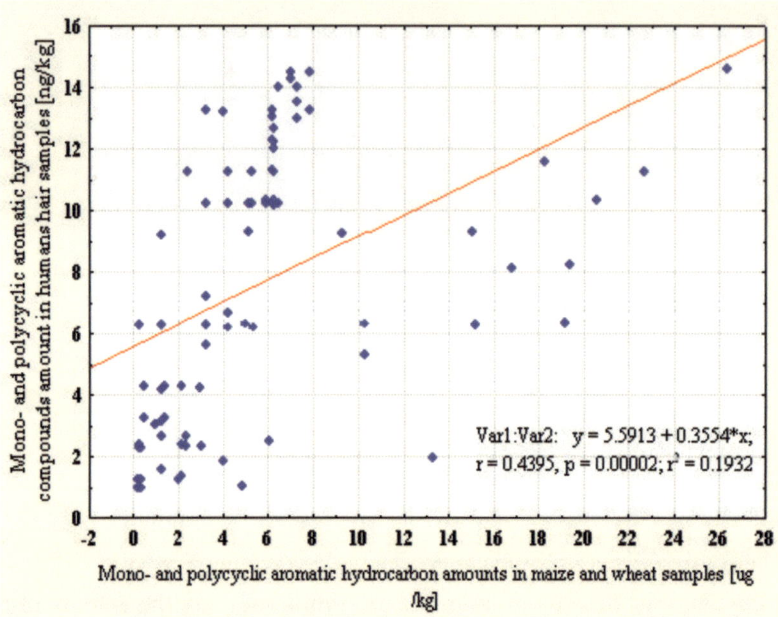

Fig. 6. Mono- and polycyclic aromatic hydrocarbons amount from human hair samples dependence by the amount of these compounds in the consumed foodstuffs.

Comparing the uptake levels of mono- and polycyclic aromatic hydrocarbons it was observed that polycyclic aromatic hydrocarbons are better retained in living organisms than monoaromatic hydrocarbons.

5.3 Organometallic compounds fingerprint on biological samples

Methylmercury was the most prevalent organometallic compound that was detected in environmental samples. This was a special case which is attributed to banned and improper managed old chemical factory. The amount of this compound founded in soil and water samples are listed through table 4.

Environmental contaminant	Environmental compartment	Average value during time period	
		2009	2010
Organometallic compounds	Soil	0.5 – 13.5 $\mu g \cdot kg^{-1}$	0.5 – 15.1 $\mu g \cdot kg^{-1}$
	Water	0.6 – 10.8 $\mu g \cdot L^{-1}$	0.6 – 13.9 $\mu g \cdot L^{-1}$

Table 4. Methylmercury range values in soil and water samples in the last two years.
Methylmercury amounts in vegetables samples grown in this region was between 0.5 - 12.1 $\mu g \cdot kg^{-1}$ while in human and animal hair samples it's amount was between 0.2 - 19.1 and 0.5 - 10.3 $ng \cdot kg^{-1}$, respectively. Once again higher values being detected in case of human subjects.

6. Advantages and limitation in eco-toxicological profiling based on biological samples collected non-invasively

Main disadvantages of these techniques are the difficulty in making difference between external and internal contamination - as regards hair samples. Also in case of hair samples the results vary intensively once with color, age and gender without existing a good correlation between the same parameters. The lack of a standardized procedure of analysis make more difficult to establish the reliability of results in order to estimate the method uncertainty and making real laboratory intercomparisons. Also a large number of studies showed that there are no correlation between the contaminant level from hair samples and blood samples in case of the same person.

As regards the study performed by us using of such sampling matrixes made possible to evaluate the impact of certain environmental contaminants on humans and animals (grown for consumption purposes) which are living in such contaminated areas. Also through the significant number of subjects that were monitored we was able to conclude that the environmental contaminants are uptake by humans mainly through food web chains (as a result of consumption of contaminated vegetables and animals). This is the same in case of home grown animals which are similarly exposed to these environmental contaminants mainly through the consumed contaminated food.

7. Conclusions

Major environmental pollutants that were detected in different environmental compartments of the studied regions were monoaryomatic hydrocarbons and chlorinated solvents, which were also the most prevalent as presence as well as amount. Methylmercury presence in environment is considered as a special case - environmental contamination in this case coming from the old banned chemical factory on which sites no remediation was applied. In almost all environmental contaminants cases, their presences are attributed to the industrial activities through these regions. All these contaminants leave an impact on the living things that live in the surrounded environment. Vegetables growing on a soil contaminated by former industrial activities may contain chlorinated or hydrocarbon pollutants in their tissues at significant concentrations.

Uses of noninvasive matrices in both animals as well humans' biomonitoring purposes have been showed to be suitable.

These compounds were detected in all humans and animals' hair samples. Compounds with higher molecular weight masses were easily accumulated by human bodies than the easier compounds.

8. References

Bhandari, A.; Surampalli, Y.P.; Adams, D.G.; Champagne, P.; Ong, S.K.; Tyagi, R.D.; Zhang, T.Ch. (2009). *Contaminants of emerging environmental concern*. pp. 1-6, American Society of Civil Engineers (ASCE) ISBN 978-0-7844-1014-1, Reston, Virginia USA.

Hansen, A. (1993). *The mass media and environmental issues – Studies in communication and society*. pp. 12-28, ISBN 978-0718520533 Leichesters University Press, UK.

Neuzil, M.; Kovaric, W. (1996). *Mass media & environmental conflict – America's green crusades.* pp. 105-125, ISBN 076190333X Sage Publication Inc., USA.

Harrison, M.R. (2001). *Pollution: causes, effects and control.* 4th edition, pp. 12-28, ISBN 0-85404-621-6 The Royal Society of Chemistry, UK.

Pierce, J.J.; Weiner, F.R.; Vesilind, P.A. (1998). *Environmental pollution and control.* 4th edition, pp. 15-29, ISBN 978-0-7506-9899-3 Butterworth-Heinemann, Wobum USA.

Byers, P.J. (2006). *Metalworking fluids.* 2nd edition, pp. 386, ISBN 978-1-57444-689-0 Taylor & Francis Group, CRC Press, USA.

Walker, C.H. (2009). *Organic pollutants: an ecotoxicological perspective.* 2nd edition, pp. 94-98, ISBN 978-1-4200-6258-8 Taylor & Francis Group, CRC Press, USA.

World Health Organization. (2006). *Principles for evaluating health risks in children associated with exposure to chemicals.* pp 7-15, in Environmental Health Criteria 237, IPCS – International Programme on Chemical Safety, ISBN 92-4-157-237-X Geneva, Switzerland.

Watson, S. (2009). *Animal Testing: Issues and Ethics.* 5th edition, pp. 5-12, ISBN 978-1-4358-5022-4 Rosen Publishing Group, Inc., New York, USA.

Beebee, T.J.C., Rowe, G. (2008). *An introduction to molecular ecology.* 2nd edition, pp. 92-101, ISBN 978-0-19-929205-9 Oxford University Press Inc., New York, USA.

Angerer, J., Ewers, U., Wilhelm, M. (2007). Human biomonitoring: state of the art. International Journal of Hygiene and Environmental Health, vol. 210, pp. 201-228, ISSN 1438-4639.

Liu, J.F., Li, N., Jiang, G.B., Liu, J.M., Jonssin, A.J., Wen, M.J. (2005). *Disposable ionic liquid coating for headspace solid-phase microextraction of benzene, toluene, and xylenes in paints followed by gas chromatography-flame ionization detection.* Journal of chromatography A, Vol. 1066 (1-2), Pag. 27-32, ISSN 0021-9673.

Johansen, P., Muir, d., Asmund, G., Riget, F. (2004). *Human exposure to contaminants in the traditional Greenland diet.* Science of the total environment, Vol. 331 (1-3), pp. 189-206, ISSN 0048-9697.

Behrooz, R.D., Sari, A.S., Bahramifar, N., Ghasempouri, S.M., (2009). *Organochlorine pesticide and polychlorinated biphenyl residues in human milk from the Southern Coast of Caspian Sea, Iran.* Chemosphere, Vol. 74 (7), pp. 931-937, ISSN 0045-6535.

Skrbic, B., Szrwinska, K., Durisic-Mladenovic, N., Nowicki, P., Lulek, J. (2010). *Principal component analysis of indicator PCB profiles in breast milk from Poland.* Environmental international, Vol. 36 (8). pp. 862-872, 0160-4120.

Quan, X., Chen, S., Platzer., Chen, J., GFrerer, M. (2002). *Simultaneous determination of chlorinated organic compounds from environmental samples using gas chromatography coupled with a micro electron capture detector and micro plasma atomic emission detector.* Spectrochimica acta part B: Atomic spectroscopy, Vol. 57 (1), Pag. 189-199, ISSN 0584-8547.

Chen, E.C.M., Chen, E.S. (2004). *Molecular electron affinities and the calculation of the temperature dependence of the electron-capture detector response.* Journal of chromatography A, Vol. 1037 (1-2), Pag. 83-106, ISSN 0021-9673.

Khajeh, M., Yamini, Y., Hassan, J. (2006). *Trace analysis of chlorobenzenes in water samples using headspace solvent microextraction and gas chromatography/electron capture detection.* Talanta, Vol. 69 (5), Pag. 1088-1094, ISSN 0039-9140.

10

Polysystemic Approach to Risk Assessment

Mikhail Karganov, Irina Alchinova and Nadezhda Khlebnikova
Institute of General Pathology and Pathophysiology, Russian Academy of Medical Sciences, Moscow Russia

1. Introduction

Health hazard from natural and anthropogenic sources has begun to be analyzed 2-3 decades ago. Until recently this analysis was primarily applied for human safety in incidents. However, due to stochastic features of risk assessment its applicability to the diagnosis, prevention, and protective or compensatory measures in some cases can be very limited.

The proposed concept combining approaches based on risk analysis and sanogenetic analysis is aimed at increasing the efficiency of both approaches in addressing issues of risk prediction.

There are several reasons necessitating integrated approach to risk analysis:

1. Identifying of risk groups in the population subjected to anthropogenic influences is associated with the formation of groups in which the frequency of fixation of certain pathological processes is dramatically increased. However, modern approaches of epidemiological analysis are based on statement of the accomplished fact (prevalence of cancer, hematological diseases, congenital and hereditary abnormalities, reduced life span, *etc.*). Most these consequences are long-lasting (sometimes for many years) and state the hazard rather than predict the risk.
2. In cases when anthropogenic influences are evaluated on the basis of detection of undesirable biological responses (e.g. chromosome aberrations, accumulation of lipid peroxidation products, changes in activity of some enzymes etc.) the recorded shifts usually ambiguously determine the expected consequences. Biological aftereffects of the detected shifts can to a certain degree reflect both desirable (development of resistance) and undesirable (fixation of certain pathologies) outcomes.
3. In most cases, the above-listed approaches are informative only in relatively high doses and concentrations of this or that anthropogenic factors and are almost irreproducible in case of low-dose and low-concentration influences.
4. Many effects (mortality, pathologies) as consequences of anthropogenic risk factors are unspecific, i.e. can be caused by other factors.
5. The data on the effect of some harmful factors suggest that the probability of this effect is proportional to the probability of spontaneous appearance of the same effect, caused by a complex of other factors of human life. We can speak about synergism of various factors.
6. Modern industrial conditions practically exclude single-factor exposure. Existing methods of epidemiological risk assessment practically cannot differentiate between

chemical, radiation, climatic, noise, vibration, and other impacts. Hence, they do not provide a reliable argument for choosing optimal preventive measures.

7. The method of documentation of pathological consequences rules out the possibility of early prophylactic protection of the population contacting with hazardous industries, which is a priori more efficient and less costly than treatment of the realized risks.

By the wide range of effects, all existing real dangers of radiation, physical, chemical, and biological nature can be divided into two categories: 1) risks in the deterministic range of doses and concentrations (doses and concentrations far surpassing than the established thresholds), 2) risks in a stochastic range of doses and concentrations (doses and concentration near the established thresholds). In the deterministic range of doses and concentrations, the biological effects strictly depend on doses and concentrations of anthropogenic factors and can be detected by existing methods of epidemiological analysis. In the stochastic range of doses and concentrations of anthropogenic factors, the consequences strictly depend on individual sensitivity of biological objects, including humans.

In the human body, many regulatory systems operating at different levels of organization provide sensitivity or resistance to both external and internal factors. On the results of analysis of functional adequacy of sanogenesis systems predicts the level of resistance or sensitivity to relatively tolerable doses and concentrations of anthropogenic influences. Hence, sanogenetic monitoring is more informative for the range of doses and concentrations that cause stochastic effects.

Due to individual variability of functioning of sanogenetic processes, the same anthropogenic factor in equal doses and concentrations will cause certain effects in some organisms (sensitive), will not cause in others, and will induce resistance in the third. This implies that at the population level three subpopulations should be determined at relatively low-dose and low-concentration exposures: sensitive, neutral, and super resistant. The ratio between these subpopulations can eventually serve as a criterion of population risk from this exposure.

Any stable fixation of the pathological trace is preceded by processes of dysregulation of the corresponding functions. The most probable pathological outcomes can be predicted on the basis of the results of polysystemic sanogenetic monitoring by detecting dysregulation in certain systems of the organism (cardiorespiratory, psychomotor, and metabolism systems). Monitoring is carried out using computerized measurement instrumentation and data processing systems, which provides the basis for strict quantitative assessment of the dynamics of risk for the studied populations. The risks assessment goes from the instrument of control to the rank of controlled processes, which is the basis for successful operation of potentially hazardous industries.

2. Methods

Hardware base of the sanogenetic monitoring complex includes three major appliances adapted to non-invasive screening survey:

- spiroarteriocardiorhythmograph for continuous non-invasive recording of blood pressure, expiration and inspiration air flows with a highly sensitive ultrasonic transducer, and electrocardiogram;

- computer-aided device for express-evaluation of psychomotor activity from motor tests;
- laser correlation spectrometer intended for identification of the pattern of regulation of metabolic and immune processes.

All tests were performed with strict adherence of general bioethical standards.

Individual *functional sufficiency of the cardiorespiratory system* was evaluated using a Spiroarteriocardiorhythmograph instrument complex (SACR, recommended by Ministry of Health Care and Social Development of the Russian Federation for clinical use; registration certificate #29/03020703/5869-04, St. Petersburg) allowing simultaneous recording of the heart, vascular, and respiratory rhythms. The method makes it possible to calculate the relative contribution of sympathetic and parasympathetic autonomic nervous system (ANS) into heart rate and BP regulation, integrated values of cardiogram intervals, parameters of lung ventilation, baroreflex parameters, etc.

Electrocardiogram (ECG) was recorded in standard lead I over 2 minutes. The time-amplitude parameters of PQRST complex and heart rhythm variability (HRV) were evaluated using statistic, geometric, and spectral parameters. HRV power in different frequency bands determined using Fourier-transform analysis characterizes ANS activity and the function of the central mechanisms of heart rate regulation. Three frequency bands can be distinguished in spectra: very low frequency (VLF, 0-0.04 Hz), low frequency (LF, 0.04-0.15 Hz), and high frequency (HF 0.15-0.4 Hz), which are measured in absolute values of power ($msec^2$). These values can also be presented in standardized units (LFn, HFn) calculated as the ratio of each spectral component to their sum. Index of autonomic balance (AB=LF/HF) and index of centralization (C=(VLF+LF)/HF) were calculated from HRV spectral parameters.

Peripheral systolic and diastolic blood pressure (SBP and DBP, respectively) and their variability were measured on middle phalanx using the method of Penaz. From the parameters of BP pulse wave, hemodynamic parameters, stroke volume, and cardiac output were calculated using phase analysis of cardiac cycle and BP. Spontaneous arterial baroreflex sensitivity ($BRS=LF_M/LF_{SBP}$) was also evaluated. From geometric parameters of HRV (mode, mode amplitude, amplitude of oscillations, etc.), autonomic balance index, parameter of adequacy of regulation processes, autonomic rhythm index, and regulatory system strain index (SI) were calculated.

For evaluation of functional reserves of the cardiovascular system, a functional test with increased "dead space" was used (Trukhanov et al., 2007). Reactivity of the cardiovascular system was evaluated by changes in the parameters describing its function (in %) during ECG recording in spirometric mask in comparison to ECG recorded without the mask. The time of inspiration and expiration, volume rate of inspiration and expiration, and respiratory volume of quiet breathing in an averaged cycle were evaluated. Parameters of forced expiration (vital capacity of the lungs and volume of forced expiration) were also measured.

The study of the *latent period of simple sensorimotor reaction and other psychomotor parameters* was performed using a specific instrument called "CMM" (computer movement meter), Registration Certificate # 29/03041202/5085-03. The accuracy of measuring the time of simple sensorimotor reaction was 1 ms. A subject was placed in a comfortable chair, while his/her hand was placed on a special handle and lever, which, in turn, may revolve around a vertical axis. The fulcrum of the rotating segment was treated as the upper third of the forearm. In this position the forearm and wrist could commit abduction and adduction. Mechanical resistance to rotational movement of the forearm was insignificant and therefore ignored in the calculations. Initially, the lever with forearm and wrist resting on it was fixed on the zero position by electromagnetic stoppers. Participants were instructed to focus on the cross in the centre of the screen, and to adduct the handle of the lever with their forearm and wrist as quickly as possible in response to the visual signals started at random intervals (4–8 seconds). The visual signals (vivid light) were slightly peripheral to the central visual field, in order to potentially speed up sensor motor reaction. At the instance of switching on light the electromagnetic stoppers were simultaneously removed and the lever was able to move freely. The latent period of the motor response (reaction time - RT) was measured from the moment of switching on light until the angular displacement of the the handle with forearm at 1 degree was recorded by a computer. The subjects were not limited in the amplitude of translation of the lever. The study was performed for both dominant and subdominant hands. Each subject received 16 signals for reaction with each hand. Training before the experiment included reaction for 10 visual stimuli which were randomly distributed in time.

Subfractional composition of blood serum was analyzed using laser correlation spectrometer (LCS, certificate of Committee on New Medical Instrumentation, Ministry of Health Care and Social Development of the Russian Federation, RU.C. 39.003.A N 5381, St. Petersburg). The method is based on changes in spectral characteristics of monochromatic coherent helium-neon laser radiation due to light scatter in disperse system (blood serum, urine, and other biological fluids) (Karganov et al., 2011). The degree of this scatter is proportional to particle speed, which depends on its hydrodynamic radius. The spectra of blood serum samples (0.2 ml) were recorded and processed routinely (Karganov et al., 2011b).

Evaluation of individual sensitivity to ionizing radiation was carried out on mitogen-stimulated peripheral blood lymphocytes. The cells were cultured in glass flasks in a medium containing sterile embryonic calf serum ("Perbio-HyClone", USA), RPMI-1640 medium with 25 mM HEPES and sodium bicarbonate, phytohemagglutinin ("Sigma", USA), glutamine, and antibiotics.

The cells were irradiated on a "Luch-1" γ-apparatus (Medical Radiological Research Center, Russian Academy of Medical Science, Obninsk) at 0.25 Gy/min radiation power and 65 cm distance. The adapting dose (AdD) was 0.05 Gy and the damaging dose (DD) was 0.5 Gy. Three samples for each examinee were irradiated according to the following schemes: 1) AdD at G0 stage of the cell cycle without DD (for evaluation of the effect of AdD); 2) DD at G2 stage of the cell cycle 48 hours after AdD; 3) DD without AdD (for evaluation of DD). The cells were incubated and fixed using standard methods (Hungerford D.A., 1965). The preparations for routine analysis were stained with azure and eosin and examined under a microscope in transmitted light under oil immersion at ×1000. The following chromosome aberrations were counted: chromatid and isochromatid fragments and symmetrical and asymmetrical chromatid exchanges. For each donor, 100 metaphase plates per term were analyzed. The RAR coefficient at different irradiation doses was calculated by the ratio of the number of chromosome aberrations: RAR=control + DD/(AdD+DD); RAR≥1.5 indicates the presence of the adaptive response.

Statistical processing of experimental data was performed using nonparametric tests (Kruskal-Wallis, Mann-Whitney U test, Spearman correlation coefficient, and Fisher exact test), because empirical data did not conform normal distribution according to Kolmogorov-Smirnov test. The differences between the parameters within the group were evaluated using paired Wilcoxon T test. The data are presented as M±SEM. The significance level was 5%.

3. Results

3.1 Screening examination of workers of the nuclear fuel plant

Device complex and methodological approaches have been tested during screening examination of workers of the nuclear fuel plant.

Laser correlation spectroscopy was used for the analysis of blood serum from individuals exposed to repeated or single irradiation. Changes in spectral characteristics associated with shifts in the homeostatic system were revealed (Akleyev, Kisselyov, 2000). Further studies in this field showed that even single radiation exposure leads to metabolic shifts in the organism towards predominance of catabolic processes. Redistribution in the blood serum spectrum was also directed towards accumulation of the low-molecular fraction (Akleyev & Kisselyov, 2000).

We examined workers (n=328) employed at nuclear fuel cycle plant in Electrostal' town contacting with open (shop #1, uranium and its products, MPC level) and sealed (shop #2) sources of radiation, or exposed to combined influence of radiation (shop #3, uranium and its products, above MPC level), chemical, and other factors. Blood serum samples from workers not contacting with radioactive materials were used as the control (n=16).

Comparison of the percentage of metabolic shifts in different shops showed that the percentage of normological spectra was low in shop #1; the percentage of anabolic shifts in all shops was decreased in comparison with the control (Fig. 1).

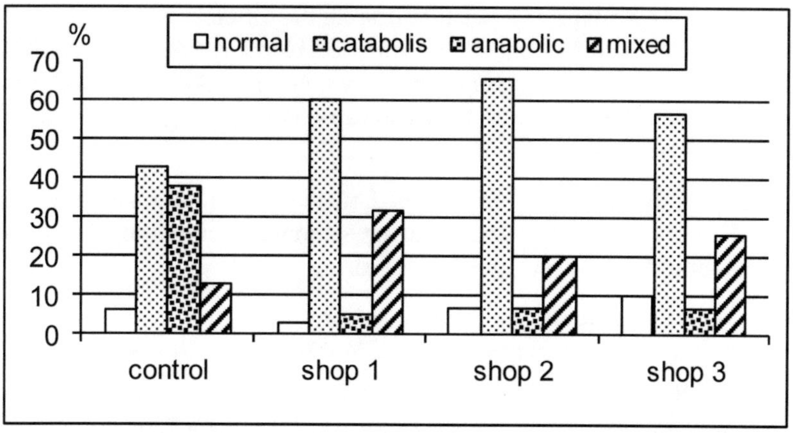

Fig. 1. Incidence of different types of metabolic shifts at nuclear fuel cycle plant of Electrostal' town.

High percentage of catabolic shifts is typical of all shops, which reflects similar life conditions and influence of similar environmental factors.

Detailed analysis of integral LC histograms with evaluation of the contribution of particles of a certain size into light scattering allows identification of the fraction determining maximum differences between the shops (Fig. 2). Significant differences of averaged LC histograms in different shops from summary LC histograms of blood serum in the control group were noted at various points.

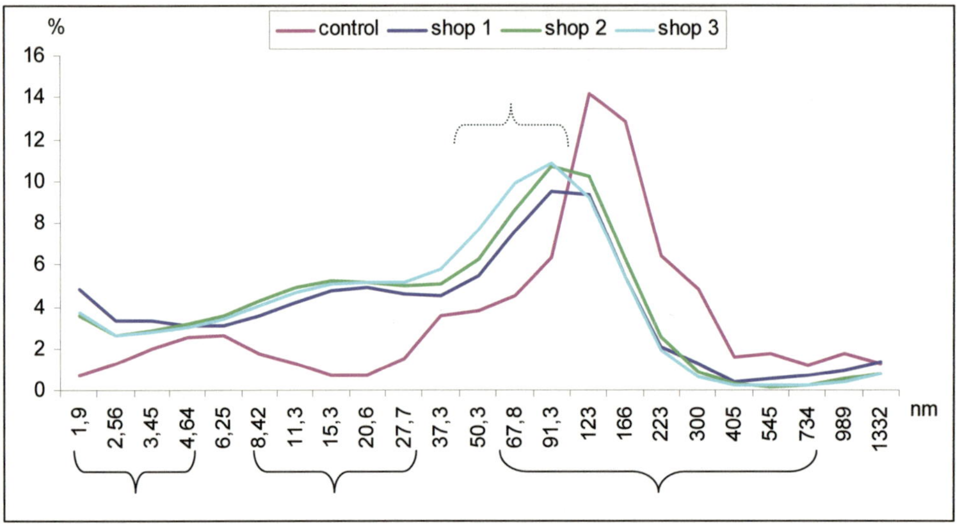

Fig. 2. Integral LC histogram of blood serum in different shops of nuclear fuel cycle plant of Electrostal' town. Points significantly differing from the control group are shown with solid bracket, dashed bracket shows difference of LC histogram of shop # 1 from that of shop # 2 and # 3; $p<0.05$.

This method of data presentation shows that despite the differences of all three shops from the control group, the subfraction composition of blood serum in shop # 1 workers significantly differs from that in two other shops by many points. This situation makes us to conclude that technological factors are the main cause determining different distribution of metabolic shifts in this sample.

One of the leading technological factors is ionizing radiation (primarily γ-radiation). According to the current sanitary and hygienic standards, occupational irradiation from artificial sources should not exceed 50mZv per year.

Shop #1 workers receive 2 times higher annual dose of irradiation than workers of other shops. Therefore, plant management is expected to improve medical monitoring and safety control for workers of this shop.

Twofold increase in the dose may cause changes in metabolic shifts; therefore it is necessary to analyze the dynamics of the ratio of predominant directions of metabolic shifts in 2002-2003 (Fig. 3).

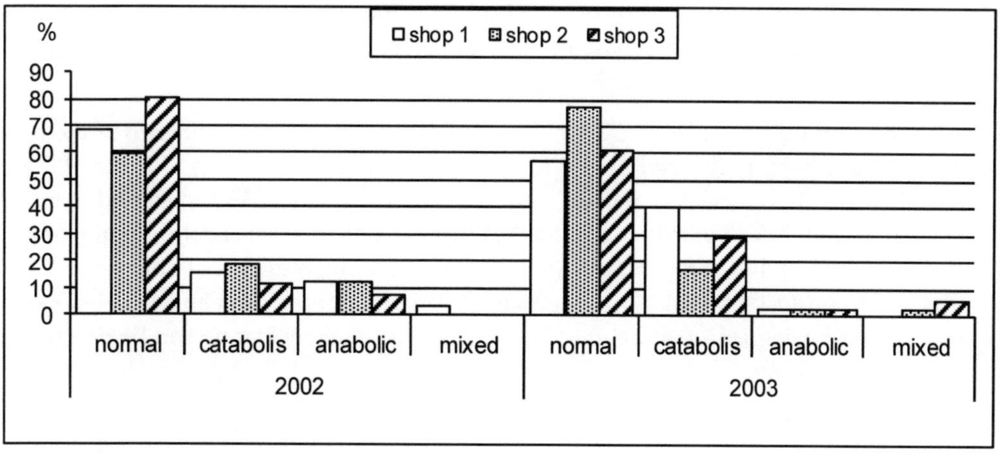

Fig. 3. Ratio of metabolic shifts in different shops

Evaluation of the dynamics of metabolic shifts of plasma homeostasis in different shops revealed the same character of differences: increased contribution of catabolic shifts. Questions arise whether this increase is related to accumulation of the irradiation dose and in which shop the strain of adaptive resources in workers was maximum. To this end, the ratio of metabolic shifts in the same workers after repeatedly evaluated after 12 months (Fig. 4).

Fig. 4. Changes in the ratio of metabolic shifts over 12 months. Ordinate: differences between the percentage of the spectra in 2002 and 2003.

It is obvious that more pronounced changes were found in shop #55 (work with sealed sources of radiation): significant increase in catabolic shifts after exposure to apparently less damaging influences.

The studied parameters were expressed in absolute values (msec, ml, mm Hg, %) and also scored. This method of analysis is based on comparison of the real parameter with averaged data for conventionally normal sample with consideration for age and gender.

Centile rating for the parameters used in our studies implies that the incidence of the most balanced values (within ±0.5 centiles) is about 50%, moderate strain (from -0.5 to -1.5 and from +0.5 to +1.5) is detected in 40% observations (of them 20% are shifted towards hypofunction and 20% towards hyperfunction) and more pronounced (pre-pathological) strain is detected in 10% observations (<-1.5 centiles for hypofunctional shifts and >+1.5 centiles for hyperfunctional shifts). Thus, each measured parameter can be attributed to one of the three above degrees of strain and the direction of functional strain can be designated with plus or minus sign.

Apart from parameter evaluation, we used integral systemic rating; to this end, modal values of all parameters of the system were summed up. If the system included 7 parameters (e.g. the system characterizing myocardial contractility includes the following parameters: HR and P, PQ, QR, QRS, QT, and ST intervals), the corresponding ratings of the integral strain of the system were determined as follows: 1, balanced, modal sum <4.5 centiles; 2, moderately strained, modal sum from 4.51 to 6.5 centiles; 3, strained, modal sum >6.51.

Apart from modal summation, algebraic sum reflecting predominant direction of functional shifts for the whole system was calculated: negative values attested to hypofunction and positive to hyperfunction.

According to general polysystemic monitoring concept, individual functional reserve of the organism to certain working conditions can be determined only on the basis of dynamic observation. Since in our study we performed only single functional scanning, interpretation of the results can be performed using centile distribution of the parameters in populations not burdened by verified pathologies (conventionally healthy individuals).

The examinees were primarily male individuals: from 93% (shop 3) to 61% (shop 2) and 63% (shop 1).

The analyzed groups differed by the age: individuals aged 40.1-50 years predominated, while the percentage of individuals of other age groups was similar (19 and 24% for the groups of 30.1-40 and over 50.1 years, respectively).

Analysis of variability of RR intervals of integral ECG showed that the incidence of balanced states (level 1) was maximum in shop 2 (70%) and minimum in shop 3 (54%). The percent of individuals with strained functional state (level 3) was similar in all shops (8-9%) and did not exceed the range of frequencies characteristic of healthy population (≤10%).

The latter suggests that the myocardial contractility regulation system is not the target of unfavorable factors.

It should be noted that the incidence of strained levels of autonomic regulation of the heart rate in shop 1 (17%) and shop 3 (15%) surpassed the level for healthy population by 1.5-1.7 times and the corresponding value in shop 2 by 2.5 times.

On the basis of age-dependent relationships between the incidences of different strain levels for autonomic regulation of the heart rhythm, we can conclude that the increase in the strain in the heart rhythm regulation system is determined by aging, rather than by technological processes.

Integral evaluation of the functional state of peripheral resistance regulation system (based on measurement of systolic and diastolic pressure strain in digital arteries and baroreflex levels by low- and high-frequency components of the arterial spectrum) showed that the incidence of strained states (rating level 3) in all three groups almost 2.5 times surpassed the level observed in healthy population (10%).

Hence, occupational risks can be expected to be realized in the system of vascular regulation. This assumption is also confirmed by more strained situation in shop 3. Moreover, the incidence of detection of strained parameters practically did not depend on the age of the examinees both in the whole group and in each shop separately.

Of the parameters constituting the integral rating of the whole system, the contribution of hypertensive strain of systolic and diastolic pressures (15 and 14%, respectively) was lower than the contribution of hyperfunctional baroreflex strain (by the low- and high-frequency components of blood pressure spectrum) observed in 23 and 19% cases, respectively. Baroreflex regulation is most tightly coupled with endogeneous mechanisms of tissue metabolism regulation, which was conformed by LCS analysis of the blood plasma, oropharyngeal washout fluid, and urine.

The levels of systolic and diastolic blood pressure are regulated by different mechanisms; the strain in these mechanisms can differ in different groups. The incidences of strained states by the individual components of systolic pressure regulation in the whole examined group considerably surpassed those observed for individual components of diastolic pressure regulation (16-27% *vs.* 10-18%). However, different picture was observed in some shops. For instance, in shop 2 strained ratings for systolic pressure regulation parameters were least incident (≤10%). For systolic pressure regulation, the worst picture was revealed in shop 3 (22-31% strained states, i.e. 2-3-fold more incident than in shop 2); while shop 1 was the worst by diastolic pressure regulation, especially by variability parameter. These differences most likely reflect different degree of adaptation of the blood pressure regulation systems to the working conditions, rather than are age dependent. Thus, predominant contact with sealed sources of radiation minimizes direct damage to the vascular bed, which enables optimization of diastolic pressure regulation processes (shop 2). Predominant contact with open radiation sources (shop 1) increases the risk of strain in the regulation of both systolic and diastolic pressure; this effect was most pronounced in case of combined exposure (shop 3).

All these findings suggest that processes of neuroendocrine regulation of peripheral circulation are most susceptible to the influence of occupational risk factors. Therefore, it was interesting to study in detail functional adequacy of the processes of psychomotor functions at various levels of neuromuscular regulation.

By functional characteristics of the respiratory system (on the basis of spirometry data) and constitution (anthropometry data), the studied populations by the majority of criteria did not differ from the normological one. The third strain rating level by the vital capacity of the lungs was found in 15% cases and by airway conductance (Tiffeneau index) in 20%. In the latter case, hypofunctional shifts were detected in 57% cases and hyperfunctional in 43%. The differences between the groups were insignificant, which excludes age-related and occupational differentiation.

When studying the functional state of psychomotor regulation of employees, we revealed a fivefold increase in the incidence of strained states. In all cases, the strain had hyperfunctional direction (i.e. professional activity primarily mobilizes psychomotor functions) and its incidence was similar in all examined groups (i.e. the observed strain is not related to age).

Thus, working conditions selectively complicate neuromuscular regulation and first of all regulation of smoothness of motions and correction of motions. Both parameters are related to the central and subcortical levels of regulation. Hence, the observed strain in the regulation of locomotion is not associated with muscular fatigue, but is determined by constant need to accurately perform motor acts, which leads to mobilization of neuroendocrine regulatory mechanisms.

However, we revealed certain differentiation in the structure of psychomotor strain depending on the nature of technological process. For instance, in shop 2 (predominant contact with sealed sources of radiation) relative low incidence of strain in correction of movements and extremely high incidence of strain in smoothness of movements were observed. In shop 1 (primary contact with open radiation sources), an opposite situation was observed: maximum strain in error correction and relatively low strain in smoothness of movements. In shop 3 (combined exposure), the situation by the discussed criteria was intermediate.

These findings suggest that working conditions at the industrial plant primarily mobilize the processes of neuroendocrine regulation of psychomotor functions. It should be noted that working with sealed sources of radiation requires extreme smoothness of movements, while working with open sources requires precision of movements.

The tested approaches are aimed at detection of individual functional strain, which ensures targeted correctional and rehabilitation measures. This is the field where we the above-discussed basic methods of polysystemic monitoring can be practically applied for maintaining more stable functional reserve of the organism under the influence of technological factors.

3.2 Assessing the health of military pilots

In examination of a group of **military pilots**, radioadaptive response (RAR) of peripheral blood lymphocytes was additionally studied. The health status of flight personnel is one of the most important components of flight safety requiring realization of complex measures aimed at prevention of diseases, prolongation of career longevity, and improvement of tolerance of unfavorable flight factors. In modern aviation medicine and physiology, the primary attention is paid to the control of health reserves, rather than transition of health to disease. The focus is thus shifted to individual evaluation of the functioning of the main organs and systems in pilots. Flight personnel is exposed to various potentially dangerous factors during flights. Evaluation of biological risks in modern aviation is complicated by the fact that these influences do not exceed or only slightly exceed the maximum permissible level. This is the range where organism's response to the studied factors is maximally individual.

Long-term predictions of the state of organism's functional reserves and evaluation of the probability of negative biological effects during flights should be based on the results of not only standard physiological tests, but also detection of pathological markers at the cellular and biochemical levels. Complex examination of passenger aircrafts crews carried out by Italian researchers revealed changes in a number of cytogenetic parameters of buccal epithelial cells and lymphocytes (e.g. elevated chromosome aberrations frequency, oxidative damages in DNA, etc.) (Cavallo et al., 2006). Changes in cell functioning lead to activation of DNA reparation, on the one hand, and induce apoptosis of damaged cells, on the other, thus promoting adaptation of the organism to stressors. However, chronic exposure to minor influences leads to exhaustion of adaptive reserves, which creates prerequisites for the development of organism hypersensitivity.

When discussing delayed consequences of the effect of occupational factors on morbidity in aviators, ionizing radiation is considered to be the most important genotoxic factor (Cavallo et al., 2006). The compliance of radioactivity situation during flights with modern radiation safety standards is now intensively discussed. It was suggested to consider pilots as professionals working with ionizing radiation sources (Bhatti et al., 2010; Yong et al., 2009). During flights the pilots are exposed to low-dose radiation; therefore the individual reactions of the organism at the cellular and systemic levels are difficult to predict. The capability of a biological object to respond to ionizing radiation in this or that way, i.e. individual radiosensitivity, is determined by genetic and environmental factors and depends on various processes such as postradiation reparation capacity, intensity of metabolism, rate of redox, physicochemical, and biochemical reactions in cells, etc.

Irradiation of peripheral blood lymphocytes is a convenient tool for evaluation of individual radiosensititvity. Low doses of radiation trigger nonspecific defense mechanisms, i.e. the results can be extrapolated to the effects of other damaging agents. One of the above-mentioned effects of low-dose radiation is radioadaptive response (RAR): cells exposed to low-dose radiation become more resistant to ionizing radiation (or other agents) in high doses (Pelevina et al., 2003). RAR can be used as a measure of individual radiosensitivity.

Our aim was to study the correlation between individual radiosensetivity and changes in molecular composition of blood serum and function of the cardiorespiratory system in pilots with different flight time.

The examinees were divided into three groups: control group A included people who had no flight hours, but their work is related to aviation (flight operations officer, technical staff, and parachutists; n=9, mean age 23.1±2.6); group B comprised pilots with flight time <1000 hours (n=17, mean age 33.1±1.5); group C includes pilots with flight time >1000 hours (n=12, mean age 39.7±1.1).

All examined pilots were in one age group by the level of chromosome aberrations. The incidence of RAR of PHA-stimulated lymphocytes decreased proportionally to the flight time: 78%, 59% and 33% in groups A, B and C, respectively ($p<0.05$ for groups A and C, Fisher's exact test). A negative correlation was revealed in the examined cohort between the flight time and RAR coefficients (r=-0.37, p=0.021). RAR coefficient decreased with increasing flight time: 1.68±0.15, 1.08±0.10, and 0.93±0.17 in groups A, B. and C, respectively (H(2, n=36)=6.32, p<0.05, ANOVA Kruskal-Wallis test). Post-hoc analysis revealed significant differences in group C in comparison with group A (p<0.05; U test). Previous studies on lymphocytes isolated from animals and healthy donors showed that RAR develops in not all individuals and is not detected in 30-60% cases. The intensity of RAR depends on genetic predisposition and physiological status of the organism (Weissenbok et al., 2000).

Comparison of averaged LC spectra using the method of trapezoidal integration revealed a pronounced tendency towards a decrease in the contribution of small particles (hydrodynamic radius <11 nm) into light scatter in group B in comparison with the control (p=0.073, U test). In group C this decrease attained the level of statistical significance. The contribution of large particles (>165 nm) into light scatter increased in both groups. The increase in the relative number of 300-400-nm particles in group B and 500-900-nm particles in group C in comparison with the control is worthy of note (Fig. 5, A, B). Since the observed

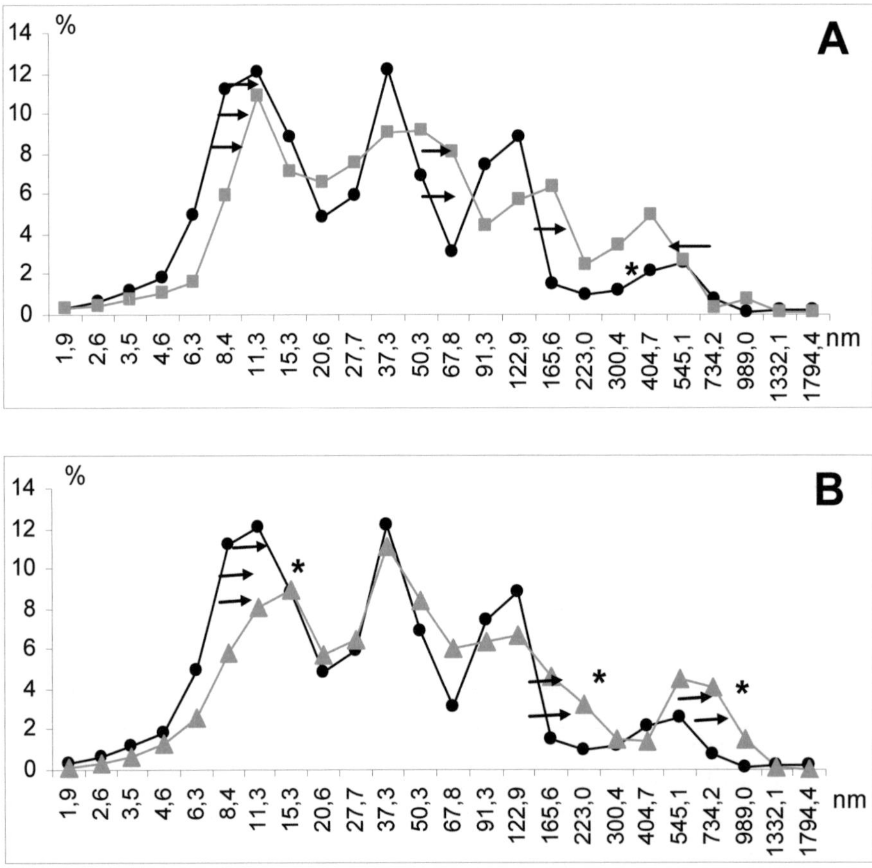

Fig. 5. Changes in the LC spectra of blood serum of pilots with a touch of up to 1000 hours (A) and more than 1000 hours (B). Abscissa - particle size (nm), the vertical axis - the contribution of particles of different hydrodynamic radius of the light scattering (%).

differences were detected in healthy individuals undergoing regular medical examinations, it can be hypothesized that they are determined by the action of flight factors (including radiation) and various mechanisms of adaptation. Low radiation doses are considered to produce an activating effect on the immune system, stimulate phagocytosis and antibody production, increase lysozyme activity, and improve general immunological status.

According to the data obtained on a large sampling of conventionally healthy individuals, the total light scatter spectrum for blood serum can be divided into 5 discrete informative zones. Zone I (0-10 nm) corresponds to monomer albumins and glycolipid complexes; zone II (11-30 nm) contains globular proteins and low-molecular-weight lipoprotein complexes; zone III (31-70 nm) includes high-molecular-weight lipoprotein complexes and low-molecular-weight immune complexes; zone IV (71-150 nm) comprises medium-size immune complexes, ribonucleoproteins (RNP), and deoxyribonucleoproteins (DNP). Larger particles (>150 nm, zone V) appear in cases of induction of immunopoiesis with the formation of high-molecular-weight complexes; this process usually accompanies allergization and autoimmune sensitization of the organism. Depending on the increase (or decrease) in the percent contribution of particles of this or that fraction into light scatter, we can speak about the direction of shifts in homeostasis and humoral immunity.

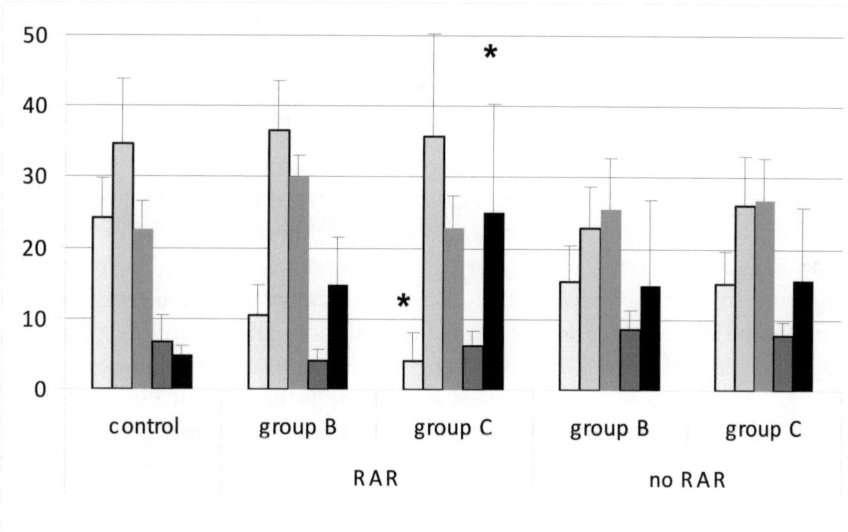

Fig. 6. The distribution of light-scattering particles in the standard zones in the serum of pilots with a flight time up to 1000 h (group B) and 1000 hours (group C), depending on the presence or absence of RAR.

Analysis of LC spectrometry data with consideration for the presence or absence of RAR showed that in pilots with RAR the contribution of small particles (<10 nm) into light scatter decreased and the contribution of large particles (>150 nm) increased in comparison with the control values (Fig. 6). In pilots without RAR, no significant differences in subfraction composition of blood serum from the control were revealed. Apart from the tendency towards a decrease in the contribution of zone I particles and the increase in the contribution of zone V particles into light scatter, a decrease in the contribution of zone II particles was noted, which led to changes in the general pattern of spectra (Fig. 6). It should be noted that we failed to collect a representative control subgroup without RAR in this study, therefore pilots with and without RAR were compared only with controls with RAR. The observed differences for the subgroups with RAR can be explained by epigenetic influences of flight factors on metabolic processes in the organism. At the same time, changes in subfraction composition of blood serum in pilots without RAR in comparison with that in pilots with RAR in groups B and C can be genetically determined.

The development of RAR is probably related to changes in cell metabolism. Comparison of LC spectra of lymphocytes-conditioned culture medium showed that the contribution of particles with a radius of 91.3–223.0 nm was lower in samples conditioned by cells with RAR in comparison with samples conditioned by cells without RAR ($p<0.05$, U test). It can be hypothesized that irradiation in the absence of RAR leads to accumulation of a factor with a size of about 100 nm in the medium, e.g. RNP and DNP particles (80-100 nm).

Induction of RAR in peripheral blood lymphocytes is a loading test that allows evaluation of the adaptation reserves at the cellular level by complex parameters. Here we evaluated the quality and the intensity of reparation systems. Irradiation of interphase cells with low AdD induces stimulation of reparative mechanisms in cells. DD was applied after the cells passed the main reparation checkpoints (pre- and postreplicative reparation). It can be hypothesized that prolongation of the exposure to flight factors produces a negative effect on the adaptation capacities of the organism at the cellular level.

The cell response to stimulation aimed at conservation of genetic material is a particular case of stress reaction development. The function of the genome determines the relative role of mechanisms of cell reparation, proliferation, and death; these processes, in turn, determine the function of enzymes, cells, tissues, organs, and systems responsible for adaptation of human organism to the environment. The efficiency of this process is also determined by the intensity of the influence, duration of exposure, and adaptation capacity of the organism depending on functional resources and the corresponding strain of regulatory mechanisms. Thus, the rate and quality of adaptation depend on both the state of the genetic apparatus and physiological characteristics of organism's systems. The state and function of the cardiorespiratory system can be an indicator of adaptive reactions of the organism (Baevsky et al., 2004).

In our study, the parameters of PQRST complex, heart rate, SBP and DBP at rest and during functional tests were similar in all groups and did not exceed the normal limits for male individuals of the corresponding age. However, stability of the functioning of the cardiovascular system in groups was provided by different mechanisms.

The total power (TP) of HRV spectrum was different in groups A, B, and C at rest ($H(2, N=30)=6.944$ $p=0.031$) and during functional tests ($H(2, N=29)=8.740$ $p=0.013$). In both cases, this parameter in groups B and C was considerably lower than in the control (Table 1). This drop of TP was due to a decrease in absolute values of frequency components in all frequency bands in groups B and C. SI was maximum in group B, but the results of analysis of variance showed that pair-wise comparison of groups cannot be performed ($H(2, N=29)=5{,}360$ $p=0.067$).

	Resting state (without mask)													
	HR (1/m)	**TP (ms2)**	**VLF**	**VLF, %**	**LF**	**LF %**	**HF**	**HF %**	**LFn, n.u**	**HFn, n.u**	**AB**	**C**	**SI**	**BRS**
A	77.1 ± 2.9	6126.4 ± 1620.5	1311.7 ± 398.2	22.9 ± 5.7	2413.1 ± 1004.1	33.8 ± 7.7	2401.6 ± 741.8	43.4 ± 8.5	44.2 ± 9.6	55.8 ± 9.6	1.2 ± 0.45	2.0 ± 0.6	155.0 ± 101.7	13.1 ± 4.8
B	81.1 ± 3.1	2507.6 ± 514.2*	486.0 ± 143.9 *	21.7 ± 4.0	1224.4 ± 319.2^	44.3 ± 4.1	797.1 ± 184.3^	34.0 ± 4.5	57.0 ± 5.0	43.0 ± 5.00	1.9 ± 0.54	2.9 ± 0.8	183.8 ± 41.0	8.6 ± 0.9
C	78.4 ± 4.3	1745.8 ± 484.5*	428.5 ± 120.6^	32.9 ± 4.9	943.1 ± 449.1^	41.5 ± 6.7	374.2 ± 93.7*	25.6 ± 5.3	61.4 ± 6.9	38.6 ± 6.90	3.6 ± 1.60	5.4 ± 1.6	513.7 ± 196.9	7.5 ± 1.5
	Functional test with hypoxic stress (in mask)													
	HR2	**TP2**	**VLF2**	**VLF%2**	**LF2**	**LF%2**	**HF2**	**HF%2**	**LFn2**	**HFn2**	**AB2**	**C2**	**SI2**	**BRS2**
A	77.1 ± 2.4	8201.1 ± 2584.9+	1412.0 ± 238.9	20.9 ± 5.5	4872.3 ± 1875.1	58.7 ± 7.4+	1916.8 ± 1082.1	20.3 ± 6.8+	74.4 ± 7.4+	25.6 ± 7.4+	5.1 ± 1.50+	3.8 ± 1.3+	76.6 ± 40.1	15.9 ± 2.7
B	82.5 ± 3.8	1707.3 ± 272.9*	342.0 ± 489.1*	25.3 ± 4.2	915.7 ± 199.2*+	49.0 ± 5.4	449.6 ± 122.7^+	25.6 ± 4.8	65.4 ± 5.7	34.6 ± 5.80	5.1 ± 2.30	3.2 ± 0.9	212.9 ± 65.0	9.1 ± 0.9*
C	79.8 ± 4.3	2105.5 ± 763.9*	641.8 ± 192.2^	33.2 ± 3.6	1041.6 ± 462.9*	46.9 ± 3.9	422.2 ± 153.6^	19.9 ± 4.3	70.8 ± 4.9	29.2 ± 4.9	3.6 ± 0.90	5.2 ± 1.2	412.7 ± 88.5*	8.3 ± 1.4*

*$p < 0.05$; ^$p = 0.07$ - compared with the value of the indicator in the control group (Mann - Whitney); + $p < 0.05$ - compared to the value of the index at rest in the appropriate group (Wilcoxon test).

Table 1. Indices of heart rate variability in pilots with different number of flying hours

Functional tests revealed no significant changes in TP and absolute values of power in individual frequency bands in comparison with the corresponding values at rest in all the groups. However, we observed an increase in the percent contribution of low frequencies (LF%) and a decrease in the contribution of high frequencies (HF%) in control group A, which led to a shift in AB and CI towards a decrease ($p<0.05$, Wilcoxon test). In groups B and C, the contribution of very low frequencies increased in comparison with that at rest (VLF%). AB and SI in these groups remained practically unchanged. Moreover, we revealed negative correlations between the total number of flight hours and the total power of HRV spectrum (r=-0.55, p=0.006) and powers of low-frequency (LF) and high-frequency (HF) components of the spectrum (r=-0.62, p=0.001; r=-0.48, p=0.021, respectively). SI after functional load increased with increasing the total flight time (r=0.64, p=0.001) and in groups C this parameter significantly differed from the control (H (2, N=21)=6.019, p=0.049). BRS in the control groups was higher than in groups B and C (tendency at rest and significant differences during functional tests H (2, N=27)=6.142, p=0.046).

The total power of SBP rhythm variability spectrum in the control group remained unchanged during functional test (in mask) in comparison with that at rest, while the power of VLF band significantly decreased (Fig. 7) and LF component of the spectrum increased (pronounced tendency p=0.087 Wilcoxon test) under these conditions. In group B, the total spectral power decreased during functional test due to reduced contribution of VLF and HF bands. In group C, no significant changes in SBP rhythm variability were revealed. Groups A, B, and C did not differ by DBP parameters and parameters characterizing the state and function of the respiratory system at rest and during functional tests.

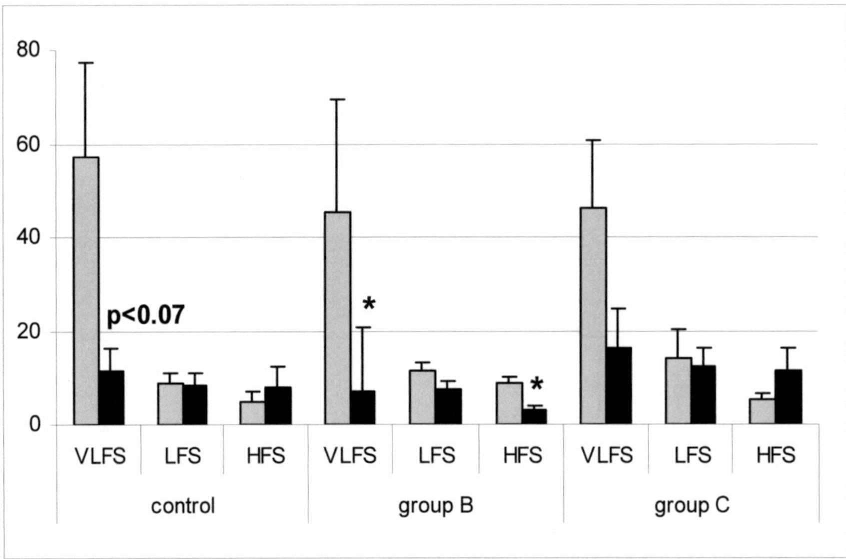

Fig. 7. Frequency ranges of systolic blood pressure variability changes when performing a functional test in pilots with different total time of the raid. Gray columns - measurement at rest, black columns - functional test (spirometric measurements in a mask) (hypoxic stress). * p <0.05 compared with the value of the index at rest (Mann - Whitney)

Comparison of the parameters characterizing functional activity of the cardiorespiratory system in pilots with and without RAR revealed no significant differences between these groups at rest by all analyzed characteristics. After functional test, peripheral SBP and DBP and total and specific peripheral resistance increased in the group without RAR in comparison with the corresponding values at rest (Table 2). In individuals with RAR these parameters little changed. HRV and SBPV also underwent different changes in groups with and without RAR during functional tests. In pilots without RAR, the total spectral power of SBPV and the contribution of very low frequency band considerably decreased in comparison with the corresponding values at rest. In the group with RAR, the total spectral power of SBP did not change, but the relative contribution of low frequencies into HRV and SBPV increased and the contribution of high frequencies decreased under these conditions, which led to an increase in their ratio and CI in comparison with those at rest.

Intergroup comparison showed that in the group with RAR, the contribution of low frequencies into HRV was greater and the contribution of high frequencies was lower than in the group without RAR. The relation between these values and CI were also higher, while SBP was lower in the group with RAR in comparison with the corresponding values in the group without RAR (Table 2). Parameters of regulation of the respiratory system after functional test were similar in these groups.

Resting state (without mask)														
	LFn, %	HFn, %	AB	C	SBP, mmHg	DBP, mmHg	TPS, ms^2	VLFS, ms^2	LFS, ms^2	HFS, ms^2	LFSn, %.	HFSn, %	ABS	TPVR
No RAR	56.1 ± 6.2	43.9 ± 6.2	2.7 ± 1.2	4.0 ± 1.3	126.6 ± 6.6	79.7 ± 4.5	66.1 ± 15.27	44.2 ± 11.5	13.8 ± 4.8	8.1 ± 1.7	59.6 ± 5.5	40.4 ± 5.5	3.0 ± 1.1	1327.1 ± 84.8
RAR	55.0 ± 5.2	45.0 ± 5.2	1.9 ± 0.4	3.0 ± 0.6	132.2 ± 5.9	88.3 ± 5.6	68.3 ± 21.0	53.0 ± 19.7	10.0 ± 1.4	5.4 ± 1.1	67.7 ± 2.6	32.3 ± 2.6	2.5 ± 0.3	1576.0 ± 130.3
Functional test (the mask)														
	LFn	HFn	AB	C	SBS	DBD	TPS	VLFS	LFS	HFS	LFSn	HFSn	ABS	TPVR 2
No RAR	61.4 ± 5.0	38.6 ± 5.0	2.4 ± 0.6	2.8 ± 0.5	143.4 ± 8.9+	88.6 ± 6.2+	40.8 ± 10.7+	19.6 ± 6.3 +	9.6 ± 3.2	11.6 ± 4.2	60.7 ± 6.5	43.7 ± 7.1	1.8 ± 0.4	1581.7 ± 160.0 +
RAR	75.1 ± 4.1 +	24.9 ± 4.1 *+	6.1 ± 1.6 *+	5.2 ± 1.0 *+	132.9 ± 4.6 *	87.1 ± 6.0	33.2 ± 7.3	20.5 ± 6.2	9.5 ± 1.6	3.2 ± 0.7	76.0 ± 2.6 *	24.0 ± 2.6 *	3.7 ± 0.5*	1542.3 ± 116.1

*$p < 0.05$ - compared with the value of the index in the group without RAR (Mann - Whitney); + $p < 0.05$ - compared to the value of the index at rest in the appropriate group (Wilcoxon test)

Table 2. Indices of heart rate variability and peripheral blood pressure to the presence and absence of lymphocyte radioadaptive response in pilots

The total power of HRV spectrum and its constituents were reduced in groups with individuals with flying practice. In contrast, SI increased with increasing flight time; the difference from the control attained the level of statistical significance in the group with long flight time. It is currently accepted that the total power of HRV spectrum reflects the influence of the humoral system on myocardial functions. The decrease in HRV is associated with exhaustion of functional reserves of the organism (Baevsky et al., 2004). It can be hypothesized that flight factors produce a negative impact on the humoral component of the cardiovascular system regulation. The data of LC spectrometry confirm this assumption.

In the control group, the contribution of LF component into the total spectral power of HRV and SBPV increased in response to functional test. Calculated parameters AB and SI also increased under these conditions. In pilots, functional test elevated the percent contribution of VLF band into HRV and increased SI. According to averaged data obtained on adult individuals not employed in aviation, a decrease in relative LF power and an increase in relative HF power were observed during testing in spirometric mask, the total spectral power being unchanged. Hence, activation of the respiratory system leads to relative strengthening of the influences of the autonomic regulatory contour. Reduced HF power and/or increased LF power together with the increase in HF/LF ratio are related to activation of the sympathetic ANS, which is considered to be an adaptive response to stress load. Increased contribution of VLF band into HRV spectrum in pilots can attest to both metabolic changes and psychoemotional factors. Elevated SI attest to mobilization of functional reserves leading to their exhaustion with increasing the degree of sympathetic activation. An unfamiliar test included into aviation medical expertise can act as significant stress factor and the response of pilots and examinees of the control group to this stress can differ.

Functional test led to an increase in peripheral vascular resistance and BP in examinees without RAR. In this subgroup, no changes in HRV spectra and no significant differences in subfraction composition of blood serum from the control were revealed. In examinees with RAR, BP and peripheral vascular resistance remained unchanged during the functional test. However, a redistribution of the HRV and SBP spectral power was observed: increase in LF power and decrease in HF power. In pilots with RAR, pronounced metabolic shifts in blood serum in comparison with the control were detected.

It is now proven that activation of signal pathways (cytokines, protein kinase C, etc.) plays an important role in induction of RAR. Adaptation in cell culture is associated with an increase in the content of reactive oxygen species and NO, which can act as the key signal molecules (Coates et al., 2004). According to the hypothesis of V. N. Titov, in cell pools regulated by paracrine mechanisms, stress induces metabolic changes accompanied by weakening of the dilatation effect of NO, functional shutdown of peripheral peristaltic pumps (muscular arteries), and increase in total peripheral vascular resistance (TPVR). The compensatory reaction consists in participation of the myocardium as the central pump in the regulation of homeostasis and increase in systemic BP.

4. Conclusion

Our findings suggest that in individuals with and without RAR, different regulatory mechanisms are involved into adaptation to varying environmental conditions, due to which they can respond to physical and psychoemotional loads in different ways at both the organism level and metabolic level. The pattern of systemic responses for humans with and without RAR are apparently genetically determined. The intensity of RAR, in turn, decreases under the action of flight factors. It can be hypothesized that individual radiosensitivity reflects general resistance of the organism to negative environmental influences.

Authors are grateful to M. M. Antoshchina, Chief Researcher of Medical Radiological Research Center, Ministry of Health Care and Social Development of the Russian Federation, Obninsk, for her help in the experiments.

5. References

Akleyev A.V., Kisselyov M.F. eds. Medical-biological and ecological impacts of radioactive contamination of Techa river. Moscow. 2000. 431 P.

Baevsky RM, Chernicova AG, Funtova II, Pashenko AV & Tank J. The autonomous regulation system functional reserves evaluation in 7-day head down bedrest. -J Gravit Physiol., 2004 V.11, #2, P.91-92.

Bhatti P., Yong L.C., Doody M.M. et al. X-ray examinations and increased chromosome translocations: evidence from three studies // Radiat Environ Biophys. 2010. V. 49. #4. P. 685-692

Cavallo D., Ursini C.L., Carelli G. et al. Occupational exposure in airport personnel: characterization and evaluation of genotoxic and oxidative effects // Toxicology. 2006. V.223. #1-2. P. 26-35.

Coates PJ, Lorimore SA & Wright EG. Damaging and protective cell signalling in the untargeted effects of ionizing radiation. // Mutat Res. 2004. V.568. #1. P. 5-20.

Hungerford, D. A. Leukocytes cultured from small inocula of whole blood and the preparation of metaphase chromosomes by treatment with hypotonic KC1 // Stain Technology. 1965. V.40. P.333-338.

Karganov M., Alchinova I., Arkhipova E. & Skalny A. Laser correlation spectroscopy approach to bioelementology problems: nutritional, ecological and toxic aspects. - "Biophysics", ISBN 979-953-307-290-5, InTech, 2011, in press

Karganov M., Skalny A., Alchinova I., Khlebnikova N., Grabeklis A., Lakarova E. & Eisazadeh S. Combined use of laser correlation spectroscopy and ICP-AES, ICP-MS determination of macro- and trace elements in human biosubstrates for intoxication risk assessment. // Trace elements and electrolytes. 2011. V.28, N.2. - P.124-127

Pelevina II, Aleshchenko AV, Antoshchina MM, Gotlib VIa, Kudriashova OV, Semenova LP & Serebrianyĭ AM. The reaction of cell population to low level of irradiation //Radiats Biol Radioecol. 2003. V.43. #2, P.161-166.

Trukhanov A. I., Pankova N. B., Khlebnikova N. N. & Karganov M. Yu. The Use of Spiroarteriocardiorhythmography as a Functional Test for Estimating the State of the Cardiorespiratory System in Adults and Children. // Human Physiology. 2007. Vol. 33, # 5. P. 585–594.

Weissenbok R.H., Pfandler S. & Fleck C.M. Radioadaptive response in human lymphocytes. In: «The Effects of Low and Very Low Doses of Ionizing Radiation on Human Health», ed. by WONUC. 2000. Elsevier Science B.V. P. 112–120.

Yong L.C., Sigurdson A.J., Ward E.M. et al. Increased frequency of chromosome translocations in airline pilots with long-term flying experience .// Occup Environ Med. 2009. V. 66. #1. P. 56–62.

11

Risk Assessment in the Anaesthesia Process

Valérie Neyns, Ophélie Carreras,
Laurie Planes and Jean-Marie Cellier
University of Toulouse le Mirail
France

1. Introduction

Typically, the patient considers the anaesthesia process as risky (Marty, 2003). Indeed, the anaesthetist has to understand risks related to the patient and also to the surgery. There are many ways to define risks according to the point of view adopted.

In medical setting, risk can be defined using the ISO 12000-1 and the OSHAS 18001 standards. So, risk is defined as a measure of threat expressed in terms of the occurrence of an adverse event (*i.e.* its probability and its frequency) and a measure of its effects or its consequences. In anaesthesia, three criteria are commonly used to describe risk: the event gravity, the occurrence frequency and the level of acceptability. The first one, the event gravity, can be seen as a qualitative scale with 4 major steps: minor risk (*i.e.* error without prejudice for the patient), significant risk (*i.e.* self–limiting prejudice), major risk (*i.e.* error needing a recovery action) and risk evaluated as serious to critical (*i.e.* permanent damage). The second scale, the occurrence frequency, contains 5 major steps: highly unlikely (*i.e.* frequency≤10-5), very unlikely (*i.e.* 10-5<frequency≤10-4), unlikely (*i.e.* 10-4<frequency≤10-3), probable (*i.e.* 10-2<frequency≤10-1) and very likely to sure (*i.e.* >10%). The last one, the level of acceptability, is divided into 3 parts: non-critical risks (*i.e.* acceptable risky situations), risks to be monitored (*i.e.* acceptable risky situations but actions are needed to identify and monitor them) and rush through risks (*i.e.* not acceptable risky situations requiring actions to reduce risks or to monitor them).

In France, there is a step entirely devoted to anaesthesia risk assessment: the pre-anaesthesia consultation. But this is a French uniqueness. Indeed, in other countries (*e.g.* Quebec), the anaesthetist will see the patient at the entrance to the operating room. However, the anaesthetist may not assess all risks during the anaesthesia consultation. In this chapter, we will study how does an anaesthetist assess risks linked to a patient who must have a surgical operation before and during this one.

We will present this chapter as follow. First, we will describe the anaesthesia process in France and some epidemiological studies on risks in anaesthesia. Then, we will present some cognitive psychology concepts related to planning, information gathering, resilience engineering and management (*i.e.* error detection, identification and recovery). Two studies will be presented by the method used and results obtained. The first one concerns a card sorting experimentation (with patient records) to understand how anaesthetists gather patient's files according to their risk. While the second one consists in semi-structured interviews revolved on simulated cases using the information on request technique. Finally, our results will be discussed in regard of theories used.

2. The anaesthesia situation

In this first section, we will present the anaesthesia situation according to the temporal organisation of the anaesthetist's activity situation characteristics and constraints. Finally, risks linked to anaesthesia will be presented through epidemiological studies.

2.1 The anaesthesia process

Anaesthesia is commonly defined as the administration of medications, called anaesthetics, to control pain during a medical procedure or surgery. There are three kind of anaesthesia: the local one used for minor surgery; the regional one inducing sedation and, the general anaesthesia used for major surgeries and inducing unconsciousness and complete pain control. In this chapter, we will mainly focus on general anaesthesia.

Moreover, the anaesthetist has 3 additional roles. First, he has to work in collaboration with the surgeon to evaluate and prepare the patient. Second, he has to provide medical care to the patient during the surgery. Finally, he is directly involved in the patient's management after the surgical intervention (Chung & Lam, 1990). According to these roles, 3 phases are distinguished: the pre-operatory one, the per-operatory one and the post-operatory phase. The figure 1 presents the anaesthesia process in France.

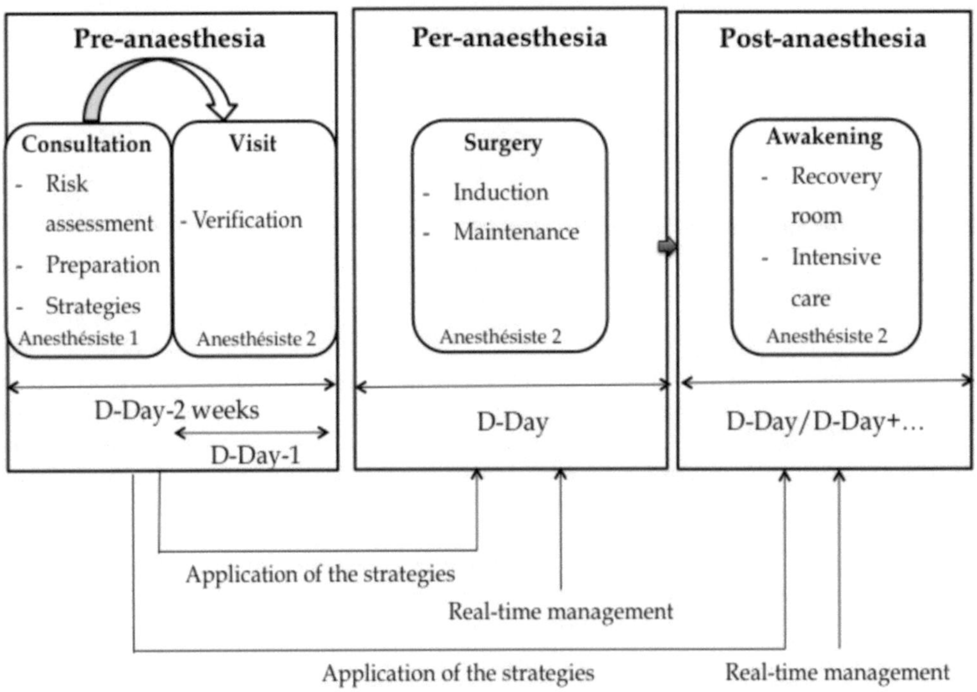

Fig. 1. The anaesthesia process in France

2.1.1 The pre-operatory phase

This phase, the first box on the left, allows the anaesthetist to prepare the patient for the surgery and to highlight important elements to consider during this one (Anceaux & Beuscart-Zéphir, 2002). Due to the French specificity (for recall, the existence of a specific consultation). This first phase is divided into two different steps: the consultation and the anaesthetic visit.

The consultation takes place 2 weeks before the surgery. During this one, the anaesthetist looks for information about the patient's disease through interviews (medical history, comorbidities, medication and allergies) and physical examination, establishes a family medical history and tries to reassure the patient. This one allows the anaesthetist to prescribe additional tests, to adapt the treatment, and if needed, to optimize the patient's health.

The second step, the anaesthetic visit, takes place the day before or the same day of the surgery. Its major aim is to verify the absence of interfering elements occurred between the consultation and the visit.

2.1.2 The per-operatory phase

This second phase starts when the anaesthetist meets the patient at the entrance to the operating room. Two different steps compose this phase. The first one is the anaesthesia induction. It means the administration of anaesthetic agents and the establishment of adequate depth anaesthesia for surgery. It is an important step physically and mentally (McDonald & Dzwonczyk, 1988; McDonald et al., 1990; Gaba & Lee, 1990). Physically because a lot of actions are needed (Xiao, 1994): preparation and injection of anaesthetic drugs, airway intubation, breathing circuit connections, programing a precise mechanical ventilation, etc. Mental workload is also observed. In fact, the patient's physiological status will change very quickly due to injected drugs. To interpret all the information relayed by the monitoring and the physical examination, the anaesthetist has to construct a specific representation of the patient's health state. Then, he has to check the data progress.

The second step of this phase is the maintenance of anaesthesia throughout the medical procedure. During this step, the anaesthetist's main task is to monitor the patient's vital signs and the progress of surgery.

A task analysis (Neyns, 2011) has highlighted 4 categories of anaesthetists' observable behaviours.

The first category relates to the anaesthetist's need to monitor the situation, assess the evolution of specific variables, understand the situation, etc. It is the **information-gathering step**. Five sources were underlined: *the patient's record* that includes both the pre-anaesthetic file (form the consultation and additional assessments) and the per-operative sheet that includes all patient data (drugs injected, vital signs, all relevant information); the *monitoring and its alarms*; the *surgery status and progress* allowing to explain changes in patient's vital signs or to restore the patient's anaesthesia; information taking on *patient* (apart from monitoring data) used to confirm hypothesis; and information related to *hour* either to check the schedule or to check the drugs effect, or for a personal need.

The second category refers to variables that are not directly controlled by the anaesthetist (*e.g.* the conduct of the intervention). He has then to refer to other members in the operating room (*e.g.* surgeon, nurses, surgical assistant, etc.) and to other health care professionals who are not directly involved in the surgical procedure (*e.g.* cardiologist, critical care services, recovery room, etc.). It is **functional communications** between the different members. These communications can go from the anaesthetist to other members but also to the anaesthetist. The communication objects can be multiple: to provide/exchange information, to help the anaesthetist to detect/identify a problem, to discuss plans for the previous/present/next patient, etc.

The third category includes the anaesthetist's **actions on the environment** (*e.g.* actions on the monitoring interfaces to facilitate data interpretation or to set to the patient's vital constants, preparation of the equipment, etc.) and **the patient** (*e.g.* drugs injections, intubation, etc.). His actions are initiated by either obligatory procedure (*e.g.* prescribed tasks), or action plans, or by environment data (*e.g.* information gathering/communication/alarms). Concerning alarms, they play a role in detecting problems. Some alarms can be the result of a failure in the patient's vital signs. The anaesthetist will act to restore the situation in a "normal way". But others can be caused by interferences (*e.g.* the electric scalpel of the surgeon can cause false cardiac dysrhythmia). These alarms are ignored because they are not significant and the anaesthetist cannot do anything against them. Finally, other alarms prompt the anaesthetist to seek for further information (*e.g.* check the surgeon' advice about the surgery, an act painful for the patient can lead to an increased heart rate).

Finally, the fourth category includes two types of actions: **non-functional actions** that are not related to the intervention (*e.g.* reading articles or emails and communications with other members, telephone, etc.) and the **anaesthetist's absence** from the operating room who delegates the process supervising to other members in the operating room.

Thus, this task analysis shows that the anaesthesia management includes extensive information and various parameters can be taken into account. It demonstrates the system complexity and its relative reliability.

2.1.3 The post-operatory phase

This last phase mainly concerns the patient's awakening. Once the surgery is completed, the anaesthetist has to reverse the effect of anaesthesia as quickly as possible.

Another characteristic of the anaesthetic process in France is that the anaesthetist involved in the anaesthesia consultation will not be necessarily the one on call during the surgery. This cooperative activity implies that the anaesthetist in charge of the consultation collects relevant information about the patient, builds a representation of this case, plans activities and finally chooses the information to convey to his colleague. This information will be provided on an anaesthetic file which is used as an external memory (Anceaux et al., 2002 ; Thuilliez et al., 2005; Neyns et al., 2010).

In summary, anaesthesia process is a dynamic and complex work environment sharing common characteristics with other dynamics work environments as aviation, car driving and nuclear power plants management (Woods, 1988; Vicente, 1999).

Such situations are highly risky due to their dynamics characteristics. In the next section, we will present epidemiological studies on anaesthetic security.

2.2 Risks in anaesthesia

As reported by the French Society of Anaesthesiology (SFAR, 2010), the rate of serious complications in the world is still 14 millions with a death rate of 2 millions. While some studies (Marty, 2003; Amalberti et al., 2005) have shown that anaesthesia was a safe system even so there are more difficult patients (older, more complicated diseases, etc.) and more complex surgeries, etc. Epidemiological studies have enabled to highlight risky factors in anaesthesia (*e.g.* Cohen et al., 1988 ; Sfez, 2002).

Classically, two kinds of risks are distinguished: risks in the operating room and out of this one. Two predominant factors were highlighted by epidemiological studies (*e.g.* Cohen et al., 1988; Arbous et al., 2001). The first one is the most important in the operatory mortality and concerns the patient's features (physical state, age, gender). The second one is related to the surgery difficulty depending on the level of complexity (minor, intermediate, major) and the status of surgery (planned or urgent). A third one is also observed: the risk linked to the anaesthesia. Three kinds of variables in risks strictly linked to anaesthesia are identified: (1) the inadequate assessment and preparation of the patient to the surgery (25%); (2) the unsuitable choice of an anaesthetic technique (15%); (3) a wrong recovery action (56%). But this last risk is not really salient. More precisely, Lagasse (2002) estimates that the risk strictly linked to the anaesthesia is less than 1/10000, while the risk linked to the surgery is 1/100 to 1/1000 and the one linked to the patient is correlated to the patient's ASA score[1] (e.g. ASA3 = 1/10000, ASA4 = 5/10000 and ASA5 =14/10000).

In summary, these results justify the relevance of anaesthetic consultation for the risk assessment and the patient's preparation to the surgery. However, this one is largely responsible of some complications (coupled with an inadequate choice of technique included in the anaesthetic evaluation). The study led by Zhang et al. (2002) has pointed out that the solution to many medical errors was to understand the operator's cognition and to adopt cognitive methodologies to prevent and avoid these errors. In the next section, we will present studies in cognitive psychology to highlight the cognitive mechanics involved in anaesthesia risks assessment (*e.g.* anticipation) and risks management (*e.g.* error detection and recovery).

3. Some psychological concepts about risk management in dynamic situation

In the management of a dynamic situation such as anaesthesia, the time management is important (*e.g.* Carreras et al., 2001). In fact, a dynamic environment can be managed before,

[1] ASA is a physical status classification system for assessing the fitness of patients before surgery. It is a six-category classification system. ASA1 for a normal healthy patient; ASA2 for a patient with mild systemic disease; ASA3 for a patient with severe systemic disease; ASA4 for a patient with severe systemic disease that is a constant threat to life; ASA5 for a moribund patient who is not expected to survive without the surgery and ASA 6 for a declared brain-dead patient whose organs are being removed for donor purposes.

during and even after the process in itself (Morel et al., 2008a, 2008b). The operator has to learn to anticipate his actions and especially their effects. However, the different interactions between multiple variables do not always allow anticipating all the effects of operator's actions and the system evolution. So the operator must deal with uncertain situations in real time and adapt to variables involved in the process and their interactions. This adaptation refers to resilience that is defined as the ability of systems to find a stable state after a major disturbance and/or in a stressful situation (Hollnagel et al., 2006). According to Cook and Woods (1994), the first source of resilience is the human operator. Indeed Reason (2008) describes the human operator as *"an hero, a system element whose adaptations and compensations have brought troubled systems back from the brink of disaster on a significant number of occasions"* (p.3). In this chapter we will mainly discuss the risk management before and during the process.

3.1 Risks management before the process

As said earlier, the human operator can perform his task on a reactive mode but also on a proactive one. This last one permits to view the situation in the near future (limited by operator's knowledge and the rapid evolution of the process and its uncertainty) and makes possible the selection of information (filtering) (Hoc, 1995). We can observe such mode before the process with anticipatory activity (plans) but also during this one with the implementation of plans built, the preparation of the next action or a sequence of actions (Hoc, 2006). According to Cellier et al. (1996), anticipation in a dynamic work environment consists in (1) evaluating the future state of a dynamic process, (2) determining the type of actions to undertake and the moment when it must be done, and (3) mentally evaluating the impact of those actions. Denecker (1999) describes the anticipation on two different modes. The first one is considered as an explicit prediction or forecasting. The second one is implicit and expressed in a selective preparation to situations or an expectation of events. In this chapter, we will mainly focus on the explicit mode.

Regarding the anaesthesia situation, several studies (Xiao et al., 1997; Neyns et al., 2010) have shown that some strategies were built in advance during the pre-anaesthesia phase. Xiao et al. (1997) revealed that anaesthetists identified during this phase a list of "points for consideration". This list includes the specific conditions in the patient's physiology, anticipated events, and risks. This one functions then as a set of "warning" that guides the anaesthetist's attention in a dynamic, multi-tasking situation. However, Neyns (et al., 2010) showed that this list reflected a wide anticipation range (Boudes & Cellier, 1998). It means that this list does not contain planning elements per se, but they can guide attention in the anaesthesia activity where timing constraints are important. Thus the information recorded by anaesthetists concerns the plans adjustment and little information concerns the development or the selection of a plan (Anceaux, et al., 2002). The information contained in the consultation files has two distinct purposes: (1) to assess the risk factors and to prevent them by the development of strategies (related to the patient and the surgery), and (2) the identification of specific problems (Neyns et al., 2010).

Finally, other studies (Anceaux et al., 2002; Anceaux et al., 2001; Beuscart-Zéphir et al., 2001; Thuilliez, et al., 2005) have pointed out that the way to anticipate during the first phase depended on the anaesthetist's expertise and the case complexity.

However the limits of projection into the future described above (the knowledge, the rapid evolution and the level of uncertainty) also require a reactive mode. This reactive mode corresponds to the resilience (Hollnagel et al., 2006)

3.2 Risks management during the process

All the risks cannot be anticipated. As we have seen before, certain risks are taken into account and actions are proposed (prediction, strategies of avoidance or management) and implemented, while others are ignored or not perceived, and their management is postponed (if they appear) in the process management. This type of management suggests a certain "allostasis risk" (Fuller et al., 2008). This second section aims to understand how operators manage risks postponed in the process management.

Although it is widely accepted that the operator commits many errors (Reason, 1990), he also detects the majority of them (at least 60% according to Allwood, 1984; Rizzo et al., 1987). Error detection is the first step in the risk management during the process. During this step, there is still no explanation of the error (Zapf & Reason, 1994). The operator is aware of the existence of a gap between what he obtained and what he expected.

Error detection can be done automatically, spontaneously and unconsciously or it is the result of a more aware comparative treatment with the initial objective (in terms of results, performance) or when the operator can no longer act on the system (Allwood & Montgmery, 1982; Allwood, 1984; Sellen, 1990; Sellen & Norman, 1992; Sellen, 1994; Rizzo et al., 1995;). Finally, the intervention of a third person is also a key factor in errors detection (Woods, 1984; Nyssen, 1997).

Sellen (1990, 1994; et al., 1992) suggests that operator, before recovering the error, would be able to identify it. This identification process consists in a comparison between results obtained and those expected. Nevertheless, it is also possible that the operator detects an error without necessarily identify it or to identify it without necessarily recover it. Indeed, a significant part of detected but also not detected errors does not have any consequences (or little) on the process. Error recovery is often considered as the last step in risk management. It allows the operator to remove or reduce the gap between what he gets and what he wanted to get. Once the error is detected, recovery actions can take many forms (Sellen, 1990). But the recovery process will not be detailed in this chapter.

Regarding the anaesthesia situation, little attention has been paid to the mechanisms of error detection. Indeed the majority of research focuses on the study of the errors prevention (*e.g.* Cooper, et al., 1984; De Keyser & Nyssen, 1993; Cooper, et al., 1982; Finley & Cohen, 1991). In 2006, Nyssen & Blavier have shown that the majority of incidents were detected during the regular monitoring of parameters and on the basis of external signs, suggesting an "automatic" mode. The expertise seems also to play a role in error detection (Nyssen & De Keyser, 1998). The most experienced ones diagnose the problem with significantly less time than novices.

Finally, most studies about risk management (see Cooper et al., 1984) showed that anaesthetists do not taken seriously into account infrequent risks. Risk management depends thus mainly on the anaesthetist's ability to respond quickly and appropriately when a problem occurs.

In summary, various studies have focused on highlighting either prevention strategies in advance or management strategies in real time. But none of them has yet made the connection between what is evaluated during the consultation and in the operating room. The French anaesthesia situation permits to study this assessment at both phases. More specifically, we want to answer the following questions: (1) what kind of risk is used when an anaesthetist has to assess a patient's file? Some studies have emphasized the effect of certain variables on the process of risk assessment. For example, Anceaux et al. (2002) showed that the case complexity could influence how to proceed with the patient during the consultation. Epidemiological studies (*e.g.* Lagasse, 2002) have also identified three kinds of risks as the cause of problems: those related to the patient, surgery and anaesthesia itself. It would be interesting to know how anaesthetists assess risks involved according to their complexity. (2) Epidemiological studies have shown on the one hand that infrequent risks are not taken into account by an anaesthetist (*e.g.* Cooper et al., 1982) and on the other hand that there are often problems when the anaesthetist can assess and prepare a patient (*e.g.* Arbous et al., 2001). How important are the risk frequency and the risk predictability in the assessment and the management in real time?

The next two studies tend to answer these questions and to understand the link between consultation and the anaesthesia in itself.

4. Study one: What kind of criteria does anaesthetist use to assess risks linked to a patient based on consultation information?

This first study aimed to understand how anaesthetists gather patient's files according to their risks. We supposed that information gathering and plans production would depend on operators' level of experience and risks level, without indication about the kind of risks used (patient, surgery or anaesthetic). First, we will present the method used to answer our question and the method used to treat our data. Finally, we will present results obtained. A brief discussion will be proposed to introduce the next study.

4.1 Method

We proposed a card-sorting task using patient's records. All the patient's files were constructed using real patients with the help of an experimented anaesthetist. Four kinds of patient's files were constructed according to the level of complexity defined by 2 combined variables: the surgery complexity (easy and complex surgery) and the complexity related to the patient's health (healthy and unhealthy patient).

An easy surgery is defined as a short-term surgery, minimally invasive (*e.g.* surgery for a breast cancer). A difficult surgery is defined as a long duration surgery (6 to 12hours) that is invasive or mutilating (*e.g.* surgery of the larynx cancer). A healthy patient is described as a normal healthy patient (*e.g.* ASA1) or with mild systemic disease (*e.g.* ASA2). The patient with a complex health status is a patient with important comorbidities and taking many drugs (*e.g.* ASA3 or 4).

Combining these 2 variables, 4 kinds of cases were obtained: easy surgery with healthy patient (called OSPS); easy surgery with unhealthy patient (called OSPC); complex surgery with healthy patient (called OCPS) and, complex surgery with unhealthy patient (called OCPC).

Each patient's file contained the following information in the same order: surgery information (name, operating position, surgery duration, type of intervention, risk of infection, latex allergy); patient's information (gender, age, weight/height, medications, comorbidities, other allergies, dentures, medical and surgical histories); data form the clinical examination (cardiovascular, pulmonary, intubation, risk of post-operative nausea and vomiting, remarks); conclusion made according to information from consultation (ASA score).

Sixteen patient's records (4 records by kind of cases) were constructed and presented in a random order. Anaesthetists were asked to sort them into several distinct groups such as they "go well together" using the think aloud method. At the end of the task, they had to explain their choices of categories giving them a title.

Twenty anaesthetists volunteers participated: 10 experienced anaesthetists (average years of experience=21.2, SD=6.23) and 10 novices (average years of experience=4.2, SD=0.4). They all came from different departments and different medical structures. They were recruited by mail or by phone. The entire experience lasted up to 30 minutes.

Data will be first analysed according to the produced categories: number of categories and titles given to each category. Then, information used to produce categories will be recorded and compared according to the level of experience. Finally, verbalizations will be analysed as a material language and exploited through an approach of natural language processing (NPL). This use of NPL is in line with works on texts profiling (see Planes, 2011). We wanted to highlight experts' and novices' profiles answering the following question: What kind of information the two groups use when they assess a patient's file? Lexico, a language tool for analysing corpus, will be used (see Lebart & Salem, 1994). This tool allows to manipulate some statistical calculation on corpus (more particularly frequencies) and to calculate specificities (in other words, "are some terms more specific to experimented anaesthetists than novices?").

4.2 Results

In this second subsection, we will firstly present categories produced by the two groups of anaesthetists according to their number and their titles. Finally, information used from their verbalisation data will be explained with the Natural Language Processing (NLP).

4.2.1 Produced categories

Anaesthetists construct different categories according to their level of experience. The more experienced anaesthetists build more categories (M=4.4, SD=0.52) than novices (M=3.5, SD=0.53). The t-test applied to the data confirms that the level of experience influences the differentiation between patient's records ($t(18)=3.86$, $p<0.001$, $d=1.8$).

Regarding the titles of categories, a difference is observed according to the level of experience. The table 1 and 2 show labels given to the categories formed by the two groups, the participants' number giving this label and the average number of reports in each category formed.

Looking to the first table, categories made by experimented anaesthetists, it seems that they mainly use the surgery difficulty and the risk level associated to patient's health. Some other categories can highlight specific problems (*e.g.* neurological problems, attention to the induction time).

Categories Label	Participants' number	Average number of reports
Easy surgery, healthy patient	10	5.8
Easy surgery but patient with mild systemic disease	8	3.4
Surgery moderately complex with minor risks	8	2.1
Complex surgery with major risks	10	4.4
Neurological problems	1	4
Very difficult case	2	2.5
Attention to the induction time	2	2.5

Table 1. Categories made by experimented anaesthetists

Categories Label	Participants' number	Average number of reports
Not at risk	8	5.5
Low level of risk	6	4.5
Long surgery, low-risk patient	2	3
Very high risk	10	4.7
Intermediate risks	4	4.25
Additional tests	1	2
Post-operative management	1	1
Particular intubation	1	3

Table 2. Categories made by novices

Regarding the second table, categories made by novices, it seems that they manly used the overall risks and do not really distinguish what is related to the surgery and to the patient. There is just one category that includes this differentiation, the category named "long surgery and low-risk patient", but it is poorly used (just 2 novices with an average of 3 reports). Finally, they also use some categories highlighting specific problems (*e.g.* additional tests, post-operative management, particular intubation).

In summary, 90% of experimented anaesthetists distinguish the difficulty of the surgery and risks linked to the patient's health state, while 60% of novices suggest the overall risk associated with patient's files. According to the number of categories formed and qualitative analysis of labels, experimented anaesthetists use more differentiations between records than do novices. In the next subsection, we will present how anaesthetists analyse a patient's file through the natural language processing.

4.2.2 Information used by the two groups

This subsection concerns the information use to assess patient's files.

The table 3 shows the average number of information used according to the cases and the anaesthetists' level of experience. Even if there is an augmentation of the information used according to the level of complexity (F(3,54)=66.09, *p*<0.0001, *f*=*1.92*), there is no effect of the level of experience on the number of information used (F(1,18)=1,09, NS, *f*=*0.25*).

		OSPS		OSPC		OCPS		OCPC		Average
	N	M	SD	M	SD	M	SD	M	SD	
Experimented	10	19.8	3.08	30.3	2.94	31	3.22	37.3	4.95	27.1
Novices	10	24.9	4.68	36.4	5.03	26.7	4.24	33.5	3.82	30.38

Table 3. Average number of information used according to the level of complexity and the anaesthetists' level of experience

According to the NLP, the figure 2 shows frequencies of criteria used by the two groups (experimented and novices) in their verbalization data. Results show that experimented anaesthetists use 4 different criteria. They use more frequently the patient criteria (frequency=406), followed by the surgery criteria (frequency=217), the risk criteria (frequency=138) and finally, the problem criteria (frequency=118). Novices use also 4 different criteria. They use more frequently the surgery criteria (frequency=373), directly followed by the patient criteria (frequency=321) and less often the risk (frequency=100) and the problems criteria (frequency=57).

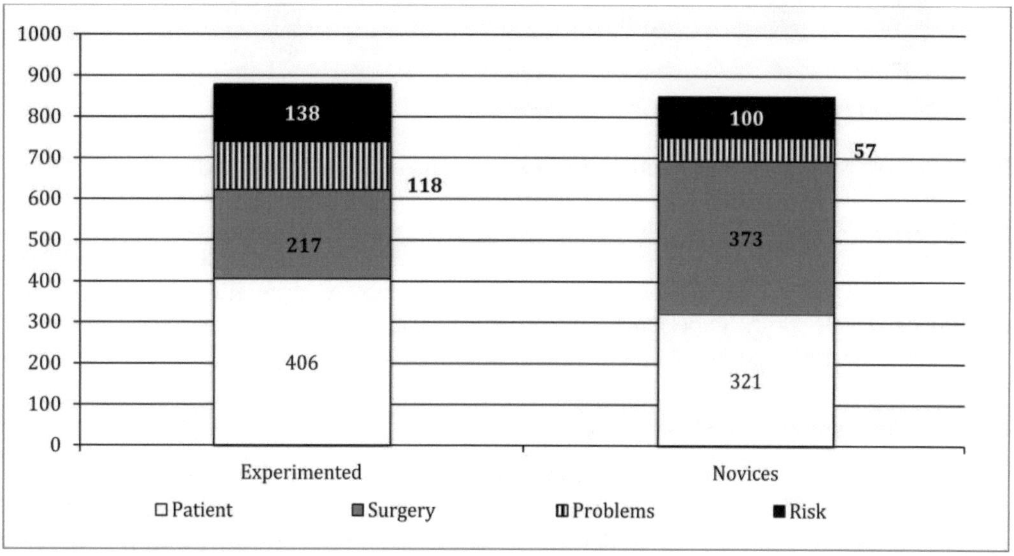

Fig. 2. Frequencies of criteria used to assess patient's files according to the level of experience (verbalization data)

As showed on the figure 1, two major differences appear between the two groups. Experimented anaesthetists mainly use the patient criteria when they have to assess a case, while novices use the surgery and the patient criteria without significant distinction. Finally, experimented anaesthetists and novices assess not only risks but they also highlight some specific problems to consider, even if the experimented ones use it more often. This last difference has a particular interest for us: either novices do not look at these particular problem as much as experts, or they use other words to describe these elements due to a difference in the vocabulary used.

The figure 3, called problems frequencies underlined by experimented anaesthetists and novices, shows that the two groups highlight all the problems. But the experimented ones underline problems more often than do novices (4 kinds of problems are more often cited by experimented anaesthetists while novices identified only two kinds most frequently).

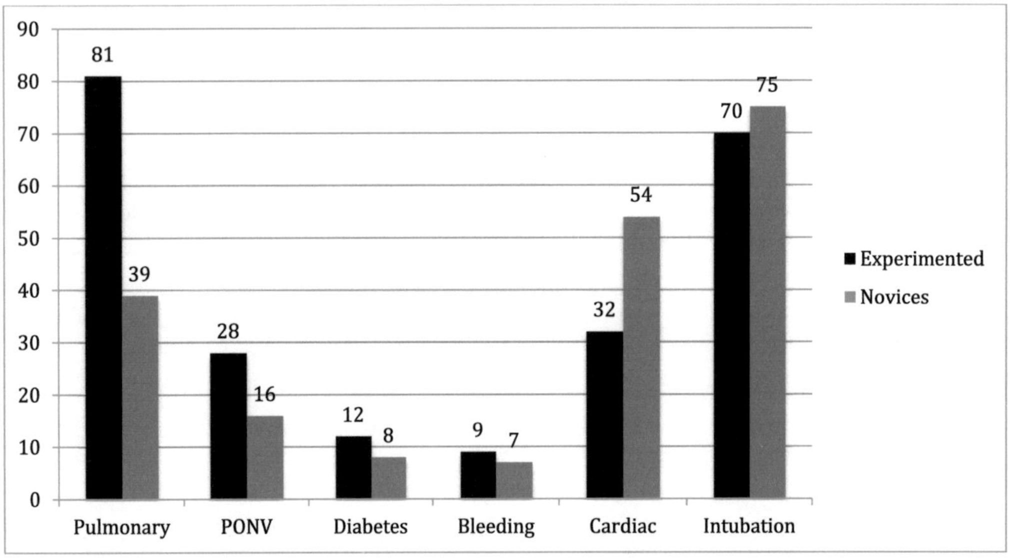

Fig. 3. Problems frequencies underlined by experimented anaesthetists and novices

In summary, it appears that novices look the same information in the records than experimented anaesthetists. However, novices consider them less frequently as problems and analyse it as a whole. This suggests the existence of different strategies among experience level. With experience, anaesthetists can focus only on the potentially problematic aspects of patient's files.

4.3 Brief discussion of the first study

This first study permits to put in light differences in the risk assessment. In fact, the level of experience seems to be a relevant factor in it. As developed through epidemiological studies, our results insist on difficulties linked to the patient's health state and the surgery. Anaesthetists, whatever their level of experience, use these two criteria more frequently. However, the more experimented anaesthetists seem to insist on patient's risks factors than novices. These last ones seem to use both of these criteria to assess risk linked to one case. Finally, risk assessment is also a way to put some warning flags on points for consideration. That confirms Xiao's results (1996).

5. Study two: How does an anaesthetist assess risks during a surgical intervention?

The first study was only focused on the risk assessment that can occur during the consultation. This second study aims to investigate it during all the anaesthesia process. We wanted to see how anaesthetists assess a predictable risk and how they do when they face an unpredictable risk. We supposed that anaesthetists would pay more attention to a predictable risk than to an unpredictable one. This attention would be clearly observed in the risk assessment during the consultation and also, in the risk management (during the surgery) by a better detection/identification. First, we will present the method chosen to answer our question and finally, we will present and discuss briefly a part of our results.

5.1 Method

This second study consists of semi-structured interviews revolved on simulated cases (already validated in Sfez, et al., 2008) using the technique of information on request (see Rimoldi, 1963). These cases were presented on vignettes that described an anaesthetic situation that deviated form its normal course.

Cases were selected based on two criteria: the frequency of occurrence and the predictability. Four cases were used in the experiment: the case of difficult intubation (predictable or not) and the case of malignant hyperthermia (predictable or not). The first case, difficult intubation, is a common situation. Moreover anaesthetists have to evaluate the intubation criteria during the consultation through patient's body inspection (mouth opening, length of upper incisors, relation of maxillary and mandibular incisors during normal jaw closure and during voluntary protrusion of mandible, intercisor distance, visibility of uvula, shape of palate, compliance of mandibular space, thyromental distance, length and thickness of neck, and range of motion of head and neck) and patient's interview (familial history). They also have a difficult airway algorithm (developed by the French Society of Anaesthesia) allowing them to choose an adequate technique according to the patient's criteria. In this simulation, the case of difficult intubation was coupled with gastric oesophageal reflux, which occurs if the anaesthetist does not practice preventive manoeuvre.

The second case, malignant hyperthermia, was chosen because it is a rare life threatening condition in anaesthesia but it can be detected during the consultation. It is usually triggered by exposure to certain drugs used for general anaesthesia. There are also algorithms that manage this risk (strategies for prevention, avoidance and recovery actions). In the case presented, the anaesthetist can either detect the problem during the consultation, either during the anaesthetic induction with warning signs as muscle stiffness, difficulty of intubation due to spasms, breathing problems, increasing of the patient's body temperature and arrhythmias.

Each vignette was divided into 3 phases. The first one was related to the preoperative phase during which the anaesthetist has to be aware of the patient's physical conditions before to put him to sleep and has to determine an action plan to intervene. This phase corresponded to the anaesthesia consultation and to the risk that a problem could appear.

After a brief description of the case including age, weight, height, the surgery planned and the position required, four questions were asked to the anaesthetist. (1) What information do you need to assess risks linked to this patient? (2) On a scale of 0 to 10 (where 0 means "very easy" and 10 "very difficult") can you estimate the difficulty level for this case? (3) What are the difficulties envisaged? (4) Do you recommend some specific actions for this patient? (Additional strategies: surveillance, specific preparation, etc.). In summary, the anaesthetist was asked to assess risk level associated to this patient and to highlight problems expected.

The second phase was related to the anaesthesia induction. During this phase, the problem really appeared. After reading the description of the situation, three questions were asked to the anaesthetist. (1) Do you need additional information? (2) Are you concerned about some elements of the current patient's state and the surgery? Could you explain the reason? (3) Are there specific actions to implement? In summary, the anaesthetist had to provide details about his situation awareness, to formulate hypotheses and to suggest recovery actions.

The last phase concerned the anaesthesia maintenance. This one is related to the recovery situation. During this third phase, if the anaesthetist did not make anything, the situation will not be anymore recoverable. After reading the vignette, four questions were asked to the anaesthetist. (1) Do you need more information? (2) Do you need to do something in this situation? (3) On a scale of 0 to 10 (where 0 means "very easy" and 10 means "very difficult") can you estimate the level of difficulty of the case you just faced? (4) Do you think you could change the course of this event? If so, could you explain how?

The figure 4 presents an example of vignette about an infrequent but predictable problem, the malignant hyperthermia case.

All the interviews were recorded. Each anaesthetist was faced with a vignette presented in a random order. Thirty-four French anaesthetists participated to our study (average years of experience = 17.12, SD = 11.02). They all came from different departments and different medical structures. They were recruited by mail or by phone. The duration of the entire experience lasted up to 10 minutes or 15 minutes.

Data will be analysed as follow. First we will present the anaesthetists risk assessment (the estimation scale of 0 to 10 at the beginning and at the end of the simulation) of the patient according to the two variables: predictability and frequency. There will be also a qualitative treatment of this assessment to demonstrate if the anaesthetist had taken into account or not difficulties presented in the vignette (*e.g.* in the case of difficult intubation, if the anaesthetist diagnosed it in the first stage, he has adequately understood the problem. It is noted 1. Conversely, if the real risk is not taken into account, it is noted 0). Finally, the risk management will be apprehended by 2 kinds of variables: the quality of the detection, identification and recovery actions proposed by anaesthetists and, the number of hypotheses generated. It tends to answer the following questions: "Does the anaesthetist identify correctly the problem?" and "Does the anaesthetist use the correct recovery actions according to specific algorithms?". It means that if the proposed diagnosis and recovery actions are correct, they are rated 1. In the opposite, if they are false, they are rated 0.

Step 1: A 42 years patient, weighing 65kg for 160cm, comes for an excision of ovarian cyst. The surgeon plans to perform this procedure laparoscopically. The patient is very anxious.

Information on request related to the Step 1:
Marital status: Single
Occupation: no
Functional capacity: 8 MET, dynamic
Smoke: no
Alcohol consumption: 1 to 2 glasses of wine during the weekend
Drugs: no
Neurological Evaluation: RAS
Surgical history: curettage after a spontaneous abortion 20 years ago with episodes of ventricular tachycardia during surgery. Seen by a cardiologist who told her that everything was back to normal and that this was probably due to an allergy to halothane. Stayed 3 days in hospital. Ached everywhere (myalgia) with very dark urine.
Cardiac system: RAS Heart Rate: 70/minutes; Blood pressure: 120/75 mmHg
Allergies: unknown
INR rate: 1.04; PT: 12, PTT: 30
Criteria for intubation: Sizes of the upper incisors normal; position of the incisors smoothly during the closing of the jaw, easy to make the movement of voluntary protrusion of about 1 cm, mouth opening more than 3 cm, good visibility uvula (Mallampati 1), thyro-mandibular distance of more than three fingers, length and thickness of the neck normal, the patient can touch her chest with her chin, no dental prothesis, no known sleep apnea, no history of difficult intubation, no gastroesophageal reflux

Step 2: The induction is performed with 10 mg of sufentanil and 150 mg of propofol with loss of consciousness. 50 mg of rocuronium were also injected. The patient is pre-oxygened properly and the ventilation is also easy. But the intubation seems to be more difficult than expected, the second test with a larynx manipulation is successful. Anesthesia is then maintained with O2 (FiO2 40%) and desflurane (ET to 4.5%).
The PETCO2 (end-expiratory pressure) is 35mHg. Respiratory rate is 10/min.
Everything is stable until the PETCO2 increased from 35 to 40 mHg in 3 minutes.

Information on request related to the Step 2:
Ventilation:
Verification of the soda lime: Normal
One-way valves: Normal
Tidal volume: 600 ml
Respiratory rate: 10/min
Auscultation:
Pneumothorax: Absence
Neck-thorax palpation for subcutaneous emphysema: Absence

Step 3: The operation begins but within 3 minutes, the heart rate increases from 80/min to 115/min. It seems that the patient had tachycardia.

Information on request related to the Step 3:
Muscle stiffness
Maximum inspiratory pressure: increased from 18 to 30 cm with a tidal volume of 600 ml
PETCO2: Increased 48mHg

Fig. 4. The infrequent but predictable problem: the malignant hyperthemia vignette

5.2 Results

We will present in this chapter only the results concerning risk assessment before and during the anaesthesia.

5.2.1 Risk assessment during the patient's consultation

The table 4 shows the average level of difficulty estimated by the anaesthetists according to the frequency and the predictability. As expected, a predictable event is always considered as more risky (M=4.06, SD=1.52) than if it was not predictable (M=2.18, SD=1.19). In fact, the t-test shows that this difference is extremely statistically significant (t(32)=4.02, *p<0.0005*, *d*=0.468). In general, the same is observed for the frequency: a common risk is assessed as more difficult (M=3.35, SD=1.9) than an infrequent one (M=2.88, SD=1.36). But by conventional criteria, this difference is not statistically significant according to the t-test (t(32)=0.83, *NS*, *d*=0.57).

	Frequent	Infrequent	**General mean (SD)**
Predictable	4.88 (SD=1.25)	3.33 (SD=1.41)	4.06 (SD=1.52)
Unpredictable	2 (SD=1.22)	2.38 (SD=1.19)	2.18 (SD=1.19)
General Mean (SD)	3.35 (SD=1.9)	2.88 (SD=1.36)	

Table 4. Average level of difficulty according to the frequency and the predictability

Specifically, two major differences appear. Predictable risks are considered as more difficult when they are frequent (M=4.88, SD=1.25) than when they are infrequent (M=3.33, SD=1.41). This difference is statically significant by the t-test applied (t(15)=2.38, *p<0.05*, *d=0.65*). The same is observed concerning the frequent risk. They are considered as more difficult when they are predictable (M=4.88, SD=1.25) than the reverse (M=2, SD=1.22). By conventional criteria, this difference is considered to be extremely statistically significant (t(15)=4.8, *p<0.0005*, *d*=0.6).

Regarding the quality of their risk assessments, only 35.3 % of the anaesthetists correctly assess the risk they face and mainly when the risk is frequent and predictable (7 anaesthetists have a correct assessment against 1 who is wrong). Typically, risks identified are mostly distorted by other problems than those presented in vignettes.

5.2.2 Risk assessment during the surgery

Although risks are not correctly identified at the consultation, when problem arises, the majority of the anaesthetists (97%) identified that there was a problem but only 79.4% of the anaesthetists identified correctly the problem they faced.

In general, all the risks are correctly identified. As shown on the figure 5, frequent risks are always identified correctly (100%). Anaesthetists frequently recognized predictable and unpredictable risks (respectively, 83,33% and 75%). The most difficult risk to identify is the one that is infrequent (58,33%) and mainly when it could not be foreseen during the consultation (25%).

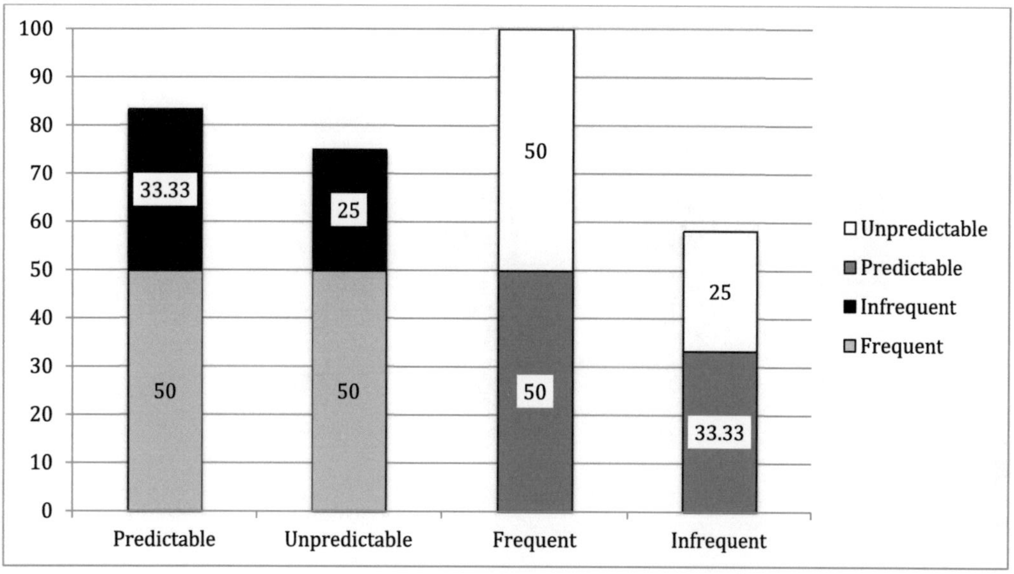

Fig. 5. Percentage of risks correctly identified according to their predictability and their frequency

In average, anaesthetists have advanced 2.24 hypotheses before giving the correct diagnosis (SD=1.16). As shown in the table 5, whatever the frequency or the predictability, there is no difference in the number of hypotheses.

	Frequent	Infrequent	Mean (SD)
Predictable	2 (SD=0.93)	2.89 (SD=1.36)	2.47 (SD=1.23)
Unpredictable	1.89 (SD=0.93)	2.13 (SD=1.25)	2 (SD=1.06)
Mean (SD)	1.94 (SD=0.9)	2.53 (SD=1.33)	

Table 5. Means of hypotheses generated according to the predictability and the frequency (SD)

Concerning the recovery, three results are interesting to quote. First, 3 anaesthetists did not attempt to recover the problem. Second, even if 97% of the anaesthetists identified the risks, only 67.6% of them were able to recover the problem presented. As shown on the figure 6, anaesthetists recovered correctly 94,44% of frequent risks and more often when the risk was predictable (50%). Finally, the less managed risk is the infrequent one (40,97% of anaesthetists were able to provide adequate recovery actions) regardless of its predictability (22,22%) or not (18,75%).

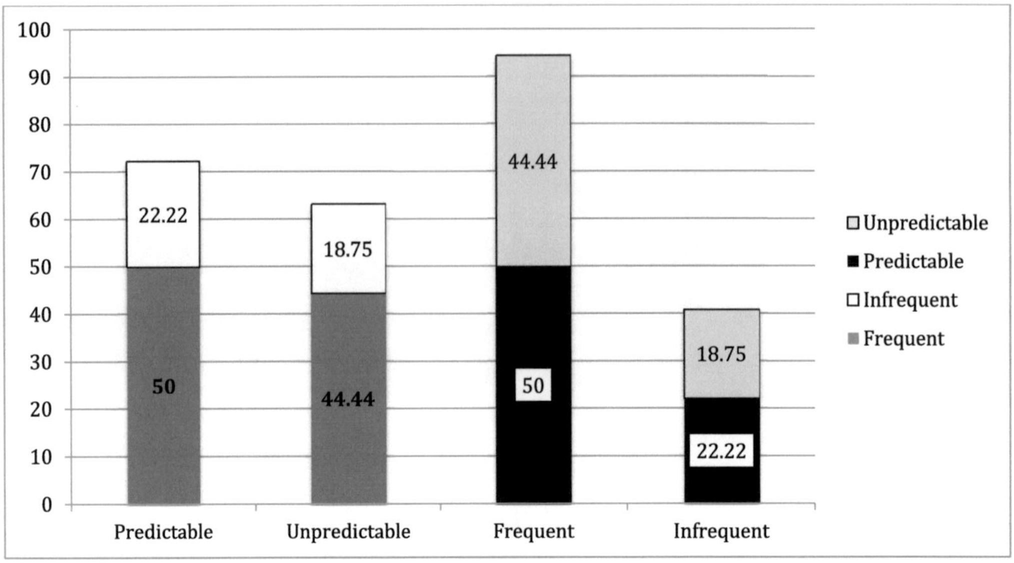

Fig. 6. Percentage of correct recovery actions according to the predicatbility and the frequency

5.3 Brief discussion of the second study

In this second study, we wanted to see how anaesthetists assessed risks before and during the surgery. We were interested by two kinds of variables: the predictability and the occurrence frequency. Our results showed that anaesthetists assessed problems as more risky when they can anticipate it than when they are unforeseeable. However, few anaesthetists were able to correctly assess the risks during the consultation. In fact, their representation were often distorted by others problems. It means that they put a lot of warning flags for each case. One variable seem relevant in the risk assessment and management: the frequency. A frequent risk is always correctly identified and recovered.

In the following section, all the results will be explained regarding the theories used.

6. General discussion

The main objective pursued in this article was to emphasize the risk assessment during the consultation (phase 1 of the process) and during the anaesthesia (phase 2). More specifically, 2 questions have structured our work: (1) what kind of risk is used when anaesthetist has to assess a patient's file? (2) How important are the risk frequency and the risk predictability in the assessment and the management in real time?

The first research permits us to answer the first question and to study the effect of the experience level on the risk assessment. Our analyses show that the most experimented anaesthetists differentiate more patients' files regarding to the number of categories formed, the labels given and their verbalizations. Concerning the kinds of risks used, the experimented anaesthetists use most often the complexity linked to the patient to assess a file whereas novices use a more general level of risk without distinction between patient, surgery and anaesthesia risk, even if their verbalizations show that they use the patient and the surgery criteria with almost the same frequency. Both of them use more information according to the files complexity. Finally, the two groups of anaesthetists use some warning flags to underline some specific problems.

Through this study, our results show that the more experimented anaesthetists build more functional representations because they analyse the situation more broadly. This confirms the results of Cellier et al. (1997) and Hoc (1989). In other words, experimented anaesthetists consider two levels of risk (the patient and the surgery) even if they use more often the criteria related to the patient. That should enable them to anticipate potential incidents associated with the patient and the surgery.

Our results also confirm that two variables can influence the consultation (Anceaux et al., 2001, 2002, 2005): the level of experience and the case complexity. Finally, regarding information used by anaesthetists (all experience level combined) to assess a file, they all point out specific problems. This last point confirms Xiao's results (et al., 1997).

All these elements show that the risk assessment during the consultation leads to a schematic representation that can be then specified later during the surgery (Hoc, 1987).

The second research permits to answer the second question (How important are the risk frequency and the risk predictability in the assessment and the management in real time?). Our results indicate that the risk predictability increases the perceived difficulty level associated with it. Moreover this perceived difficulty is higher when the risk is both frequent and predictable. The risk frequency seems to be important for the identification during the surgery and also for the recovery actions.

Overall, few anaesthetists assess the problem correctly at the end of the consultation (for recall: only 35.9% of anaesthetists assess the real risk). Anaesthetists correctly assess only frequent and predictable risks at this early stage. Concerning the other risks, the anaesthetist's representation is distorted by other problems. Even if the risk is not really perceived during the consultation, the identification is almost correct for all the anaesthetists. We also find that anaesthetists identified and recovered less correctly the infrequent risks.

Finally our results point out that anaesthetists suggest that anaesthetists propose several hypotheses before reaching the right result whatever the nature of the risk according to its frequency and its predictability.

These results confirm epidemiological and psychological studies previously cited. In fact, several explanations can be given to our results. Firstly, previsability permits to assess (consultation) and to identify (surgery) correctly the risks. Secondly, as Cooper's results (et al., 1982) showed, anaesthetists not taken into account infrequent risks. These ones are not correctly assessed, identified and recovered because there is a misunderstanding. Indeed the majority of anaesthetists faced with a infrequent risk explained that they have been rarely confronted with this kind of risk and most often through simulations. Thus, when the patient presented signs of this problem, they tended to minimize the facts (Amalberti, et al., 2005). Moreover, their representation seems distorted by highlighting other problems (points for consideration, Xiao, et al., 1997). Finally, frequent and predictable risks are most often correctly assessed, dectected and recovered. The explanation lies in algorithms developed and the current practice that allow the anaesthetist to identify problems quickly (by information filtering).

7. Conclusion

Two questions were asked at the beginning of this chapter and one aim was pursued. Regarding anticipation performed by anaesthetists, we wanted first to highlight the explicit elements of anticipation. However, the results show that the anaesthetist has on the one hand, explicit predictions about specific events and on the other hand, builds expectations that are not clearly communicated (implicit expectations) but known as points for consideration. This result confirms a certain "risk allostasis" (Fuller, et al., 2008). Further analysis (Neyns, 2011) enables us to explain this risk allostasis through the availability of algorithms developed by scientific societies such as SFAR (French Society of Anaesthesia) or the American Society of Anesthesiologists (ASA). These algorithms allow the anaesthetist to manage promptly when a problem occurs. For example, the management algorithm of malignant hyperthermia describs all the symptoms and strategies to use for stopping the crisis.

Thus, during the consultation all the risks are not assessed but it allows the anesthetist to put warning on specific problems that may arise. The anaesthetist therefore has a schematic representation and uses generic and abstract plans with a particularisation in real time (Anceaux et al., 2002; Thuilliez et al., 2005; Van Daele & Carpinelli, 2001). This means that during the intervention, this representation is the result of filtering information and a highlighting of risks. Throughout the surgery, the anesthetist will complete this representation with contextual information. Finally, when a problem occurs, anaesthetists tend to put routines in place to assess the situation (algorithms, hypothesis generation). We also observed, whatever the risks they face (frequent/infrequent, predictable/unpredictable) anaesthetists tend to explain the situation using several hypotheses. This result demonstrates that the situation of anesthesia is a complex situation where multiple variables interact and can be the source of several problems. Moreover, epidemiological studies pointed out that there were problems arising from inadequate evaluation and a bad patient's preparation for surgery. Our results tend to show that it can only be part of a highlighting of several risks as those actually involved. The anaesthetists' representation is distorted by some points for consideration when they have to assess the situation. Another study (Neyns, 2011) shows that certain events (such as difficult intubation) could be due to a misidentification according to qualitative judgments on criteria.

But these first results cannot attest to the necessity or not of a preanesthetic consultation but allow to emphase the importance of the anaesthetists' experience and the need to develop habits of action (Norros & Klemola, 1999) to recognise and manage some cases. For example, the infrequent risks. Because rare does not mean impossible, it is important to establish some specific training through simulations, conference, seminar, etc. In another study (Neyns, 2011) we compared the risk management between France (with consultation) and Quebec (without specific anaesthesia consultation). This study showed that in Quebec they have developed specific patterns of actions that allow them to quickly manage the problem, even if they took longer time to identify the problem.

This chapter, in line with the work on resilience, contributes to a positive view of risk management in anaesthesia. The operator is a central key to the system resilience, not only in terms of preparation but also in real-time management. It points out adaptation strategies to the system variabilities by a proactive identification of risk factors and reactive strategies in response to changes of the patient's health conditions (Patterson et al., 2010).

Finally, the use of different approaches to address resilience is relevant, it permits to obtain and confront additional information. It is interesting to use several techniques to obtain additional information. However, methods used are subject to numerous biases. The categorization of files can not really be considered as a consultation. The patient was not present, the anesthetist has to build his representation on written data, not physical or verbal ones. In the simulation, the anesthetist is confronted alone to the case but it is a team-work where detection by a third person is very important. Thus, detection strategies could not be identified. Moreover, in this second study, we focused on the risks occurring in the operating room. It is clear that these risks also require increased monitoring after surgery because they can affect the patient's health. However, for purposes of the study, the simulation did not take into account the latter period.

8. Acknowledgment

This research project would not have been possible without the support of many people. The lead author wishes to express her gratitude to her three colleagues, Prof. Dr. Cellier, Dr. Carreras and Ms Planes who offered invaluable assistance, support and guidance in these two studies. She also whishes to thank all the anaesthetists who were abundantly helpful to understand their work. Deepest gratitude are also due to all the members of the Laboratory of Cognition for sharing the literature and invaluable assisstance, and the members of the French Society of Anaesthetists whithout whose knwoledge and assistance these studies would not have been successful. The authour would also like to convey the Faculty for providing the financial means and laboratory facilities.

Finally, the lead author would also like to express her love and gratitude to her beloved families for their understanding and endless love.

9. References

Allwood, C.M., & Montgomery, P. (1982). Detection of errors in statistical problem solving. *Scandinavian Journal of Psychology, 23*, 131-139.

Allwood, C.M. (1984). Error detection processes in statistical problem solving. *Cognitive Science, 8*, 415-437.

Amalberti R, Auroy Y, Berwick D, Barach P. (2005). Five system barriers to achieving ultrasafe health care. *Ann Intern Med, 142*(9), 756-764.

Anceaux, F., & Beuscart-Zéphir, M.C. (2002). La consultation préopératoire en anesthésie : gestion de la prise d'informations et rôle des données retenues dans la planification du processus d'anesthésie. *Le Travail Humain, 65*(1), 59-88.

Anceaux, F., Thuilliez, H., & Beuscart-Zéphir, M.C. (2001). Gestion de la prise d'informations pour la planification en situation dynamique : l'anesthésie. In V.

Gorsjean & E. Raufaste (Ed.), *Actes des premières journées d'études en Psychologie Ergonomique* - Epique 2001, INRIA, Rocquencourt, pp71-82.

Arbous, M.S., Grabbee, D.E., Van Kleef, J.W., de Lange, J.J., Spoormans, H.H.A.J.M, Touw, P., Werner, F.M., & Meursing, A.E.E. (2001). Mortality associated with anaesthesia: a qualitative analysis to identify risk factors. *Anesthesia, 56*, 1141-1153.

Beuscart-Zéphir, M.C., Anceaux, F., Crinquette, V., & Renard, J. (2001). Integrating users' activity modelling in the design and assessment of hospital electronic patient records: the example of anesthesia. *International Journal of Medical Informatics, 64*, 157-171.

Nyssen, A.-S., & Blavier, A. (2006). Error detection: A study in anaesthesia. *Ergonomics, 49*, (5-6), 517-525.

Boudes, N., & Cellier, J.M. (1998). Etude du champ d'anticipation dans le contrôle du trafic aérien. Le Travail Humain, 61, 29-50.

Carreras, O., Cellier, J.M., Valax, M.F., & Terrier, P. (2001). Ajustement temporel à la dynamique des situations. *Psychologie Française, 46*(2), 119-129.

Cellier, J.M., De Keyser, V., & Valot, C. (1996). *La gestion du temps dans les environnements dynamiques*. Paris, Presses Universitaires de France.

Cellier, J.M., Eyrolle, H., & Mariné, C. (1997). Expertise in dynamic environments. *Ergonomics, 40*, 28-50.

Chung, D.C., & Lam, A.M. (1990). *Essentials of Anesthesiology (2 ed.)*. Philadelphia: W.B. Sanders.

Cohen, M.M., Duncan, P.G., & Tate, R.B. (1988). Does Anesthesia Contribute to operative mortality ?. *JAMA, 260* (19), 2859-2863.

Cook, R.I., & Woods, D.D. (1994). Operating at the sharp end: the complexity of human error. In Bogner M.S. (Ed.). Human Error in Medicine. Hillsdale, NJ: Lawrence Erlbaum Associates. pp. 255-310.

Cooper, J.B., Long, C.D., Newbower, R.S., & Philip, J.H. (1982). Critical incidents associated with intraoperative exchanges of anesthesia personnel. *Anesthesiology, 56* (6), 456-461.

Cooper, J.B., Newbower, R.S., & Kitz, R.J. (1984). An analysis of major errors and equipment failures in anesthesia management. Considerations for prevention and detection. *Anesthesiology*, 60, 34-42.

De Keyser, V., & Nyssen, A.S. (1993). L'erreur humaine en anesthésie. *Le Travail Humain*, 56, 243-266.

Denecker, P. (1999). Les composantes symboliques et subsymboliques de l'anticipation dans la gestion des situations dynamiques. *Le Travail Humain, 62* (4), 362-385.

Finley, G.A. & Cohen, A.J. (1991). Perceived urgency and the anaesthetist: responses to common operating room monitor alarms. *Canadian Journal of Anaesthesia,* 38, 958-964.

Fuller, R., McHugh, C., & Pender, S. (2008). Task difficulty and risk in the determination of driver behaviour. *European Review of Applied Psychology*, 1, 13-21.

Gaba, D.M. & Lee, T. (1990). Measuring the workload of the anesthesiologist. *Anesthesia and Analgesia*. 71, 354-361.

Hoc, J.M. (1995). Planning in diagnosing a slow process: some implications on the design of human operator computer support. *Zeitschrift für Psychologie, 203*, 101-115.

Hoc, J-M. (1989). La conduite d'un processus à longs délais de réponse : une activité de diagnostic. *Le Travail Humain*, 52, 289-316.

Hoc, J.-M. (1987). *Psychologie cognitive de la planification*. Grenoble : Presses Universitaires de Grenoble.

Hoc, J.-M. (2006). Planning in dynamic situations: some fidings in complex supervisory control. In W. van Wezel, R. Jorda, & A. Meystel (Eds.) *Planning in Intelligent Systems : Aspects, Motivations, and Methods*, Hoboken, NJ: John Wiley & Sons.

Hollnagel, E., Woods, D.D., & Leveson, N., (2006). *Resilience Engineering: Concepts and Precepts*. Aldershot, UK: Ashgate Publishing.

Lagasse, R.S. (2002). Anesthesia safety : model or myth? A review of the published literature analysis of current original data. *Anesthesiology*, 97, 1609-1617.

Lebart, L. & Salem, A. (1994). *Statistique Textuelle*, Dunod, 344 p.

McDonald, J.S. & Dzwonczyk, R. (1988). A time and motion study of the anaesthetist's intraoperative time. *British Journal of Anaesthesiology*. 61, 738-742.

McDonald, J.S., Dzwonczyk, R., Gupta, B. & Dahl, M. (1990). A second time-study of the anaesthetist's intraoperative period. *British Journal of Anaesthesiology*. 64, 582-585.

Morel, G., Amalberti, R., & Chauvin, C. (2008a). Articulating the differences between safety and resilience: The decision-making process of professional sea-fishing skippers. *Human Factors*, 50(1), 1-16.

Morel, G., Amalberti, R., & Chauvin, C. (2008b). Safety and Resilience: Articulation and difference - The example of the professional fishing industry. *Human Factors*, 50 (1), 1-16.

Neyns, V., Carreras, O., & Cellier, J.M. (2010). Evaluation et gestion des risques en anesthésie : Stratégies mises en place par les médecins-anesthésistes. *Le Travail Humain, 73* (4), 319-337.

Neyns, V. (2011). *Les modalités du contrôle cognitif en situation dynamique : anticipation et gestion des dérives. Le cas de l'anesthésie.* Thèse de doctorat, Université de Toulouse.

Norros, L., & Klemola, U.M. (1999). Methodological considerations in analysing anaesthetists' habits of action in clinical situations. *Ergonomics, 42* (11), 1521-1530).

Nyssen, A.S., & De Keyser, V. (1998). Improving training in problem solving skills: Analysis of anesthetists' performance in simulated problem situations. *Le Travail Humain*, 61(4), 387-401.

Nyssen, A.S. (1997). Une nouvelle approche de l'erreur humaine dans les systèmes complexes : Exploration des systèmes complexes en anesthésie. Thèse de doctorant en Psychologie, non publiée, Université de Liège, Liège.

Patterson, E.S., Cook, R.I., Woods, D.D., & Render, M.L. (2010). Gaps and resilience. In M.S. Bogner (Ed.), *Human error in medicine*, (second edition ed.).

Planes, L. (2011). Comparaison de verbalisations d'anesthésistes experts et novices. Mémoire de Master 1: Sciences du langage/Traitement automatique des langues. Université de Toulouse.

Reason, J. (2008). *The Human Contribution: Unsafe Acts, Accidents and Heroic Recoveries.* Ashgate Editions.

Reason, J. (1990). *Human Error*. Cambridge: Cambridge University Press.

Rimoldi, H.J.A. (1963). Processus de décision et fonctions mentales complexes. *Revue européenne de psychologie appliquée, 54*, 25-32.

Rizzo, A., Bagnara, S., & Visciola, M. (1987). Human error and detection processes. *International Journal of Man-Machine Studies*, 27, 555-570.

Rizzo, A., Ferrante, D., & Bagnara, S. (1995). Handling human error. In J.M. Hoc, & P.C. Cacciabue (Eds.), *Expertise and Technology: Cognition and Human- computer Cooperation. Expertise: Research and Applications*, NJ: Lawrence Erlbaum Associates, pp. 195-212.

Sellen, A.J., & Norman, D.A. (1992). The psychology of slips. In B.J. Barrs (Eds.). *Experimental Slips: Exploring the Architecture of Volition*. New York: Plenum Press, pp. 317-339.

Sellen, A.J. (1990). Mechanisms of human error and human error detection, PhD thesis, University of California, San Diego (published 1991), University Microfilms, *Abstract in Dissertation Abstracts International*, 51, 5618.

Sellen, A.J. (1994). Detection of everyday errors. *Applied Psychology: An International Review*, 43, 475-498.

Sfez, M. (2002). Analyse et maîtrise du risque en anesthésie. *Conférences d'actualisation*, 371-385. (http://www.anesthesie-foch.org/s/article.php3?id_article=138)

Sfez, M., de Marcellis-Warin, N., Pourreau, A., Triadou, P., Courault, M. & Lienhart,A. (2008). Combiner entretiens et cas simulé pour identifier les facteurs favorisant la récupération. *Risques et Qualité*, *5*(3).

Thuilliez, H., Anceaux, F., & Hoc, J.M. (2005). Rôle de l'opérateur et du statut fonctionnel des informations lors de la prise d'informations en anesthésie. *Le Travail Humain*, 68, 225-252.

Van Daele, A., Carpinelli, F. (2001). La planification dans la gestion des environnements dynamiques : quelques apports récents de la psychologie ergonomique. *Psychologie Française*, 46 (2), 143-152.

Vicente, K.J. (1999). *Cognitive Work Analysis: Toward safe, productive, and healthy computer-based work*. Mahwah, NJ: Lawrence Erlbaum Associates.

Woods, D. (1984). Some results on operator performance in emergency events. *Institute of Chemical Engineers Symposium*, 90, pp. 21-23. Columbus, OH: Ohio State University.

Woods, D.D. (1988). Coping with complexity: The psychology of human behavior in complex systems. In L.P. Goodstein, H.B. Andersen, and S.E. Olsen, (Eds.). *Tasks, Errors and Mental Models*. New York: Tailor & Francis.

Xiao, Y. (1994). Interacting with complex work environment: A field study and a planning model. Ph.D. Dissertation, University of Toronto, Ontario, Canada.

Xiao, Y., Milgram, P., & Doyle, D.J. (1997). Planning behavior and its functional rôle in interactions with complex systems. *IEEE Transactions on Systems, Man and Cybernetics, Part A*, 27(3), 313-324.

Zapf, D., & Reason, J.T. (1994). Introduction: Human errors and error handling. *Applied Psychology: An International Review*, 43, 427-432.

Zhang, J., Patel, V.L., & Johnson, T.R. (2002). Medical Error: Is the Solution Medical or Cognitive? *Journal of the American Medical Informatics Association*, 9(6), s75-s77.

12

Effects of Wearing Gloves and Sex on Endurance Time and the Corresponding Finger Skin Temperature During a Cold Immersion

Yuh-Chuan Shih* and Yo-May Wang
Department of Logistics Management, National Defense University No. 70, Sec. 2, Zhongyang N. Rd., Taipei City, Taiwan

1. Introduction

Many workers, such as commercial fishermen, power-line workers in temperate climates, and frozen-food processing industry workers, need to perform manual work in cold environments. Exposure to cold environments and contact with cold materials have been reported impair tactile sensitivity in the hands (Enander, 1984), hand dexterity (Schiefer et al., 1984; Riley & Cochran, 1984; Enander & Hygge, 1990; Heus et al., 1995), and tracking performance (Goonetilleke & Hoffmann, 2009). Manual dexterity is frequently used to evaluate hand function and is important during hand manipulation. Hand/finger skin temperature is considered a vital factor in dexterity (Schiefer et al., 1984; Enander, 1984; Enander & Hygge, 1990; Brajkovic & Ducharme, 2003, Chen et al., 2010) and hand performance (Riley & Cochran, 1984; Havenith et al., 1995, Chen et al., 2010). More importantly, such impairment may lead to an increased number of accidents (Müller, 1982; cited by Havenith et al., 1995).

Several epidemiologic studies have shown that, in addition to heavy physical work, awkward and static postures, repetition of movements, and vibration, cold may be a risk factor for occurrence or aggravation of musculoskeletal disorders (MSDs), such as in the fish-processing industry (Chiang et al., 1993; Nordander et al., 1999) and meat-processing factories (Kurppa et al., 1991; Piedrahíta et al., 2004). A report by the European Agency for Safety and Health at Work (2010) also noted that the risk of MSDs increases with work in cold environments.

In order to protect the hands from cold, gloves are recommended as a first line of defense. Unfortunately, although wearing gloves does not affect muscular fatigue (Chang & Shih, 2007), doing so could cause a negative effect on exertion (Shih, 2007; Chang & Shih, 2007) and dexterity (Bishu & Klute, 1995; Ou, 2003). On the other hand, gloves can insulate the hands against cold. For example, in a dialogue test of 12°C-water and 5-minute immersion to assess the hand-arm vibration syndrome (ISO/CD 14835-1, 2001), researchers have

* Corresponding Author

revealed that wearing gloves can not only reduce pain but also delay a decrease in finger skin temperature (FST) (Suizu et al., 2004; Suizu & Harada, 2005). In addition, Nag and Nag (2007), who evaluated the hazards and health complaints associated with fish processing activities in India, indicated that, during a 2-hour period of work, wearing latex gloves could improve the FST and HST, raise morale in female workers, and alleviate cold-induced symptoms. Therefore, due to both the delay on FST decrease and reduction in perceived pain, one of the objectives of this study is to examine if wearing gloves can prolong the endurance time (ET) in a cold immersion, and if thicker gloves will lead to longer ET.

Geng et al. (2001) indicated that type of material and surface temperature affected the contact cooling of the finger significantly. They found that materials with high thermal conductivity and/or lower surface temperature decreased FST more rapidly. The trend of a FST decrement during cold immersion should slow gradually as the heat balance approaches. At first, rapid heat dissipation occurs between two objects in contact due to a great difference in temperatures, accompanied by a rapid decrease in FST, a noticeable sensation of cold, and associated induced pain. Wearing gloves could delay the reduction in FST and reduce the perception of pain (Suizu et al., 2004; Suizu & Harada, 2005), and it is supposed that wearing gloves could also extend ET. Consequently, it is worthwhile and interesting to identify the FST at the point of pain tolerance during cold immersion, for FST has been reported as a crucial factor affecting hand dexterity (Schiefer et al., 1984; Enander, 1984; Enander & Hygge, 1990; Brajkovic & Ducharme, 2003, Chen et al., 2010). Wolff (1984) defined pain tolerance as 'that point at which a subject will terminate or withdraw from noxious stimulation' and argued that tolerance measures in the laboratory are analogous to clinical pain.

Meanwhile, experimental pain, such as that induced by cold water, is usually rated by psychophysics, which investigates the correspondence between the magnitude of stimulus properties as assessed by both the instruments of physics and the perceptual systems of people (Baird & Noma, 1978). Over the years, many types and shapes of scales have been developed for use in psychophysical studies. One of the most commonly used scales for evaluating subjective qualities during dynamic physical work is the 15-point rated perceived exertion (RPE) scale developed by Borg (1970). Later on, Borg (1982) developed a new rating scale constructed as a category scale with ratio properties called the Category Ratio scale (CR-10 scale). It combines the positive attributes of category and ratio scales and hence allows relative comparisons as well as level estimations. Borg recommends the use of the CR-10 scale to determine subjective symptoms, such as aches and pain. Åkesson et al. (1999) and Dedering et al. (2006), for example, used this scale to rate pain. Since the CR-10 scale functions as a ratio scale, most kinds of mathematical operations are permitted.

During cold-water immersion, different levels of perceived pain are generated, gradually rising from the bottom to the top of the pain tolerance scale defined by Wolff (1984). Pain tolerance as a dependent variable is the central behavioral measure of an individual's ability to endure any given level and type of pain stimulation (Keefe & Williams, 1992). The second objective of the present study is, therefore, to examine the FST of different gloved conditions under a given level of perceived pain as classified by Borg's CR-10 scale. In addition, a linear model between Borg's CR-10 scale and ET and FST is explored.

Furthermore, the variability of experimental pain perception results from a variety of sources, such as psychosocial factors, biological factors, and experimental variables. A sex difference in response to experimental induced pain is reported to be significant, and female workforces are increasing in size and importance in some light manual operations. Even though women have generally been reported to have a lower pain threshold, a greater ability to discriminate painful sensations, higher pain ratings, and a lower tolerance for pain (Berkley, 1997), studies employing different populations and types of pain have not consistently found differences in gender and pain intensity. If the capacity of cold tolerance of the hands depends on gender, such a difference is an interesting issue and should not be ignored. The functional capacity chosen here is the endurance of the hand during immersion in cold water and its correspondence to different levels of perceived pain. The issue of interest is whether two factors, wearing gloves and gender, affect this functional capacity. Pain tolerance as a dependent variable is, therefore, the central behavioral measure of an individual's ability to endure any given level and type of pain stimulation (Keefe & Williams, 1992). Consequently, the present paper also examines the sex effect on the capacity for tolerance of cold-water-induced pain on hands and the corresponding change in FST.

In summary, the main objective of the present study is to explore the effects of gloves and sex on FST and ET under different levels of cold-induced pain cataloged according to the CR-10 scale in a cold immersion. Additionally, the linear relationship between the CR-10 scale, ET, and FST will be studied.

2. Methods

Most studies on hand cooling have been carried out with immersion into cold water (cold pressor test) (Petrofsky & Lind, 1980; Suizu & Harada, 2005; Geurts et al., 2006; Coulange et al., 2006), contact with cold materials (Havenith et al., 1992; Chen et al., 1994), or exposure in cold air (Candas & Dufour, 2007). The cold pressor test, a procedure in which subjects are instructed to immerse a limb into a cold-water bath, has been considered one of the most valid methods for inducing pain to meet the criteria of controllability, reliability, discriminability, convenience, and validity (Hirsch & Liebert, 1998). Thus, the cold pressor test was employed in the present study to cool the skin temperatures.

2.1 Participants

Fifteen men and fifteen women participated in the experiment, and all but two males were right-handed. They were free from any neuromuscular and musculoskeletal disorders, and their demographics and anthropometrics are shown Table 1.

2.2 Apparatus and materials

The apparatus and materials employed were as follows.

1. A water bath made by Firstek Co. (Model: B102) was used. It can maintain the temperature constantly at a desired level with an electronic thermo-sensor and a heater. The minimum temperature is the ambient temperature plus 5°C, and the maximum is 80°C. Its exterior (W×D×H) is 54×33×27 cm^3, and the interior is 49×29×15 cm^3.

2. A submersible cooler made by Firstek Co. (Model: HC-101) was used to cool the water from the ambient to -20°C. At 20˚C, the cooling efficiency is 750 Kcal/hr. During immersion, it was submersed in the water bath to cool the water, and the water temperature was set and regulated by the electronic thermo-sensor of the previous water bath.
3. A digital thermometer and hygrometer (TECPEL Co.; Model: DTM301) was used to monitor the ambient temperature and humidity at the same time. The temperature range measured was from -10°C to +50°C. Relative humidity measured ranged from 20% to 99%.
4. A digital 4-channel thermometer made by TECPEL Co. (Model: DTM319) with a size of 184x64x30 (WxDxH) mm^3 was used to record skin temperatures. The memory capacity is 16,000 records of data. The sampling rate was 6 data/minute, and it was connected to a personal computer with an RS-232 link.
5. Latex gloves with six available sizes made by Modern Healthcare Co. (Model: 1010) were used. Here, the latex gloves are evaluated due to their widespread use in processing and packing in the frozen foodstuff industry in Taiwan.

Item (unit)	Sex	Mean	SD	Range
Age (yr.)	Male	24.1	4.9	20~35
	Female	28.8	4.8	22~38
Height (cm)	Male	172.8	4.3	165~180
	Female	160.2	4.5	150~168
Weight (kgw)	Male	70.5	10.4	55~103
	Female	54.5	5.6	45~65
Hand length (cm)	Male	18.9	0.6	17.8~19.7
	Female	17.3	0.8	16.2~19
Palm length (cm)	Male	9.8	0.5	8.8~10.8
	Female	9.0	0.6	8.0~9.8
Palm breadth (cm)	Male	8.2	0.3	7.6~8.7
	Female	7.0	0.3	6.6~7.5

Table 1. The anthropometric data of subjects (SD: standard deviation)

2.3 Experimental procedures and data acquisition

All participants were well informed of the goals and procedures first. The mean ambient temperature (standard deviation, SD) was 19.2°C (1.1°C), and mean relative humidity (SD) was 62% (8%). Before immersion, two channels of the 4-channel digital thermometer were applied to monitor the FST on the ventral side of the distal phalanges of both the thumb (FST_T) and the little finger (FST_L) of the left hand. The thermo-sensor was first fixed by surgical tape to the fingertips, then a thick waterproof bandage was wrapped, and the fingers were covered by additional rubber coverings. Finally, adhesive tape was used to close the opening of the rubber coverings without severe tape tension that might occlude blood flow. Initial FSTs (denoted by FST_{T0} and FST_{L0}) were both controlled in a range of 30-35°C by immersing the left hand in a 30°C-water bath for a short duration to minimize the possible inference of initial FST on endurance time (ET) needed to reach a corresponding level of pain perceived. To standardize limb submersion across subjects, subjects were instructed that, when signaled, they should place their left hands into the water up to about 1/3 of the forearm above the wrist.

From the beginning of immersion to withdrawal of their left hands from the water tank, participants were instructed to self-report five perceived levels of cold-water-induced pain according to the Borg 10-point CR scale: just noticeable (0.5), weak (2), strong (5), very strong (7), and extremely strong (10) discomfort (almost unbearable). At the same time, the associated ET (ET_i, $i=1,2,3,4,5$) and corresponding FSTs (FST_{Ti} and FST_{Li}, $i=1,2,3,4,5$) were recorded.

2.4 Experimental design and data analysis

A nested-factorial design was employed, in which the factors were sex, subject (nested within sex), gloved condition, and level of perceived pain. Three gloved conditions involved bare hand (0G), wearing a single layer (1G), or wearing a double layer (2G) of latex gloves. The water temperature was set at 10°C, and each treatment was repeated twice. The level of significance (α) was set at 0.05, and the responses analyzed are as follows.

1. The endurance time (ET) was the time needed to reach the five separate perceived pain levels, denoted by ET_i ($i=1,2,3,4,5$). Each ET_i was the mean value of two replications. Here the ET_5 is the time to reach the pain tolerance defined by Wolff (1984).
2. The FST_i corresponded to ET_i. FST_i is a mean value measured on the little finger (FST_{Li}) and thumb (FST_{Ti}) of two replications ($i=1,2,3,4,5$). They are 1-min average values calculated from 30 sec before and after the time ET_i of each replication.
3. The exchange rate of FST at different pain levels from the beginning. They are defined as $Rate_{i0}=(FST_i - FST_0)/ET_i$, where $i=1,2,3,4,5$.

3. Results

3.1 Endurance Time (ET)

As shown in Table 2, which shows the ANOVA results for all responses, all main effects were significant, as well as the gloved×level. In general, males had a longer ET than females (113.9 vs. 101.9 sec). Figure 1, demonstrating the gloved effect at different levels of perceived pain, indicates that wearing gloves is able to prolong ET, and more layers is associated with longer ET. Definitely, the longer ET contributed to greater cold-water-induced pain.

Figure 1 further shows that, based on the bare hand (0G) condition, subjects were willing to immerse their hands in a 10°C water bath for about 2 and 3 times longer in the 1G and 2G conditions, respectively. In detail, Table 3 indicates that the ET of the 1G condition was from

2.5 times at ET_1 to 1.9 times at ET_5; for the 2G condition, it was from about 4.3 times to 3.1 times from ET_1 to ET_5. The ET of 2G was around 1.6-1.7 times that of 1G for all pain-perceived levels.

Sources of variation	ET				Rate			FST			
	d.f.	MS	F-value	p-value	MS	F-value	p-value	d.f.	MS	F-value	p-value
Subject (Sex)	28	58468	24.408	0.0000	4.742	19.591	0.0000	28	49.7	24.9	0.0000
Sex	1	16200	6.763	0.0097	1.427	5.894	0.0156	1	22.5	11.3	0.0009
Gloved	2	531877	222.037	0.0000	1.122	4.634	0.0103	2	298.8	149.5	0.0000
Level	4	230914	96.397	0.0000	4.596	18.988	0.0000	5	304.1	152.2	0.0000
Sex*Gloved	2	4515	1.885	0.1532	0.117	0.482	0.6181	2	1.2	0.6	0.5627
Sex* Level	4	36	0.015	0.9995	0.144	0.594	0.6671	5	0.8	0.4	0.8339
Gloved*Time	8	19600	8.182	0.0000	0.541	2.234	0.0244	10	18.1	9.0	0.0000
Sex*Gloved* Level	8	83	0.035	1.0000	0.025	0.103	0.9991	10	0.1	0.1	1.0000
Error	392	2395			0.242			476	2.0		
Total	449							539			

Table 2. The ANOVA results

Gloved \ Level of perceived pain	L1	L2	L3	L4	L5
1G/0G	2.5	2.3	2.1	2.0	1.9
2G/0G	4.3	3.6	3.4	3.3	3.1
2G/1G	1.7	1.6	1.6	1.6	1.7

Table 3. The times of ET based on 0G or 1G at different levels of perceived pain

3.2 Finger Skin Temperature (FST)

First, to examine the FST_0, which was controlled in a range of 30-35°C prior to immersion, the overall mean (SD) was 32.1 (0.56). ANOVA showed that there was no significant difference in FST_0 between genders or among gloved conditions. That is, the initial FST, FST_0, was well controlled. Table 2 indicates that all main effects and gloved×level interaction were significant on FST. Figure 2 demonstrates the gloved effect on FST at different pain-perceived levels and indicates that, given a pain level, more layers corresponded to lower FST. At ET_5, the FST was 29.1, 27.3, and 24.7°C for 0G, 1G and 2G, respectively, and it was 3.1, 4.8, and 7.5°C lower than FST_0 for 0G, 1G and 2G. In general, greater perceived pain corresponded to lower FST. The significant sex effect reveals that the FST of females was 0.4°C higher than that of males, but this minor difference seems negligible in the workplace.

3.3 The exchange rate on FST

Moreover, the ANOVA results for the exchange rate on FST in Table 2 revealed that all main effects and gloved×level interaction were significant. The gloved×level interaction plotted in Figure 3 reveals that with fewer layers the gloves, the rate was steeper. The variation in heat exchange for the bare hand condition (0G) from the beginning to the end was the steepest (-1.22~ -2.00°C/min), second for 1G (-1.14~ -1.72°C/min), and the flattest for 2G (-1.44~ -1.67°C/min). In the whole immersion process, the mean exchange rates on FST ($Rate_{50}$) were -2.00, -1.71, and -1.65°C/min for 0G, 1G, and 2G, respectively. Males had a greater reduction rate in FST than females (-1. 7 vs. -1.54°C/min).

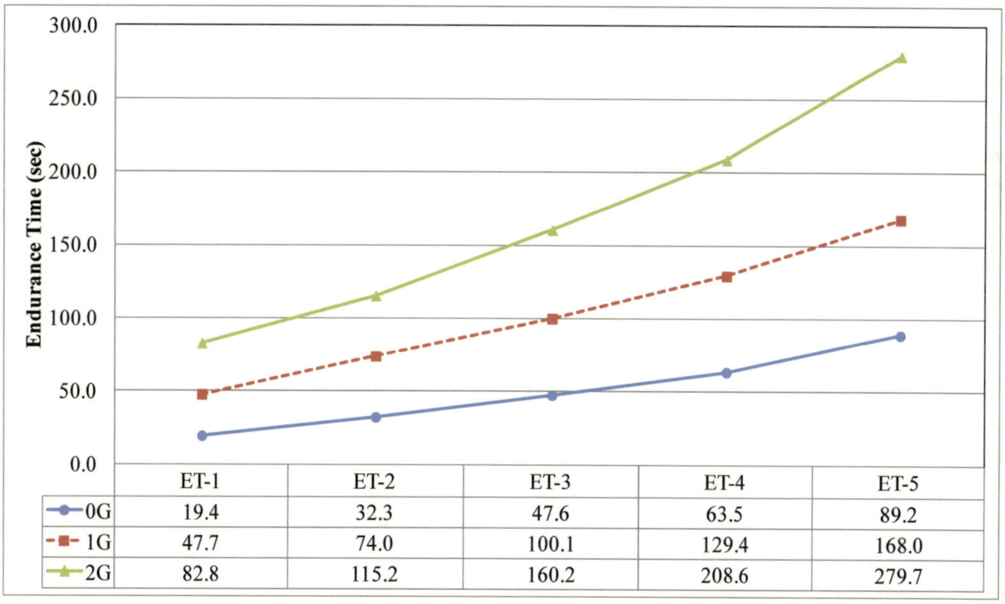

	ET-1	ET-2	ET-3	ET-4	ET-5
0G	19.4	32.3	47.6	63.5	89.2
1G	47.7	74.0	100.1	129.4	168.0
2G	82.8	115.2	160.2	208.6	279.7

Fig. 1. Gloved effect on ET at different levels of perceived pain

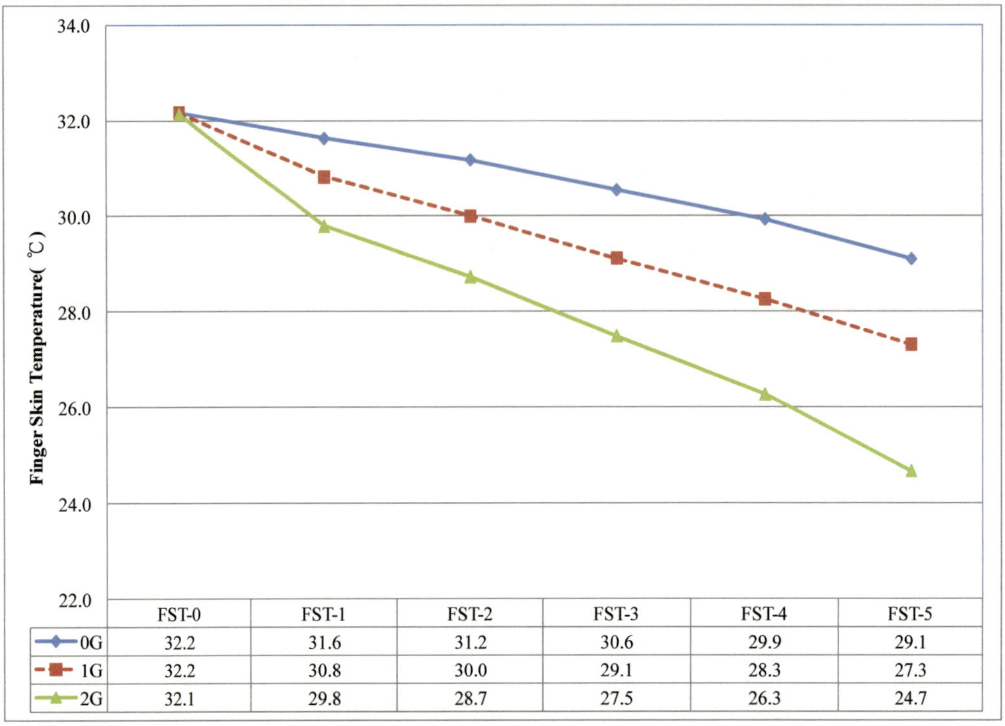

	FST-0	FST-1	FST-2	FST-3	FST-4	FST-5
0G	32.2	31.6	31.2	30.6	29.9	29.1
1G	32.2	30.8	30.0	29.1	28.3	27.3
2G	32.1	29.8	28.7	27.5	26.3	24.7

Fig. 2. Gloved effect on FST at different levels of perceived pain

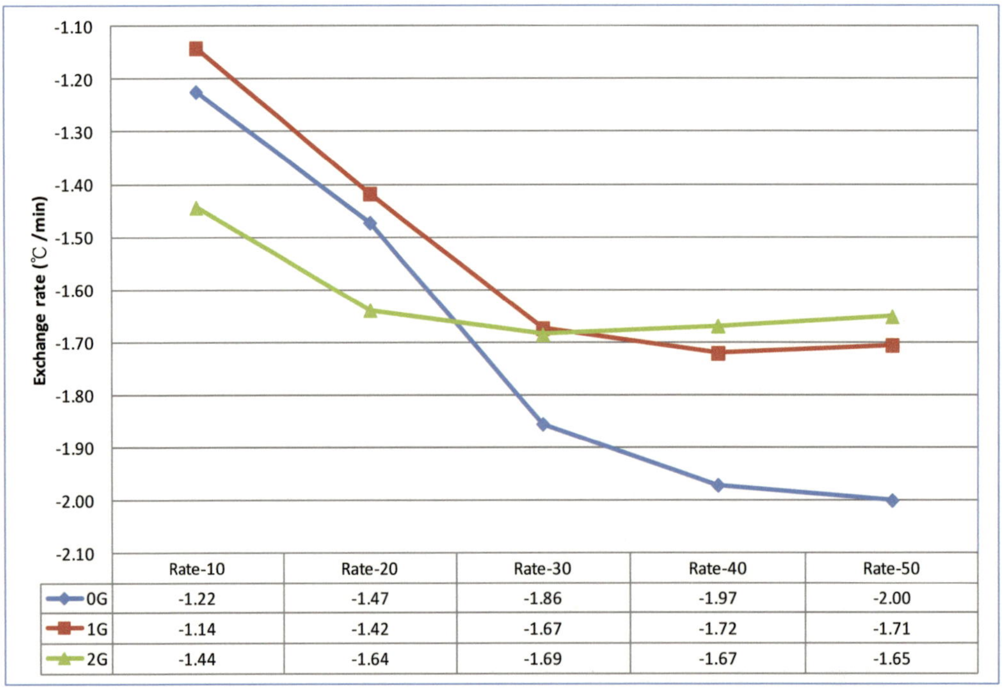

	Rate-10	Rate-20	Rate-30	Rate-40	Rate-50
0G	-1.22	-1.47	-1.86	-1.97	-2.00
1G	-1.14	-1.42	-1.67	-1.72	-1.71
2G	-1.44	-1.64	-1.69	-1.67	-1.65

Fig. 3. Gloved effect on exchange rate on FST at different levels of perceived pain

3.4 The linear relationships between Borg's CR scale, ET, and FST

Five self-reported levels of perceived pain, based on Borg's CR scale, were first assigned the number 0.5 (just noticeable), 2 (weak), 5 (strong), 7 (very strong), and 10 (extremely strong), respectively. The linear relationship between Borg's CR scale (predictor) and ET_i was examined for the bare-hand condition of each subject. For both sexes, all the intercepts (β_0) and all the slopes (β_1) differed from zero significantly. Intercepts (β_0) were in the range of -3.6 ~ 42.2 for males and 0.6 ~ 39.8 for females. The slopes (β_1) were in the range of 2.3 ~ 26.6 for males and 1.4 ~18.6 for females. The corresponding R^2 ranged from 0.70 to 0.99 for males and 0.55 to 0.97 for females. These results imply that Borg's CR scale is a good predictor for how long subjects are willing to immerse their bare hands in 10°C water.

Moreover, to compare the gloved effect on ET and ΔFST ($FST_i - FST_0$), the ET and ΔFST values among subjects were averaged. A simple regression model using Borg's CR scale as the predictor was built. Table 4 shows the intercept (β_0), slope (β_1), and R^2. It indicates the R^2 s were very high, all above 0.95.

For ET, except for β_0 under the 2G condition, the β_0 s and all β_1 s between males and females seemed not to differ pronouncedly. That could be why the ET of males was 12 sec longer than that of females (113.9 vs. 101.9 sec). In addition, β_0 s among the three gloved conditions seemed to differ dramatically, as did the β_1. The reason could be that more layers of gloves led to a longer ET. As to FST, with more layers of gloves, FST decreased more greatly when increasing by one Borg rating, and males seemed to be able to bear more reduction in FST than females. The aforementioned findings are consistent with the ANOVA results.

Gloved	Parameters	ET		ΔFST	
		Male	Female	Male	Female
Bare	β_0	16.4	14.8	-0.59	-0.23
	β_1	7.5	6.7	-0.26	-0.26
	R^2	0.98	0.99	0.99	0.99
1G	β_0	46.8	40.7	-1.55	-1.04
	β_1	12.3	12.3	-0.33	-0.39
	R^2	0.99	0.99	0.99	0.99
2G	β_0	82.2	57.3	-2.52	-1.80
	β_1	20.3	20.3	-0.50	-0.56
	R^2	0.99	0.98	0.99	0.99

Table 4. The regression parameters for using Borg's scale as predictor at different gloved conditions (ET or $\Delta FST = \beta_0 + \beta_1 \times Borg$)

In order to eliminate individual differences, normalized ET_i, say $ET_i(\%)$ ($i = 1,2,3,4,5$), was tested. It was found that neither the gloved effect nor the sex effect was significant under any given pain-level. Mean $ET_i(\%)$ was 26.5, 41.6, 58.7, 76.1, and 100 for just noticeable, weak, strong, very strong, and extremely strong, respectively. The linear model was built as $ET(\%) = 23.6 + 7.53 \times Borg$, R^2 =0.99. It indicated that to increase one point in the Borg scale, subjects could immerse their hands in the 10°C water for 7.53% more of the ET.

As expected, a longer ET (predictor) led to more reduction in FST (ΔFST), and a good linear relationship was found between the two ($\Delta FST = \beta_1 \times ET$). The slopes ($\beta_1$) of the regression model for the bare-hand, 1G, and 2G conditions were -2.04, -1.77, and -1.66 (°C/min), respectively. The R^2 s were all above 0.99, indicating that the bare-hand condition allowed greater heat dissipation on the hand than the gloved condition did.

4. Discussion

Cold may act in two ways to increase the risk of musculoskeletal disorders: directly, by its effect on body tissue, and indirectly, from the possible problems caused by the personal protective equipment used to alleviate its effect (Hagberg et al., 1995). Wearing gloves can retard the heat exchange, and, as expected, this prolongs the time of exposure of the hand in 10°C water. Unexpectedly, this longer exposure even leads to lower FST. At first, it was expected that the shorter ET of the bare-hand condition would be associated with a lower FST, but this expectation was not observed. In contrast, the shorter barehanded ET could be contributed by the impact of touching the cold water directly, which resulted in a faster rate of heat exchange. One of the goals of wearing gloves is to protect the hands from contact with cold/hot objects. According to the above results, under a given level of perceived pain, more layers of gloves extended the ET, leading to a lower corresponding FST. This trend may have been caused by the retarded heat exchange due to the wearing of gloves. Wearing gloves could prolong the ET to reach a given FST (Clark & Cohen, 1960) or delay the reduction in FST at a given endurance time (Suizu et al., 2004; Suizu & Harada, 2005). A lower FST associated with wearing gloves first numbs the hands, and gradually impairs the hand function and performance (such as dexterity, sensitivity, and strength generation), eventually leading to frostbite and damage to the hands. On the positive side, wearing gloves is at least able to prolong the exposure time in a cold environment, and this actually delays the harmful impact from the cold directly. The slower exchange rate could be one reason to encourage subjects to keep their hands in the cold water longer (with longer ET). As a result, wearing gloves seems to be a good recommendation, but adequate time to rewarm the hands is important to avoid a continuous reduction of FST.

Clark and Cohen (1960) investigated the effect of cooling rate (fast: 0°F and slow: 20°F) on manual operating time (knot tying) during cold exposure and during subsequent rewarming (75°F). During cooling, they indicated that the time needed to reduce FST to a given working temperature was longer, and more operating time was observed when the cooling rate was slow; additionally, a slow cooling rate was also associated with a longer rewarming time and longer operating time during the subsequent rewarming stage. Furthermore, Sawada et al. (2000) evaluated how repeated 10-min cooling of fingers with a 5-min rest pause schedule at work affected cold-induced vasodilatation, pain, and cold sensation in the fingers in three ambient temperature conditions (30, 25, and 20°C). They found that, at the ambient temperature of 20°C, the cold-induced vasodilatation response weakened continuously upon repetition of immersion and almost disappeared during the final immersion, and the recovery of FST during each post-immersion rest was delayed gradually upon repetition of immersion. At all three ambient temperatures, the pain and cold sensation induced by each cold-water immersion significantly decreased upon repetition of immersion. Later on, Geurts et al. (2005) additionally examined the effects of cold adaptation on the thermal response and neuromuscular function of the hand. They indicated that there was no significant change in maximal voluntary contraction strength or evoked contractile characteristics of the first dorsal interosseus after cold adaptation. Nevertheless, the mean and minimal FST decreased significantly after adaptation; the onset time was delayed and the amplitude of cold-induced vasodilatation decreased. The above findings seem to suggest that local cold adaptation of the hand does not enhance hand temperature or function, but it may put the hands at a greater risk of cold injury. Daanen et al. (1993) found impaired finger dexterity as FST fell below 14°C. Schieffer et al. (1984) found a slight reduction and a strong decrease in manual dexterity at 20-22°C and 15-16°C on FST, respectively. As a result, less heat exchange could acclimatize hands to the cold water gradually. The problem raised is that hands with lower FST and working in a cold environment for a long time could suffer reduced performance, and the lower FST could also potentially increase the possibility of musculoskeletal disorders. Unfortunately, wearing gloves for the sake of safety could enhance the local cold adaptation of the hand but lower the FST. Such a conclusion implies that when a bare hand directly contacts the cold water, FST falls more rapidly than when gloves are worn. Rapid reduction of FST causes a rapid sensation of pain, and it shortens the time for which subjects are willing to endure the pain.

The longer ET and lower FST of males seem to demonstrate that males have greater tolerance for cold-water immersion. These findings support past studies reporting a significant sex difference in response to experimental induced pain, and reports that women generally have a lower pain threshold, a greater ability to discriminate painful sensations, higher pain ratings, and a lower tolerance for pain (Berkley, 1997).

Finally, there is a good linear relationship between Borg's CR scale, ET, and reduction in FST. It is clear that the slope, β_1, is the change in response magnitude when the predictor changes by one unit. As to the intercept, β_0, it contains some unclear information, such as the difference in individuals, sex, age, and race. More studies are needed to identify and explain the meaning of the intercept.

5. Conclusions

Wearing gloves can insulate the hand in cold water and decrease the speed of heat dissipation; therefore, wearing gloves possibly allows hands to accommodate gradually to the cold environment and possibly prolongs the endurance time. With gloves, FST decreases gradually to reach a level lower than that in the bare-hand condition, and this could possibly not only impair hand performance but also encourage operators to prolong the duration of working. Such a practice could increase the probability of injury to the hands in a cold environment. In addition, males have greater tolerance for cold-water immersion. Finally, good mutual linear relationships exist among ET, ΔFST, and Borg's CR-10 scale.

6. References

Åkesson, I., Johnsson, B., Rylander, L., Moritz, U. & Skerfving, S. (1999). Musculoskeletal disorders among female dental personnel- clinical examination and a 5-year follow-up study of symptoms. *International Archives of Occupational and Environmental Health*, 72(6): 395-403.

Baird, J.C. & Noma, E. (1978). *Fundamentals of scaling and psychophysics*. Wiley, New York.

Berkley, K.J. (1997). Sex differences in pain. *Behavioral Brain Science*, 20: 371-380, 435-513

Bishu, R.R. & Klute, G., 1995. The effects of extra vehicular activity (EVA) gloves on human performance. *International Journal of Industrial Ergonomics*, 16: 165-174.

Borg, G.A.V. (1970). Perceived exertion as an indicator of somatic stress. *Scandinavian Journal of Rehabilitation Medicine*, 2(3): 137-148.

Borg, G.A.V. (1982). Psychophysical bases of perceived exertion. *Medicine and Science in Sports and Exertion*, 4(5): 377-381.

Brajkovic, D., & Ducharme, M. B. (2003). Finger dexterity, skin temperature, and blood flow during auxiliary heating in the cold. *Journal Applied Physiology*, 95: 758-770.

Candas, V., & Dufour, A. (2007). Hand skin temperature associated with local hand discomfort under whole-body cold exposure. *Journal of the Human-Environment System*, 10(1): 31-37.

Chang, C.H. & Shih, Y.C. (2007). The Effects of Glove Thickness and Work Load on Female Hand Performance and Fatigue during an Infrequent High-Intensity Gripping Task. *Applied Ergonomics*, 38(3): 317-324

Chen, W. L., Shih, Y.C., & Chi, C.F. (2010). Hand and finger dexterity as a function of skin temperature, EMG, and ambient condition. *Human Factors*, 52(3): 426-440

Chen, F., Nilsson, H., & Holmer, I. (1994). Finger cooling by contact with cold aluminum surfaces- effects of pressure, mass, and whole body thermal balance. *European Journal of Applied Physiology*, 69: 55-60.

Chiang, H.C., Ko, Y.C., Chen, S.S., Yu, H.S., Wu, T.N., & Chang, P.Y. (1993). Prevalence of shoulder and upper-limb disorders among workers in the fish-processing industry. *Scandinavian Work Environment Health*, 19(2): 126–131

Clark, R.E., & Cohen, A. (1960). Manual performance as a function of rate of change in hand temperature. *Journal of Applied Physiology*, 15: 496-498.

Coulange, M., Hug, F., Kipson, N., Robinet, C., Desruelle, A. V., Melin, B., Jimenez, C., Galland, F., & Jammes, Y. (2006). Consequences of prolong total body immersion in cold water on muscle performance and EMG activity. *Pflügers Archiv European Journal of Physiology*, 452: 91-101.

Daanen H.A.M., Wammes L.J.A., & Vrijkotte T.G.M. (1993). *Windchill and dexterity*. Report IZF A-7, TNO Institute for Perception, Soesterberg, NL.

Dedering, Å., Harms-Ringdahl, K., & Nèmeth, G. (2006). Back extensor muscle fatigue in patients with lumbar disc herniation. *European Spine Journal*, 15(5): 559-569

Enander, A. (1984). Performance and sensory aspects of work in cold environments: a review. *Ergonomics*, 27: 365-378.

Enander, A. E., & Hygge, S. (1990). Thermal stress and human performance. *Scandinavian Journal of work Environment & Health*, 16(Suppl. 1): 44-50.

European Agency for Safety and Health at Work (2010). *OSH in figures: Work-related musculoskeletal disorders in the EU – Facts and figures*. Office for Official Publications of the European Communities.

Geng, Q., Holmér, I., & Coldsurf research group (2001). Change in contact temperature of finger touching on cold surface. *International Journal of Industrial Ergonomics*, 27:387-391.

Geurts, C. L. M., Sleivert, G. G., & Cheung, S. S. (2006). Local cold acclimation during exercise and its effect on neuromuscular function of the hand. *Applied Physiology Nutrition and Metabolism*, 31: 717-725.

Geurts, C. L. M., Sleivert, G. G., & Cheung, S. S. (2005). Effect of cold-induced vasodilatation in the index finger on temperature and contractile characteristics of the first dorsal interosseus muscle during coldwater immersion. *European Journal of Applied Physiology*, 93: 524 - 529.

Goonetilleke, R. S. & Hoffmann, E. R. (2009). Hand-skin temperature and tracking performance. *International Journal of Industrial Ergonomics*, 39(4): 590-595.

Hagberg, M., Silverstein, B., Wells, R., Smith, M. J., Hendrick, H. W., Caravon P., & Perusse, M. (1995). Identification, measurement and evaluation of risk. In: Kuorinka, Forcier (Eds.), *Work Related Musculoskeletal Disorders (WMSD's), a reference book for prevention.* Taylor & Francis, London, pp. 139–145 and 162–163.

Havenith, G., Van de Linde, E. J. G., & Heus, R. (1992). Pain, thermal sensation and cooling rates of hands while touching cold materials. *European Journal of Applied Physiology*, 65: 43-51.

Havenith, G., Heus, R., & Daanen, H.A.M. (1995). The hand in the cold, performance and risk. *Arctic Medical Research*, 54(Suppl. 2): 37-47.

Heus, R., Daanen, H. A. M., & Havenith, G. (1995). Physiological criteria for functioning of hands in the cold: a review. *Applied Ergonomics*, 26(1): 5-13.

Hirsch, M.S. & Liebert, R.M. (1998). The physical and psychological experience of pain: the effects of labeling and cold pressor temperature on three pain measures in college women. *Pain*, 77: 41-48

ISO/CD 14835-1 (2001). *Mechanical vibration and shock-Cold provocation tests for the assessment of peripheral vascular function-Part 1: Measurement and evaluation of finger skin temperature.* ISO/TC 108/SC 4/WG 11.

Keefe, F.J. & Williams, D.A. (1992). Assessment of pain behaviors. In D.C. Turk and R. Melzack (Eds.), *Handbook of pain assessment*, Guilford Press, New York: 277-292.

Kurppa, K., Viikari-Juntura, E., Kuosma, E., Huuskonen, M., & Kivi, P. (1991). Incidence of tenosynovitis or peritendinitis and epicondylitis in a meat-processing factory. *Scandinavian Journal of Work and Environmental Health*, 17(1): 32–37

Nag, P.K. & Nag, A. (2007). Hazards and health complaints associated with fish processing activities in India- Evaluation of a low-cost intervention. *International Journal of Industrial Ergonomics*, 37: 125-132.

Nordander, C., Ohlsson, K., Balogh, I., Rylander, L., Palsson, B., & Skerfving, S. (1999). Fish processing work: the impact of two sex dependent exposure profile on musculoskeletal health. *Occupational and Environmental Medicine*, 56: 256-264.

Ou, Y.C. (2003). *The Influence of Operator-Equipment/tools Interface Factors on the Performances.* National Defense Management College, unpublished master thesis.

Petrofsky, J. S., & Lind, A. R. (1980). The influence of temperature on the amplitude and frequency components of the EMG during brief and sustained isometric contraction. *European Journal of Applied Physiology*, 44: 189-200.

Piedrahíta, H., Punnett, L., & Shahnavaz, H. (2004). Musculoskeletal symptoms in cold exposed and non-cold exposed workers. *International Journal of Industrial Ergonomics*, 34(4): 271-278

Riley, M. W., & Cochran, D. J. (1984). Dexterity performance and reduced ambient temperature. *Human Factors*, 26: 207-214.

Schieffer, R.E., Kok, R., Lewis, M.I., & Meese, G.B. (1984). Finger skin temperature and manual dexterity; some inter-group difference. *Applied Ergonomics*, 15(2): 135-141.

Shih, Y.C. (2007). Glove and Gender Effects on Muscular Fatigue Evaluated by Endurance and Maximal Voluntary Contraction Measures. *Human Factors*, 49(1): 110-119.

Suizu, K., Inoue, M., Fujimura, T., Morita, H., Inagaki, J., Kan, H., & Harada, N. (2004). Influence of waterproof covering on finger skin temperature and hand pain during immersion test for diagnosing hand-arm vibration syndrome. *Industrial Health*, 42: 79-82.

Suizu, K. & Harada. N. (2005). Effects of waterproof covering on hand immersion tests using water at 10℃、12℃and 15℃ for diagnosis of hand-arm vibration syndrome. *International Archives Occupation Environment Health*, 78: 311-318.

Sawada, S., Araki, S., & Yokoyama, K. (2000). Changes in cold-induced vasodilatation, pain and cold sensation in fingers caused by repeated finger cooling in a cool environment. *Industrial Health*, 38:79-86

Wolff, B.B. (1984). *Methods of testing pain mechanisms in normal man*. In P.D. Wall and R. Melzack (Eds.), Textbook of Pain, Churchill Livingstone, New York, 186-194.

Section 3

Risk Assessment in System Design

13

Risk Analysis of the Waste to Energy Pyrolysis Facility Designs for City of Konin, in Poland, Using SimLab® Toolpack

Boguslaw Bieda
AGH University of Science and Technology, Management Department, Kraków, Poland

1. Introduction

The term "*thermal treatment*" is used to describe a range of technologies that use heat to degrade the constitution of solid matter. These includes incineration and its variations, as well as advanced thermal conversion (ATC) technologies such as pyrolysis and gasification (Everard, 2004).

To ensure sustainable development in waste management, faster development and uptake of new technology is necessary. Landfills pollute the valuable underground water, incinerations emit dioxin and produce toxic ash. The solution is Integrated Waste Management, which uses all available resources for dealing with the waste problem. Novel processes utilizing pyrolysis and gasification have attracted publicity as a potential alternative to incineration. The main advantage that gasification has over incineration is its ability to conserve the chemical energy of the waste in the produced syngas rather than convert it to heat energy in hot flue gases. Therefore, gasification has greater flexibility in the recovery of energy and chemical value from waste stream (Klein et al., 2004). Gasification is by no means a novel process; in the 19th century so-called: "*town-gas*" was produced by gasification of coal and for example used for illumination purposes (Astrup & Bilitewski, 2010). Gasification (and combinations of pyrolysis plus gasification) processes are being developed in a number of countries. In Europe, there continues to be a strong desire to avoid incineration and reduce the amount of waste going to landfill in order to meet the EU landfill Directives. In the USA, low disposal costs and plenty of landfill availability in most regions have proved a significant barrier to the construction of any new thermal treatment facilities. Incinerations also increase the amount of CO2 in the carbon cycle because the have to burn fuel together with the wastes. The governments of most countries have signed a treaty to limit CO2 emissions at their 1999 levels. In Canada, a number of waste management projects are being planned based on the waste incineration technology. In Japan all leading thermal process companies now offer gasification solutions alongside incineration with financial support from the Japanese Government.

For many people, thermal treatment technologies for waste management represent an image of hell on Earth (Everard, 2004).

The main potential benefits and advantages of pyrolysis and gasification of waste with respect to incineration are (Juniper, 2001; Klein et al., 2002; Malkow, 2004 as cited in Astrup & Bilitewski, 2010):

- The possibility and flexibility to recover chemical energy in the waste as hydrogen and/or other chemicals feedstocks rather than converting this energy into flue gases.
- Potentially beter overall energy efficiency.
- Less trouble with corrosion.
- Potentially better option for CO_2 capture.
- Potentially lower emissions of dioxins.
- Improved quality of solid residues, particular for high-temperature processes.
- Gasifications units operating with a low fuel load, potentially facilitating small plants producing less than 1 MW.
- Potentially lower costs.

The main drawback of the current technology for pyrolysis and gasification are:

- Relatively homogeneous fuels are needed. Either specific material fractions can be fed to the gasifier, or mixed waste can be pretreated and homogenized.
- Although theoretically possible, the pyrolysis and gasification processes are complicated to control and troubles with slagging, tar production, and contaminants in the produced gas are not uncommon.
- Numerous waste related pyrolysis and gasification technologies exist, many of these only demonstrated in small scale and/or applicable to specific fuel types.. This requires careful review of the appropriateness of a specific technology for a particular waste mix, local conditions, etc.
- Overall energy conversion efficiencies of existing installations have been unable to compare with modern waste incinerators.

2. Market interest in gasification and pyrolysis

Gasification is a partial oxidation processes in which the majority of the carbon is converted into the gaseous form-called syngas-by partial combustion of a portion of fuel in the reactor with air, pure oxygen, oxygen-enriched air or by reaction with steam. Relatively high temperatures are employed: 900- 1100°C with air and 1000-1400°C with oxygen. Gasification as a technology underwent major development during the oil price crises of the 1970s and 1980s.

Pyrolysis is the thermal degradation of carbonaceous materials. It occurs at lower temperature than gasification (typically 400-800°C), either in the complete absence of oxygen, or with such a limited supply that gasification. Pyrolysis has been promoted for biomass applications and in the treatment of scrap tyres, bur rarely as a stand-alone application for MSW.

Energy recovery is a secondary goal of waste incineration: thermal waste treatment and energy recovery are "married" within the waste-to energy plant (Pfeiffer, 2004). From an economic point of view, a waste–to- energy plant treating MSW is an enterprise using a special fuel.

Some technologies, including gasification and pyrolysis, offers flexibility in terms of energy production and material recycling, and is an attractive technology option for Integrated Waste Management.

The main advantage that gasification has over incineration is its ability to conserve the chemical energy of waste in the produced syngas rather than convert it to heat energy in hot flue gas. Another reason for interest in gasification is the view by political decision-makers (especially in the UK) that gasification is an alternative to incineration, because which would mean that incineration would no longer be necessary.

The United States Department of Energy (DOE) sponsored the 2004 World Gasification Survey in order to accurately describe the world gasification industry as it exists today, to identify planned capacity additions, and to keep the gasification community apprised of current data and trends (National Energy Technology Laboratory [NETL], (2005).

An additional 38 plants with 66 gasifiers have been announced and are forecast to become operational between 2005 and 2010, according to the 2004 survey. The additional capacity from these new plants is 25,282 MWth, an expected increase of 56%. Worldwide capacity by 2010 is projected at 70,283 MWth of syngas output from 155 plants and 451 gasifiers.

- Regional distribution: The Africa/Middle East region will lead the world's regional growth with 43% of planned capacity growth from 2005 to 2010, all from a single gas-to-liquids (GTL) project in Qatar that will produce liquid fuels from natural gas. The Asia/Australia region has planned projects that comprise 37% of the total planned growth, with China leading in this region. By contrast, plans for new gasification plants slowed in North America due to factors such as the economy and natural gas prices.

Feedstock distribution: Coal is the feedstock of choice for new gasification projects, identified for 29 of the 38 new plants (largely on the strength of the 24 chemical plants to be built in China). However, natural gas will be used in the largest single project from 2005 to 2010 at the nearly 11,000 MWth gas-to-liquid.

3. Description of case study

Solid waste management is developing into a complex task. New or modified treatment technologies are appearing. During the past two decades, thermal wastes management followed heavily disparate trends. In the 1980s, the focus was on new market players, and then in the 1990s on new technologies, especially pyrolysis and melting processes (Bieda & Tadeusiewicz, 2008).

Novel processes utilizing pyrolysis and gasification have attracted publicity as a potential alternative to incineration. Such systems offer some benefits in terms of recycling and public acceptance. However, because they are new, they are less proven in operation than conventional technologies-and may therefore be more risky. The main advantage that gasification has over incineration is its ability to conserve the chemical energy of the waste in the produced syngas rather than convert it to heat energy in hot flue gas (Klein et al, 2004).

The new Polish environmental strategy emphasizes the principle of sustainable development and it encourages the government of Konin to develop a waste management plan for their communities based on the use the technology for a gasification with waste to energy system. One scenario has been chosen: American Gasification System (design at 200 T/D). The Capital Budget - Project Costs of the American Scenario is given in Tables 1.

The revenues were based on the Proposal to Design, Develop and Construct a Waste-to-Energy Facility for the City of Konin. The revenues include:

- the tipping fees for landfill
- the revenues from energy sales
- other revenues.

The selling prices of the marketable material, and the tipping fee for each ton of waste that is delivered to the landfill are coming from the Waste Program Revenue from the city or others. The general operating parameters of the Konin's Waste-to-Energy Facility are as follows:

- operating weeks/year - 50 weeks
- receiving days/week - 5 days
- current tons managed - 63,000 Mg/year

Municipality has been entered into a contract to supply an average of 200/250 tons of municipal waste per day with options for increased volume as the demand increases.

Capital Budget - Project Costs (USD)		
1	**Etap 1-**Construction Management	600,731.00
2	**Etap 2-**Civil& Site Design/Site Work &Building Permitting, Gasifiers System	21,120,055.27
3	**Etap 3-**Continuous Emission Control, Monitoring Systems	999,599.10
5	**Etap 4-**Automatic Loading Systems	1,687,350.23
6	**Etap 5-**Office Furniture and Computers	4,25
7	**Etap 6-**Contingency Reserve	1,167,264.40
8	**Razem-Total Project Costs (USD)**	26,000,000.00

Table 1. Capital Budget – Project Costs of the proposed American Gasification System.

4. Monte Carlo simulation with SimLab®

The first task is to create a veritable deterministic model that represents the most likely scenario.

To use the SimLab® (SimLab, 2004), we must perform the following steps (Wajs et al., 2006):

- build model the relationships
- define assumption for probabilistic variables - manufacturing costs
- define the forecast cell, that is, the output variable - Total et the number of replication
- run the simulation
- simulate the model and analyze the outputs

- report results and make decisions.

In SimLab®, the assumptions or input range for each parameter was defined by choosing a probability distribution that describes the uncertainty of the data. Input distribution may be normal, uniform, triangular, skewed, or any shape that reflects the nature of the measurement being assessed.

At the start Simlab® displays the main panel (Figure 1); this panel is divided in three frames (Saltelli et al., 2004):

1. The *Statistical Pre Processor module*: generates a sample in the space of the input.
2. The *Model Execution module*: executes the model for each point in the sample of input factors.
3. The *Statistical Post Processor module*: performs the uncertainty and sensitivity analysis.

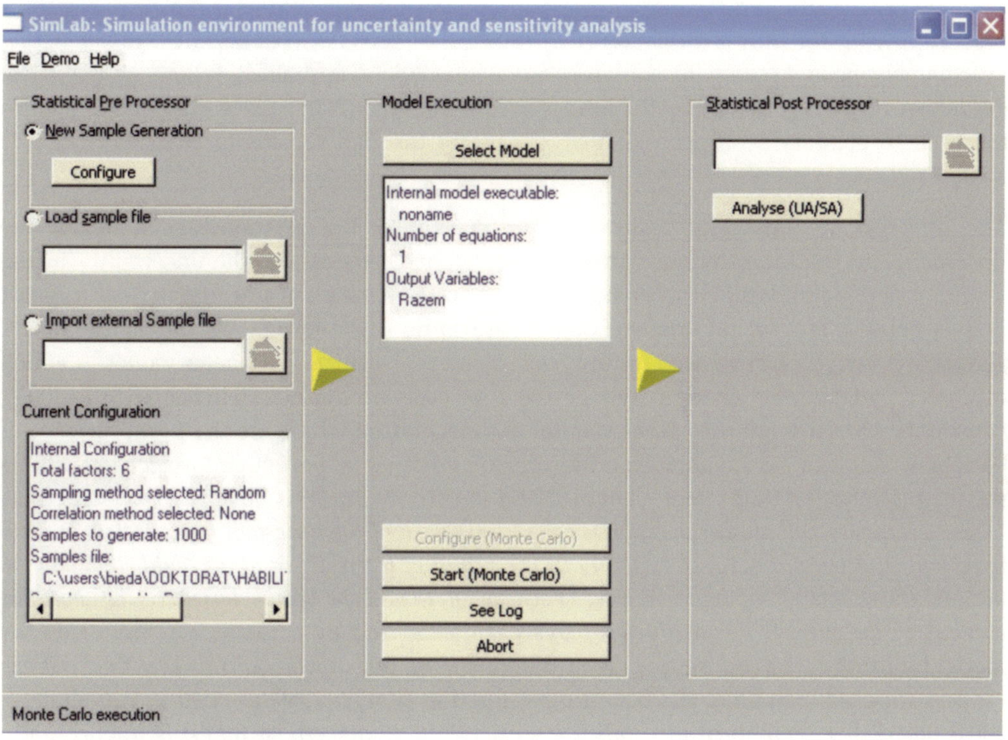

Fig. 1. Simlab® main panel.

5. Results and discussion

The deterministic project approach traditionally requires that the values for all input data be known exactly. But data in many real manufacturing projects cannot be precisely given. The stochastic approach is based on the replacement deterministic data with random variables. Important studies to stochastic variables incorporated in the data envelopment analysis can be found in (Sengupta, 1982, 1987, 1990, 1997, 1998, 2000, Cooper et al, 1998; Huang & Li, 1996; Morita & Seiford, 1999; Sueyoshi, 2000, as cited in Azadi & Saen, 2011).

Projects involve risk by nature (Zwikael & Ahn, 2011). Reducing the level of risk is extremely important in projects, and indeed results of this study suggest that project managers often use risk management planning practices, consistent with previous studies (Kerzner, 2009; Ahmed & Kayis, 2007; Voetsch, 2004; Zwikael, 2004, as cited in Zwikael & Ahn. 2011). The Project Management Body of Knowledge (PMBOK) defines risk management as one of nine project knowledge areas, alongside other topics such as scope, schedule, quality, and cost management (PMI Standards Committee [PMI], 2008). In some project contexts, risk management is perceived as a separate activity (Zwikael & Ahn, 2011).

- In countries with low levels of uncertainty avoidance, project managers place lower importance on risk management and hence do not always follow required processes.
- In industries with low levels of maturity, project managers do not frequently perform the risk management process.

When applying for European Union (EU) funding of projects, cost-benefit analysis (CBA) practitioners need to prepare comprehensive investment appraisals following the latest guidelines on CBA provided by the European Commission (2008). Since this Guide includes the need to conduct a proper risk analysis, partly through sensitivity analysis (Evans & Kula, 2011).

The purpose of sensitivity analysis is to determine the relationships between the uncertainty in the independent variables used in an analysis and the uncertainty in the resultant dependent variables. Sensitivity refers to the amount of uncertainty in a forecast that is caused by the uncertainty of an assumption as well as by the model itself. Sensitivity analysis can be used to find "*switch points*" -- critical parameter values at which estimated net benefits change sign or the low cost alternative switches (OMB, 2003). Sensitivity plots are not only fundamental to determining which are the prominent input variables, but can be invaluable indicators of whether a particular project should be pursued (Koller, 1999). In (Saltelli et al, 2004) sensitivity analysis have been presented as: "*those techniques will answer questions of the type 'which of the uncertain input factors is more important in determining the uncertainty in the output of interest?, or, if we could eliminate the uncertainty in one of the input factors, which factor should we choose to reduce the most the variance of the output?"*. Sensitivity analysis is considered by some as a prerequisite for model building in an any setting, be it diagnostic or prognostic, and in any field where models are used (Saltelli et al, 2004). Kolb quoted in (Rabitz 1989, as cited in Saltelli et al, 2004) noted that theoretical methods are sufficiently advanced, so that it is intellectually dishonest to perform modeling with sensitivity analysis. In Oreskes et al (1994) it has been shown that sensitivity analysis is not treated as a tool to build or improve a model, but it represents one of the possible licit uses that can be done of the model itself. Chapman & Ward (2004) have defined "*risk efficiency*" as the minimum risk level for a given level of expected performance.

The principal output reports provides by SimLab® are presented in Figure 2 through Figure 7 (probability distributions assigned to model input parameters), Figure 8 through Figure 13 (histograms of the output value-Razem (Total), Figure 14 (uncertainty analysis of the output value-Razem (Total), Figure 15 and Figure 16 (sensitivity analysis based on the *Standardised*

Regression Coefficients, SRC) and Figure 17 (sensitivity analysis - *Cobwebs plot* based on the *Standardised Regression Coefficients*, SRC) (Bieda, 2011). Based on the economic feasibility model presented in (Liberman, 2003), in this study used uniform distributions. Figure 15 and Figure 16 shown the results from the sensitivity analysis. The performance of the SRC is shown to be extremely satisfactory when the model output varies linearly or at least monotonically with each independent variable. MC analysis-simulation is the only acceptable approach for U.S. Environmental Protection Agency (EPA) risk assessments (Smith, 2006). Because all of the parameters of the economic model are independent, the using of the SRC is shown to be extremely satisfactory (Bieda, 2010).

SimLab® is didactical software designed for global uncertainty and sensitivity analysis, developed by the Joint Research Centre of the European Commission and downloadable for free at: http://simlab.jrc.ec.europa.eu (Simlab, 2004). The sampling techniques available in SimLab® are FAST, Extended FAST, Fixed sampling, Latin Hypercube, replicated Latin Hypercube, Morris, Quasi-random LpTau, Random and Sobol (Saltelli et al, 2004). SimLab® can also run models built in Microsoft Excel®. Using the SimLab® in order to determine the most relevant parameters sampling presented by sensitivity analysis, after selected the Monte Carlo (MC) simulation, sampling method, the optimal number of executions is considered at least 10,000 times.

There are available various commercial software packages in order to conduct the risk analysis using MC simulation. Among them, Risk® and Crystal Ball®, developed by Palisade Corporation and Decisioneering, respectively. Risk® was originally designed for business application and is easy to use without a need for extensive statistical knowledge and modeling capacity (Sonnemann et al, 2004). Crystal Ball® is a simulation program that helps analyze the uncertainties associated with Microsoft Excel® spreadsheet models by MC simulation (Sonnemann et al, 2004). Another function of the Crystal Ball® is the sensitivity analysis (Bieda, 2007).

According to Hullet (2004) the estimate of project cost risk can be made more accurate and better understood if the sources of risk are disaggregated into those that affect time and those that affect the burn tate per unit time. The schedule risk and cost risk analysis have been conducted in Microsoft Excel® and Crystal Ball®. In conclusion, Hullet (2004) muses that cost risk analysis that explicitly incorporates schedule risk analysis results, merging them with burn rate risk information in the estimates of cost risk that are more accurate than the typical approach. In the opinion of Leach (2005) anyone who is serious about realistically forecasting project schedules, in other words, truly managing projects, rather than just monitoring them, should be using MC simulation software to plan and analyze projects stochastically. Stochastic simulation (often called MC simulation) allows to capture and understand the uncertainty inherent in the project. Anderson (2005) in paper presented to the Denver Crystal Ball Conference in 2005 discusses the results of the using the MC simulation instead the analytic approach in the nuclear power plant steam generator repair/replacement cost/benefit analysis (before nuclear power plant steam generator replacement decisions have never included a MC simulation) and the strengths of the weaknesses of using Crystal Ball® and MC simulation. In conclusion, Anderson (2005) seems to think that MC approach was clearly appropriate to fully assess the impact of any decision on the ratepayer. Selection of data play a key role in application of risk analysis to project investments. In most industries the costs of raw materials and component parts constitute the major cost of a product - in some cases up to 70 per cent (Azadi & Saen, 2011). In this study the most likely Total Project Cost values are about 2.53563E+007 USD and 2.663226E+007 USD for the analyzed Scenario. Every manager has a different degree of aversion to risk In this study the most likely Total Project Cost values are about 2.53563E+007 USD and 2.663226E+007 USD for the analyzed Scenario. Every manager has a different degree of aversion to risk.

Positive coefficients indicate that an increase in assumption is associated with an increase in the forecast, negative coefficients imply the reverse (Evans & Olson, 1998). In the Sensitivity Charts (Figures 16 and 17) is presented that variables Etap2 is the most influential parameter (98.93%) in the Project Total Costs.

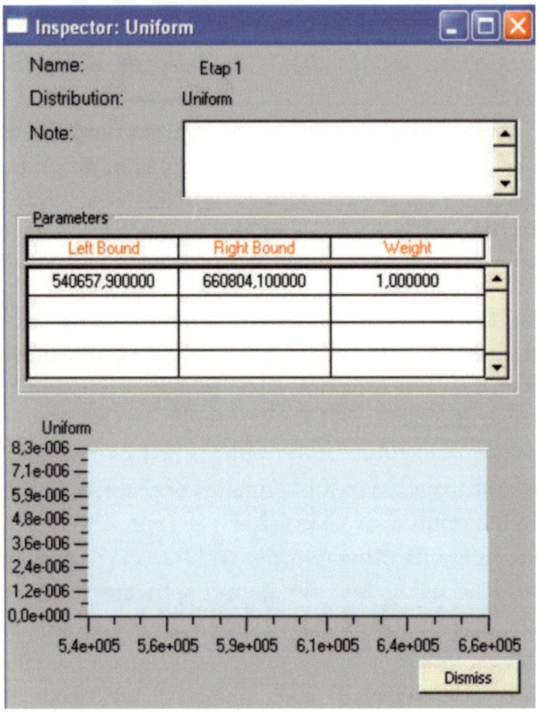

Fig. 2. Probability distributions assigned to input - Etap 1.

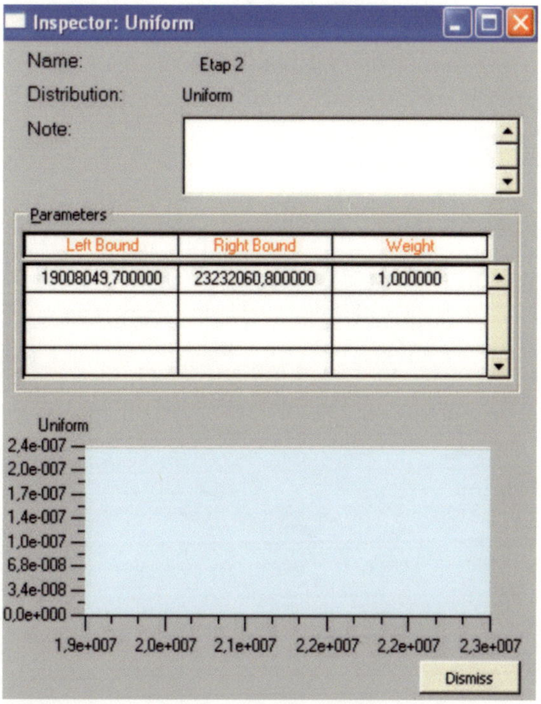

Fig. 3. Probability distributions assigned to input - Etap 2.

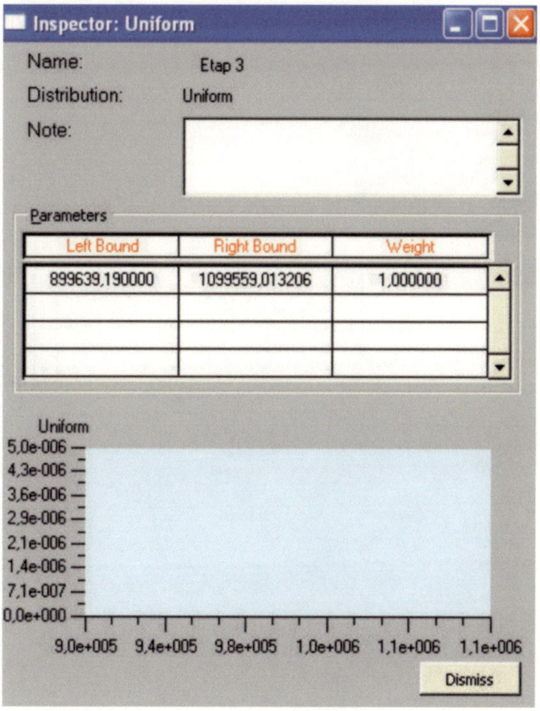

Fig. 4. Probability distributions assigned to input - Etap 3.

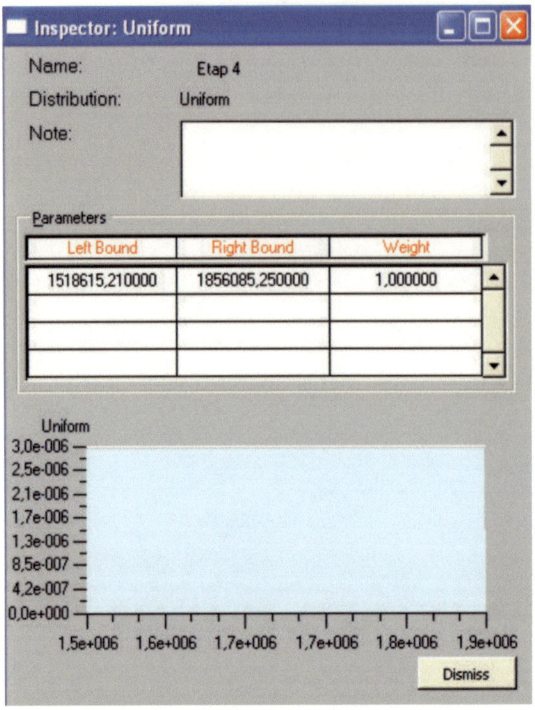

Fig. 5. Probability distributions assigned to input - Etap 4.

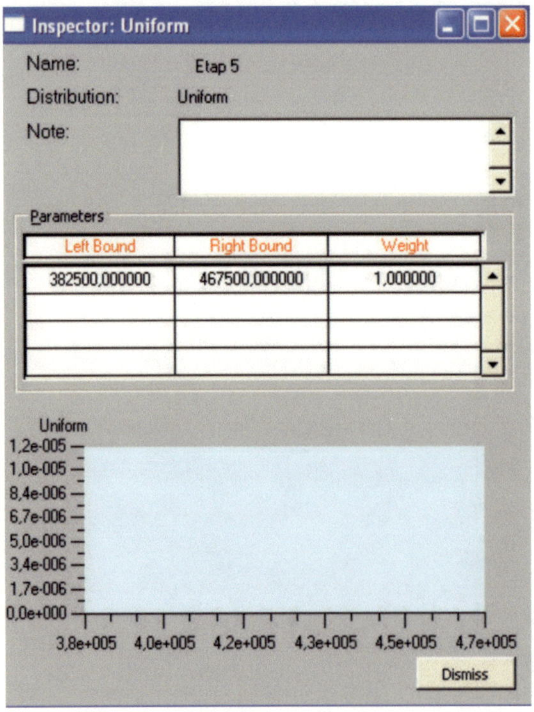

Fig. 6. Probability distributions assigned to input - Etap 5.

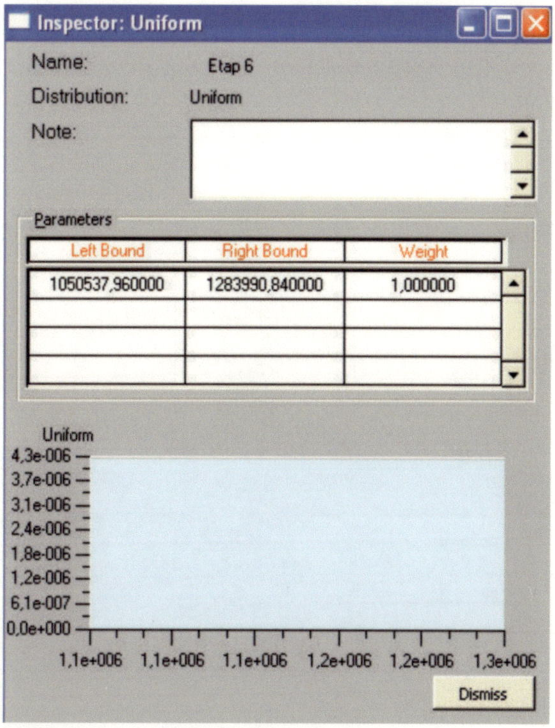

Fig. 7. Probability distributions assigned to input - Etap 6.

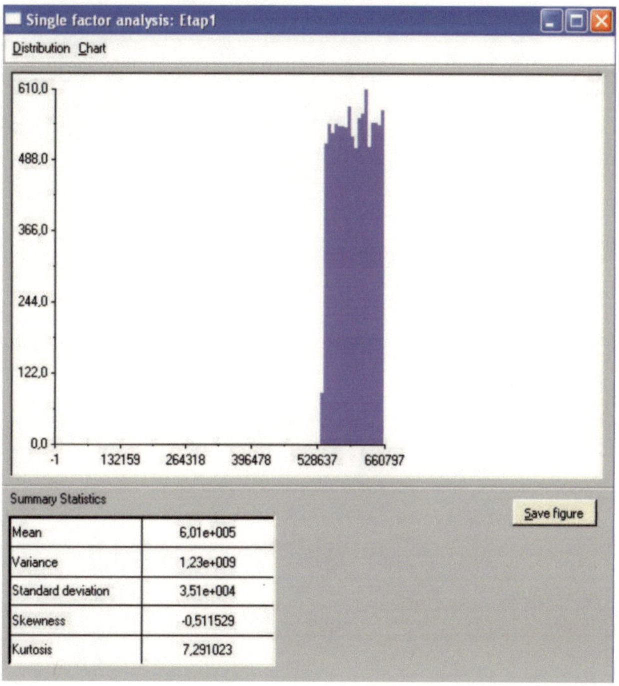

Fig. 8. Histogram results for Etap 1.

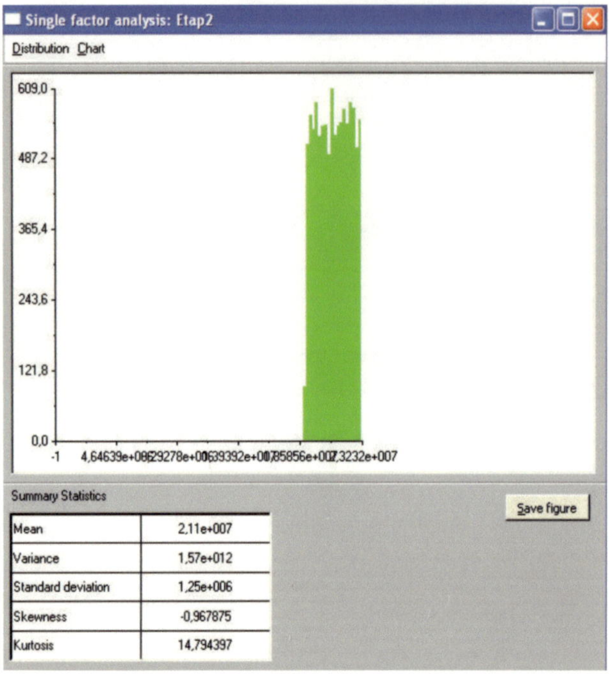

Fig. 9. Histogram results for Etap 2.

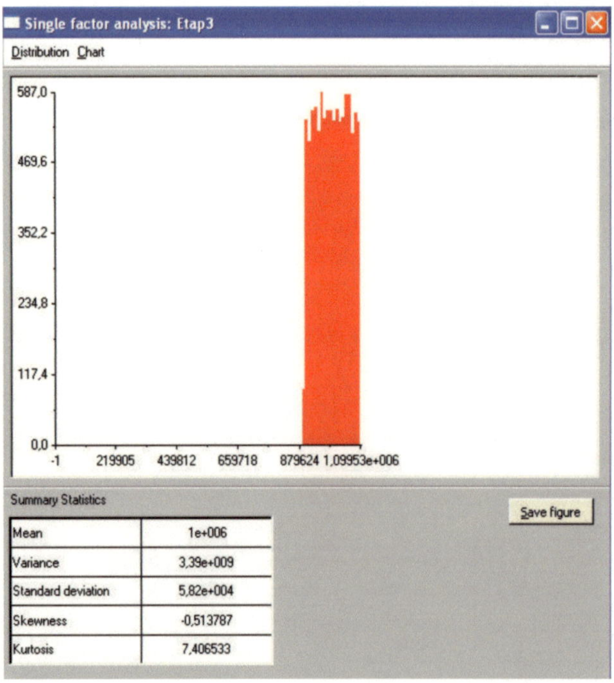

Fig. 10. Histogram results for Etap 3.

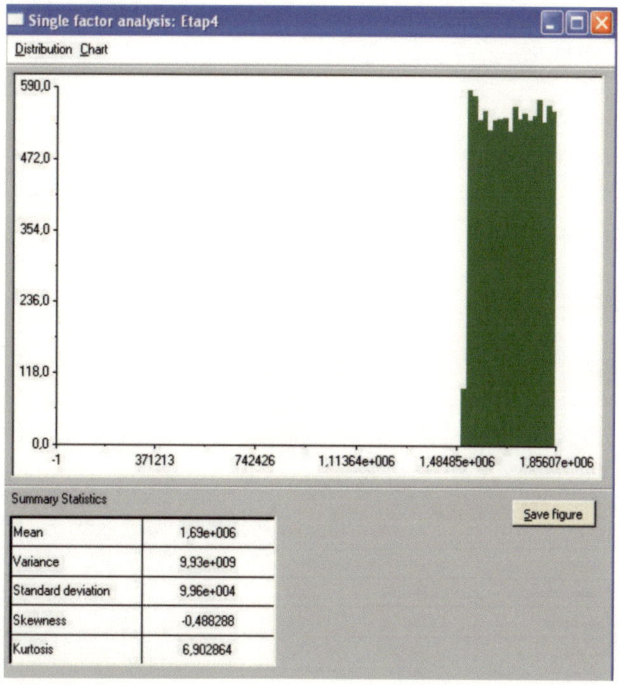

Fig. 11. Histogram results for Etap 4.

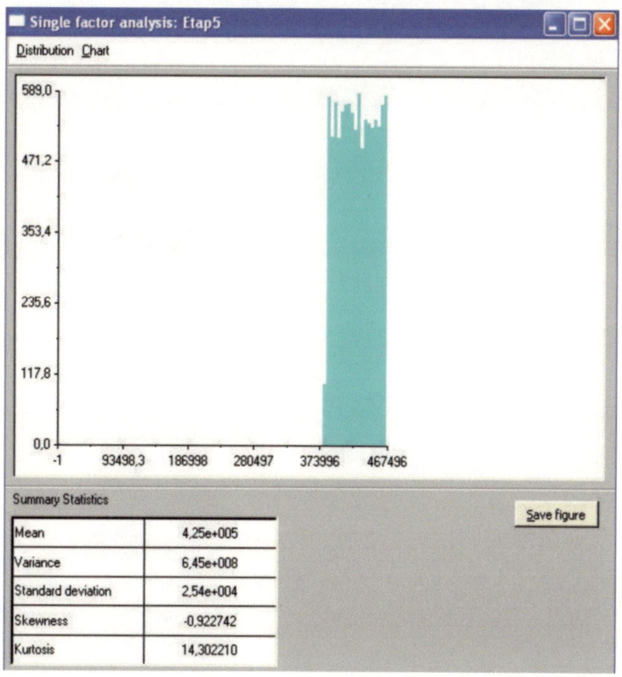

Fig. 12. Histogram results for Etap 5.

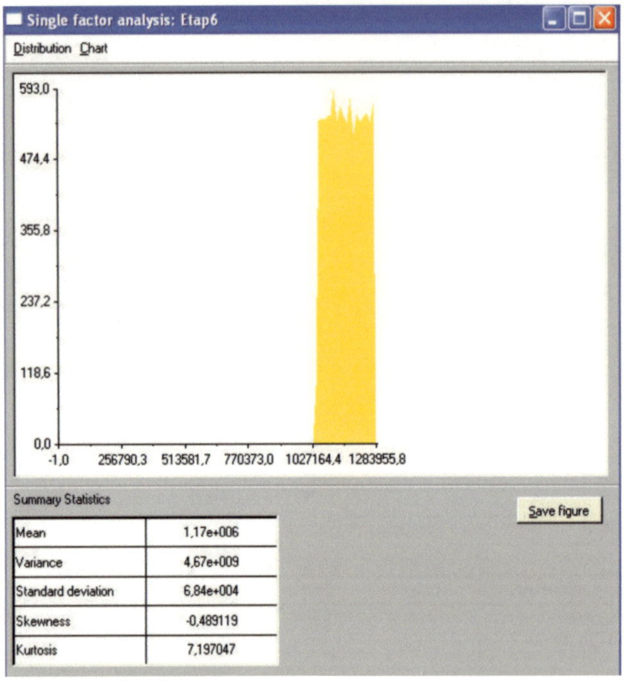

Fig. 13. Histograms results for Etap 6.

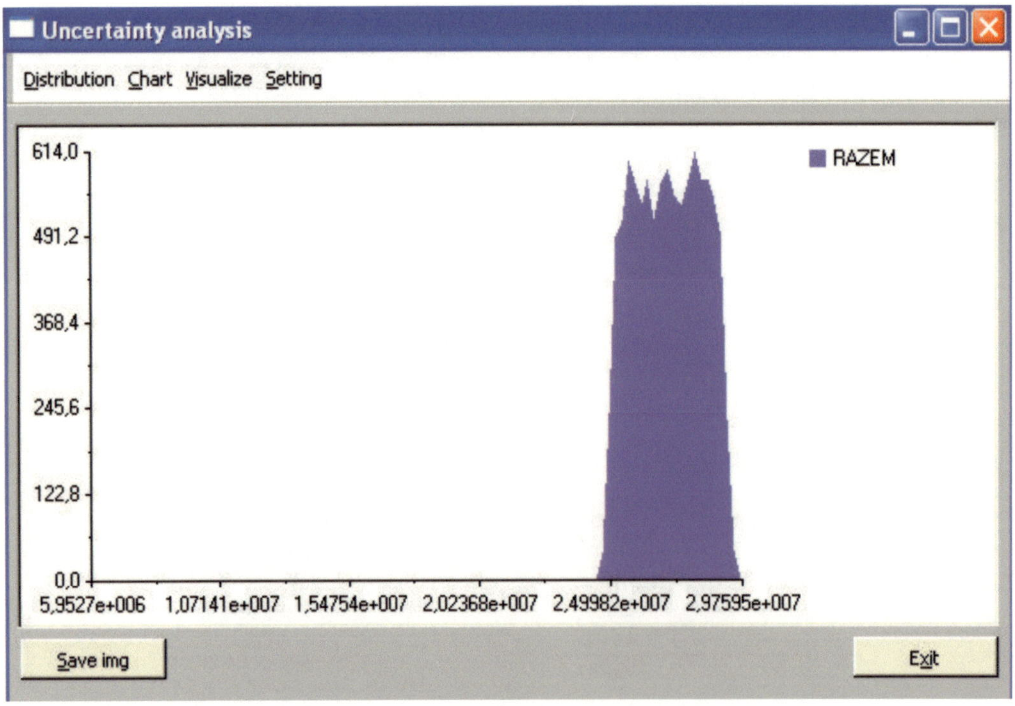

Fig. 14. SimLab® uncertainty analysis for RAZEM.

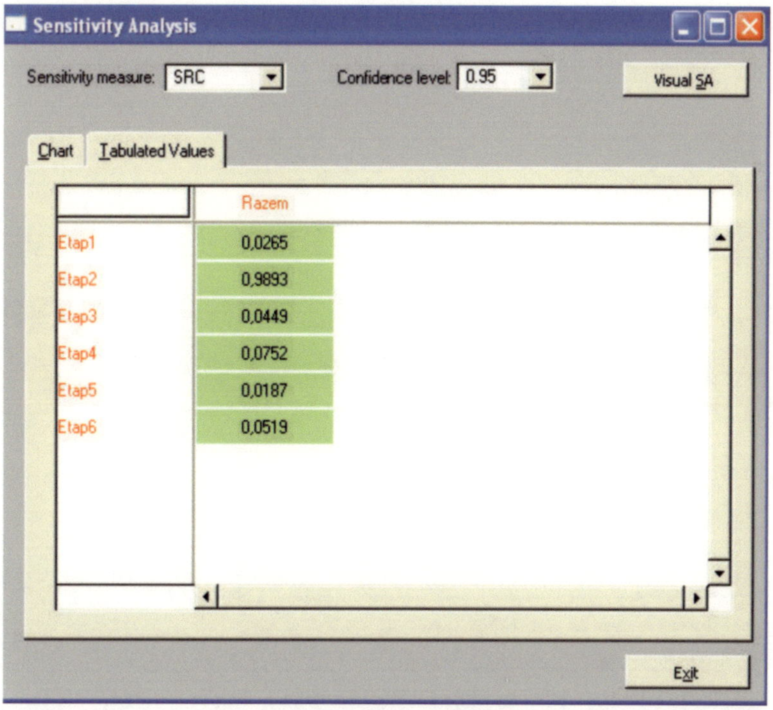

Fig. 15. Sensitivity analysis tabulated value (SRC) for the 95% confidential level.

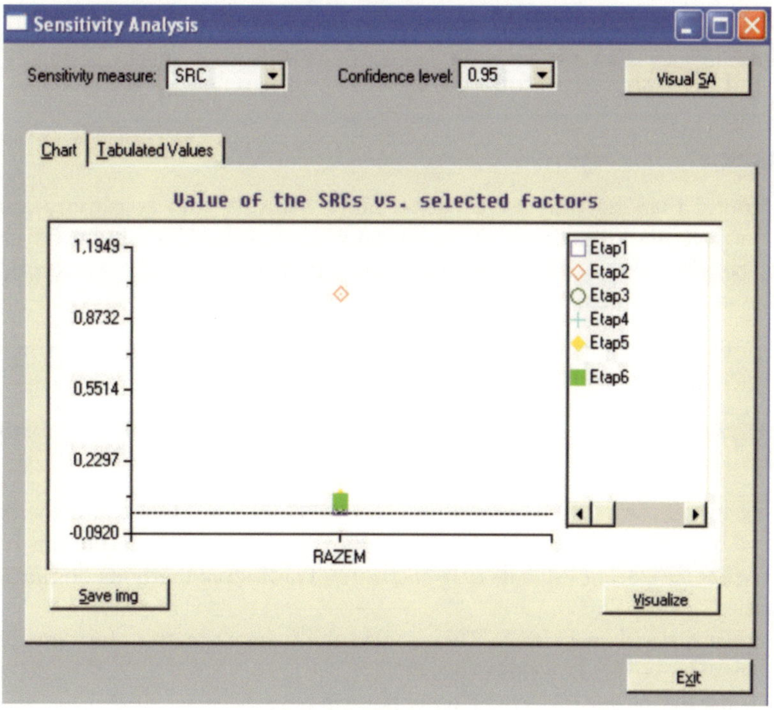

Fig. 16. Sensitivity analysis main Panel (SRC) for the 95% confidential level.

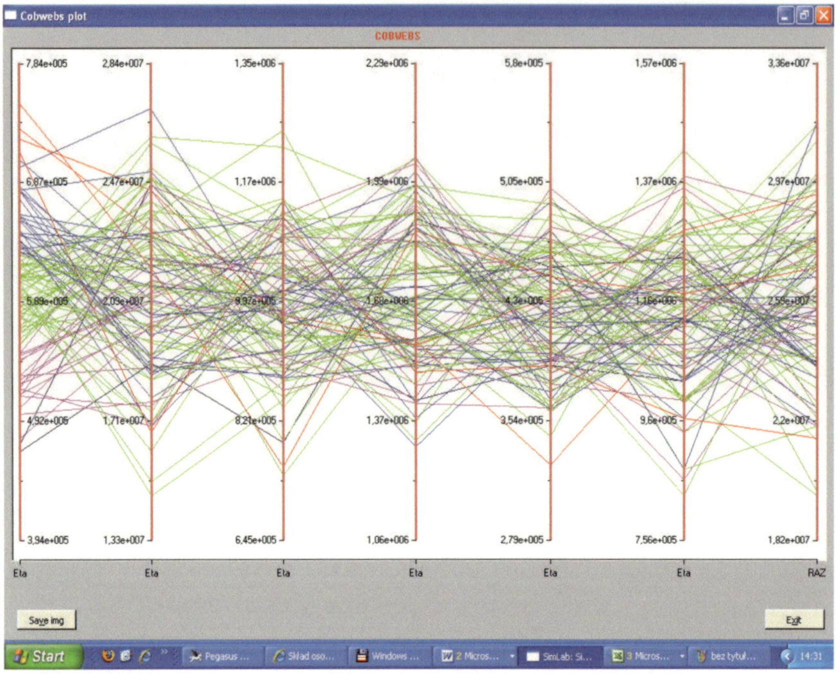

Fig. 17. Sensitivity analysis (*Cobwebs plot*) (SCR - *Spearman Rank Correlation*). (SRC) for the 95% confidential level.

When the 10,000 trials are completed, the histograms provide by SimLab®, given in Figure 9 through Figure 14, present *statistics summary.* The "*Mean*", "*Variance*", "*Standard deviation*", "*Skewness*" and "*Kurtosis*" values form the basis of starting points for the analysis.

6. Conclusions

This study found that the purpose of uncertainty analysis, and sensitivity analysis is to determine the potential directions for waste management decision support systems under uncertainty, because this technique accounts for uncertainties in the assumptions, and to introduce the sensitivity analysis

Because all of the parameters of the economic model are independent, the using of the SRC is shown to be extremely satisfactory.

Cost risk analysis can answer some questions that the traditional estimating method cannot. Included are:

- "*What is the most likely cost*?" The traditional method assumes that this is the baseline cost computed by summing the estimates of cost for the project elements, but this is not so.
- "*How likely is the baseline estimate to be overrun?*" Traditional methods do not address this problem.
- "*What is the cost risk exposure?*" This is also the answer to the question; "How much contingency do we need on this project?"
- "*Where is the risk in this project*?" This is the same as: "*Which cost elements cause the most need for the contingency*?" Risk analysis principles can be used to answer this question.

Uncertainty reduction in the project is performed during the planning phase of the project using the software package SimLab® for project risk management.

In summary, integrating risk analysis into waste to energy pyrolysis facility project management processes may be useful for the project managers. In this study the most likely Total Project Cost values are about 2.53563E+007 USD and 2.663226E+007 USD for the analyzed Scenario. Every manager has a different degree of aversion to risk.

7. Acknowledgments

This research scientific is granted by science financial support for 2011-2013.

8. References

Anderson TS. (2005). Diablo Canyon Power Plant: Steam generator repair/replacement cost/benefit analysis. *Proceedings of the 2005 Crystal Ball User Conference,* Denver, Colorado, May 2005. Date of access 25 November 2011, Available from <http://www.anarisco.com.br/gerenciador/uploads/energia_a2.pdf>

Astrup T., Bilitewski B. (2010). Pyrolysis and Gasification, In: *Solid Waste Technology & Management, Chapter 8.8,* Christensen, pp. 502-512, John Wiley & Sons, Ltd, ISBN 978-1-405-17517-3, Chichester

Azadi M, Saen RF. (2011). Developing a WPF-CCR model for selecting suppliers in the presence of stochastic data., *Operational Research Society,* Vol. 24, Nr. 1, pp. 31-48, ISSN 0953-5543 (print)

Bieda B., Tadeusiewicz R. (2008). Decision support systems based on the Life Cycle Inventory (LCI) for Municipal Solid Waste (MSW) Management under Uncertainty. *International Transactions in Operational Research.* Vol. 15, Nr 1, January 2008, pp. 103-119

Bieda B. (2011). *Metoda Monte Carlo w Ocenie Niepewności w Stochastycznej Analizie Procesów Wytwórczych i Ekologii.* Wydawnictwo Naukowe AGH, ISBN 978-83-7464-344-3, Kraków, Poland (Polish)

Bieda B. (2010). Decision Support Systems Based on the Economic Feasibility Assessment for Municipal Solid Waste (MSW) Management Under Uncertainty Using Simlab® Toolpack. *Proceedings of the Sixth International Conference on Sensitivity Analysis of Model Output, SAMO,* ISSN 1877-0428, Mediolan, October 2010

Bieda B. (2007). Assessing the Economic Feasibility of the Waste to Energy Facility Using Crystal Ball. *Proceedings of the 2007 Crystal Ball User Conference.*, May 2007, Denver, Colorado, USA. Date of Access 25 November 2011, Available from < http://www.anarisco.com.br/gerenciador/uploads/telecomunicacoes_1a.pdf>

Bieda B. (2005). The role of thermal treatment in an integrated waste management, In: *Waste Recycling,* Wzorek, Kulczycka, Fečko, Kušnierowa, pp. 104-113, ISBN 83-89174-83-9, Krakow

Chapman C, Ward S. (2004). Why risk efficiency is a key aspect of best practice projects. *International Journal of Project Management,* 22(8), pp. 619-631

European Commission (2008). *Guide to Cost-Benefits Analysis of Investment Projects,* Brussels, Directorate General Regional Policy

Evans DJ, Kula E. (2011). Social Discount Rates an Welfare Weight for Public Investment Decisions under Budgetary Restrictions: The Case of Cyprus. *Fiscal Studies,* vol. 32, no. 1, pp. 73-107, ISSN 0143-5671

Evans JR, Olson DL. (1998). *Introduction in Simulation and Risk Analysis*. Prentice Hall, New Jersey, USA.

Everard M. (2004). Hot stuff!-The role of thermal treatment in a sustainable society. *Waste Management World*, July-August, pp. 37-45

Hulett DT. (2004). Integrated Cost and Schedule Project Risk analysis. *Originally published as a part of 2004 PMI Global Congress Proceeding- Prague*. Date of access 25 November 2011, Available from <http://www.projectrisk.com/Welcome/White_Papers/Integrated_Cost_-_Schedule_Risk_Analysis.pdf>

Klein A, Whiting K, Archer E, Schwager J. (2004). Gasification and pyrolysis: what is the current situation for waste management?, *Waste Management World*, September-October, pp. 71-75

Koller G. (1999). Risk Assessment and Decision Making in Business and Industry - A Practical Guide. *CRC Press LLC*, Boca Raton, London, New York, Washington, D. C., USA

Leach P. (2005). Modeling uncertainty in project scheduling. *Proceedings of the 2005 Crystal Ball User Conference*, Denver, Colorado, May 2005. Date of access 25 November 2011, Available from <http://www.aertia.com/docs/crystalball/cbuc05-leach.pdf>

Liberman E. (2003). A Life Cycle Assessment and Economic Analysis of Wind Turbines Using Monte Carlo Simulation. *Master's thesis Air Force Institute of Technology*, Wright-Patterson Air Force Base, Ohio, AFIT/GEE/ENV/03-16

National Energy Technology Laboratory [NETL]. (2005). Gasification Robust Growth Forecast - World Survey Results. In: *A Current Perspective On the Gasification Industry - Robust Growth Forecast*, Date of access 8 November 2011, Available from: <http://www.netl.doe.gov/publications/brochures/pdfs/gasification_brochure.pdf>

Office of Management and Budget (OMB) - The Executive Office of the President. (2003). In: *Circular A-4, Regulatory Analysis*. Date of access 7 November 2011, Available from <http://www.whitehouse.gov/omb/circulars/a004/a-4.html>

Orekres N, Shrader-Frechette K, Belitz K. (1994). Verification, validation, and confirmation of nukerical models in the earth sciences. *Science*, 263, pp. 641-646

Pfeiffer E. (2004). Waste, or valuable product?. *Waste Management World*, September-October 2004, pp. 65-69

PMI Standards Committee. (2008). A guide to the Project Management Body of Knowledge, 4th ed. Newton Square, PA: Project Management Institute

Saltelli A, Tarantola S, Campolongo F, Ratto M. (2004). *Sensitivity Analysis in Practice. A Guide to Assessing Scientific Models*. John Wiley & Sons, Ltd

Simlab (2004). Software package for uncertainty and sensitivity analysis. Joint Research Centre of the European Commission

Smith, R. (2006). Use of Monte Carlo Simulation for Human Exposure Assessment at a Superfund Site. *Risk Analysis*, Vol. 14, Issue 4, pp. 433-439

Sonnemann G, Castells F, Schumacher M. (2004). *Integrated Life-Cycle and Risk Assessment for Industrial Processes*, Lewis publishers, CDD Press Company, Boca Raton, London, New York Washington, D.D

Wajs, W., Bieda, B., Tadeusiewicz, R. (2000). Project Cost Analysis for Niepolomice Municipal Solid Waste using the Monte Carlo Simulation. Int. Conference RISK ANALYSIS 2000, Bologna, Italy. In: *Wessex Institute of Technology Publishers*, pp. 225-234

Zwikael O, Ahn M. (2011). The Effectiveness of Risk Management: An Analysis of Project Risk Planning Across Industries and Countries. *Risk Analysis*, Vol. 31, No. 1, January 2011, pp. 25-37, ISSN 0272-4332

14

Methodology Applied to the Diagnosis and Monitoring of Dikes and Dams

Yannick Fargier[1,2], Cyrille Fauchard[3], Patrice Mériaux[4], Paul Royet[4], Sergio Palma-Lopes[2], Daniel François[1], Philippe Côte[2] and Fréderic Bretar[3]

1. Introduction

The recent and dramatic floods of the last years in Europe (Windstorm Xynthia, February 2010) and United-States (Hurricane Katrina, August 2005) showed the vulnerability of flood defence systems. The first key point for avoiding these dramatic damages and the high cost of a failure and its consequences lies in the conception and construction of the dams and dikes, taking into account the past flooding events. A well-designed dike with the correct height avoids failure and overtopping.

In this chapter, a dike is defined as a flood defence system, in dry condition (no contact with water). The term "levee" is often used, specially in the USA.

Many factors introduce weaknesses in the dike. Most of them are old structures. For instance, some of the French Loire River dikes were built several centuries ago. They may have been rebuilt, modified, heightened several times, with some materials that do not necessarily match the original conception of the structure. In other aspects, trees, roots, burrows or nests could modify the structure of the dike and reduce the mechanical properties.

Particular geological formation and their evolution could also threaten the dike. This is the case in the city of Orléans, France, where levees have collapsed in karstic areas. In urban context, the dikes present many other singularities, such as networks, canalisations, human constructions like houses and walls. Due to all these factors, dikes have to be considered as heterogeneous structures. Considering the social impact of a possible breach, the stretch of hundreds of kilometres and the heterogeneity of the materials, rapid, cost-effective and reliable techniques for surveying the dike must be carried out.

This chapter presents the general approach for assessing earth embankments. The first part briefly presents a synthesis of the French approach related to dike diagnosis. The second part shows the recent improvements in this geophysical area given by current applied researches and international experiments. The third part is dedicated to the airborne LiDAR (Light Detection and Ranging) technology, which provides extremely accurate topographic

[1]*EDF – R&D, Chatou, France*
[2]*Institut Français des sciences et technologies des transports, de l'aménagement et des réseaux, Nantes, France*
[3]*Laboratoire Régional des Ponts et Chaussées de Rouen, CETE NC, France*
[4]*Cemagref d'Aix-En-Provence, France*

data at a high efficient rate. The last part is a presentation of a current research work where the Electrical Resistivity Tomography (ERT) method is implemented and tested on an experimental test dam as well as real ones in order to monitor the effect of internal erosion within the structure.

2. General methodology for dike diagnosis

The management of a dike involves many stakeholders and consists in surveying, maintaining and making a diagnosis (Mériaux & Royet, 2001). The diagnosis should identify the weaknesses of the structure (zoning) and provide the degree of safety. Thus, a general methodology (Fig. 1) has been proposed by (Fauchard & Mériaux, 2007). It concerns the levees running alongside French rivers (Loire, Cher, Isère, Aggly), where the dikes are not in a permanent hydraulic heading. The diagnosis is performed in dry condition. The methodology is based on several tests carried out in the framework of the French National Project "CriTerre" and the ERINOH (Internal Erosion in Hydraulic Earthworks) project. This methodology can be applied to dams, with slight differences during in-situ inspections.

This diagnosis begins with preliminary studies, before performing geophysical surveys. It goes on with geotechnical testing, before concluding on the safety level of the dike.

2.1 Preliminary study

The preliminary study consists in gathering as much information as possible concerning the dike, the near environment and its history (Lino et al., 2000).

a. **The historical research** (Fig. 2) can establish the locations of old repaired breaches, material distribution and the way the dike was built. The study of historical archives gives clues wherefrom the materials were extracted so as to build and repair the dike.
b. The **geological study** (map and in-situ observations) of the near area gives information about materials potentially used for building the dike and on the underlying substratum.
c. The **topography** of the dike contains valuable information. From the longitudinal profile of the crest, we can assess the risk of overtopping during a flood by comparing it with the highest past flood. A map of the transverse profile is also required for stability studies and risks of piping, as well as for an accurate location of any structure (walls, crest water gates, crossing networks…) that can modify the interaction between water and dike in case of flood. Finally, the topography is useful for dike management and maintenance. It provides 3D coordinates for visual inspection, geophysical and geotechnical studies. The topographic map has usually a scale of 1:500 to 1:1000. Longitudinal profiles are performed on the crest every 20 to 25 m and transverse profiles are realized every 50 to 200 m, depending on the context. This is a critical point in the dike study, and it could be time and cost consuming for dike of long extent. In that case, **LiDAR** systems are an interesting alternative surveying technique and provide accurate 3D points along the dike with a high point density (see section 4).
d. The **visual inspection** is performed after the historical research and the topographic work. This phase confirms, completes or invalidates any information previously collected. At least three inspectors are required: one on the crest, and two at the toes dike in the riverside and landside. Any anomaly should be reported on the topographic map.

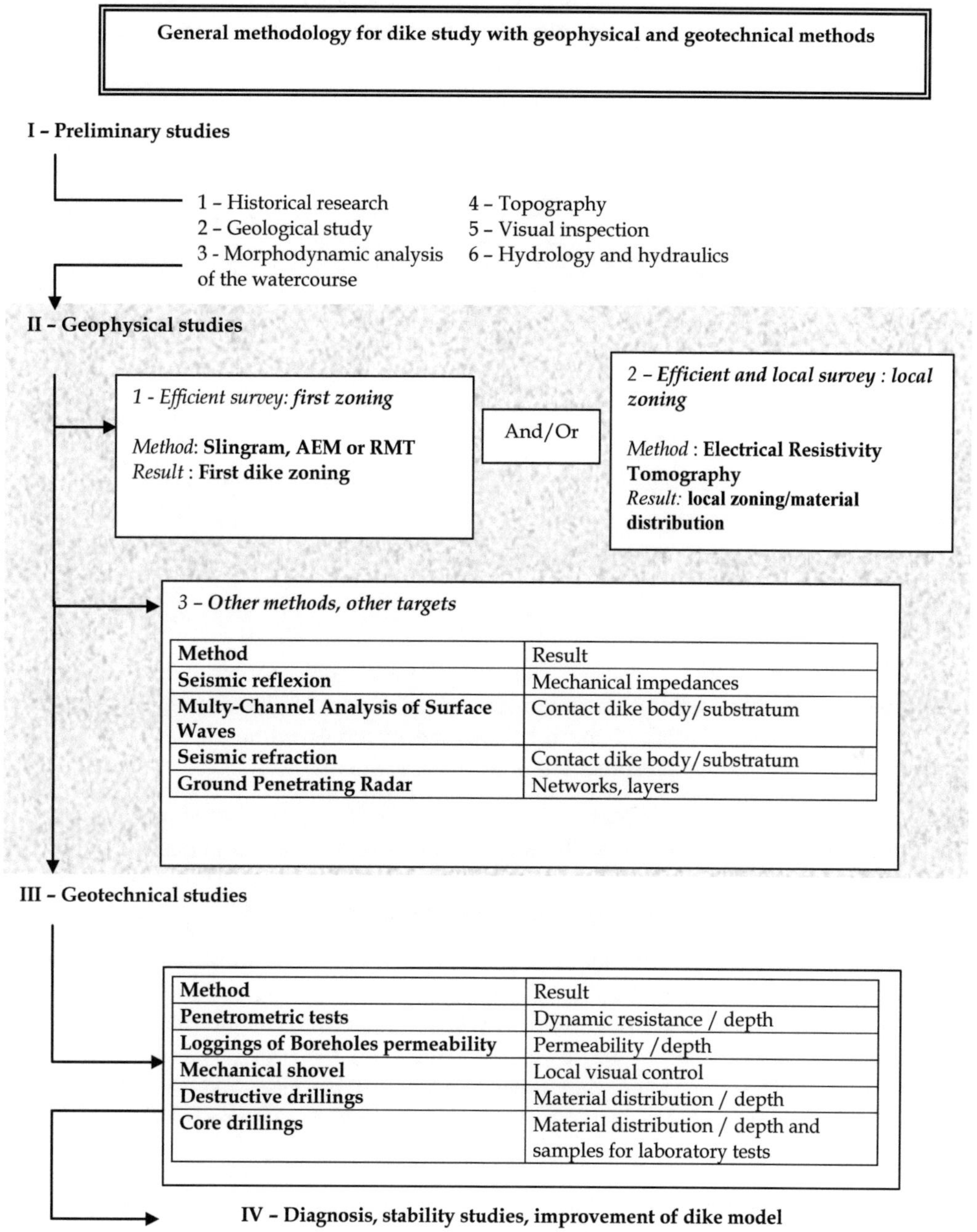

Fig. 1. General methodology proposed in 2007 by the French National Project CriTerre (source Fauchard & Mériaux, 2007)

e. The **morphodynamic study consists in** understanding the sedimentology, the hydrology and the morphometric characteristics of the waterway. It takes into account the temporal evolution of the watercourse channel. For instance, a sandy islet in the bed river modifies the water current: new parts of the dike could be threatened in case of flooding.

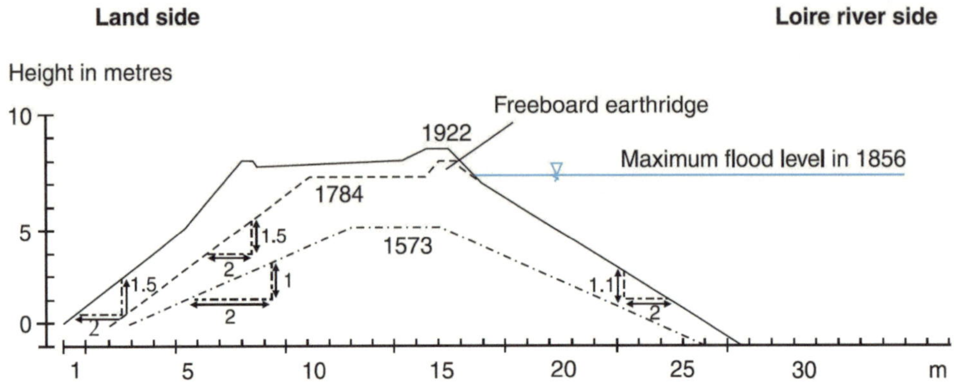

Fig. 2. Example of historical data of the dikes of the Authion river (France, Loire) (Dion, 1961)

2.2 Geophysical studies

2.2.1 Introduction

The geophysical exploration consists in mapping the dike body (nature and distribution of the material- the dike substratum is considered as a part of the dike body). Both the geometry (stretch and height) of the dike and the materials influence the choice of the methodology as well as the interpretation of the measurements.

Considering a typical study where the dike is a long structure of several kilometres, a classical approach (Fig. 3) starts with carrying out a rapid and cost-effective survey. It provides information on the homogeneity of the entire dike body. Then, heterogeneous areas that may weaken the dike body during a flood event are located.

Depending on the geophysical method, a physical parameter is measured according to different profile paths: along the crest (longitudinal profile), across the dike (transverse profile), at the toes of the dike (longitudinal profiles at the river side and the land side). The results of a geophysical survey must be **correlated** with the previous studies. This first survey helps to focus on interesting areas, which can be measured with appropriate geophysical or/and geotechnical methods. This global methodology is presented in Fig. 1.

Generally stakeholders are in charge of managing permanent critical structures like dams, multifunctional dikes in urban area or dikes in heading conditions. In this case, and regarding the potential damages that a breach could generate, the choice of a more efficient, but more time and cost consuming method may be considered. Indeed, the slightest breach during a flood event, either in urban or rural areas, leads to dramatic damages. It induces costs generally higher than the diagnosis does. As a result, some **stakeholders prefer a detailed zoning whatever the stretch length of the dike** (Fig. 3).

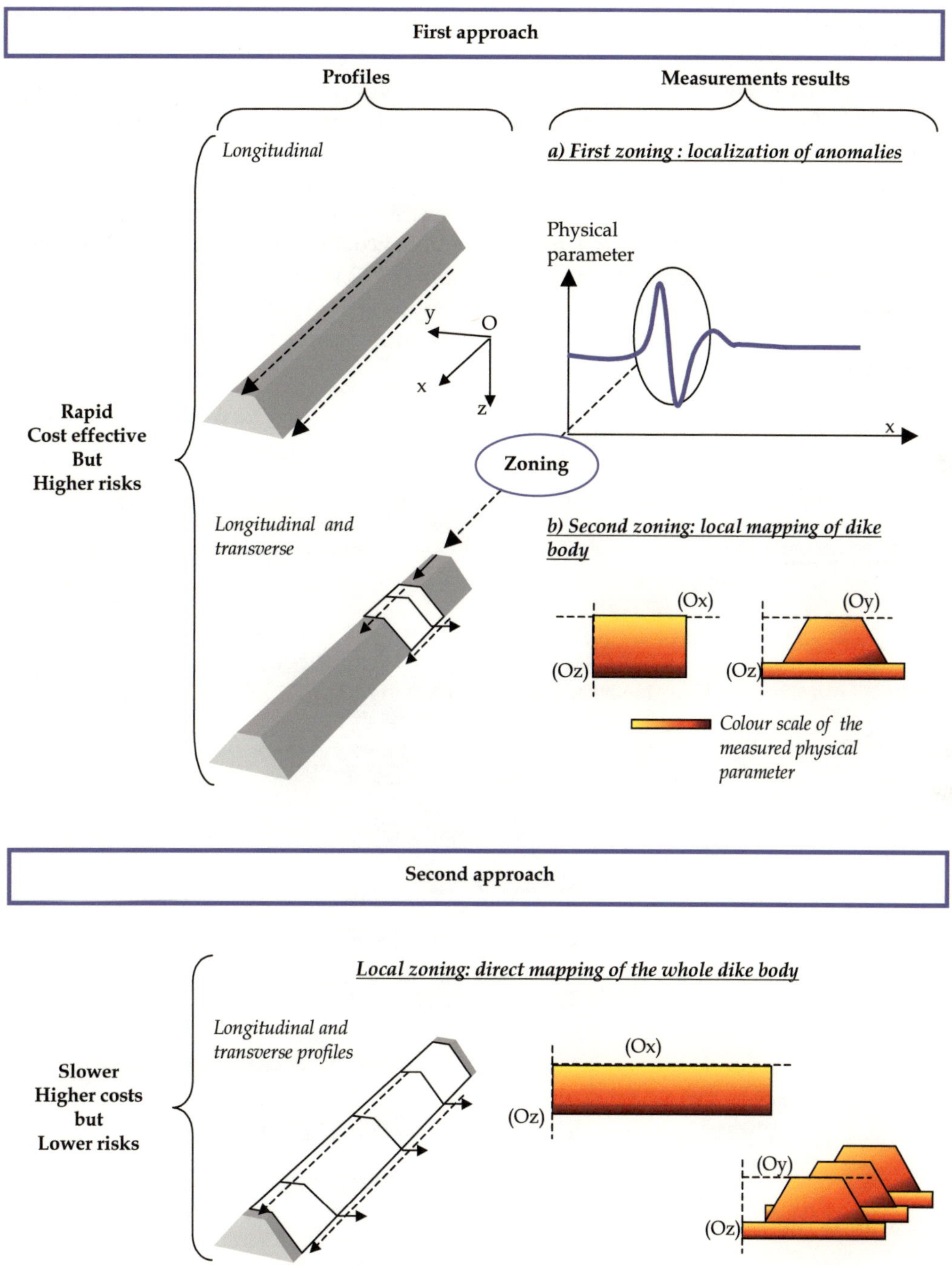

Fig. 3. Two approaches for geophysical survey on dikes

2.2.2 First zoning with geophysical methods

As discussed earlier, depending on the dike characteristics two different approaches are currently carried out. The choice is more dependent on the available time and allocated means than on the dike length.

The first approach (Fig. 3 top) consists in measuring a **physical parameter** related to the type of material of the dike body. The **apparent resistivity** (or its inverse, the conductivity) is a common physical parameter measured for this purpose. The resistivity describes how materials resist to (or conduct) electricity. It strongly depends on the nature of the studied material, its water and clay contents. Other parameters like tortuosity or water salinity of soils are also of importance. The resistivity values of encountered materials in dikes spread in a large scale: few Ω.m (ohm meter) in clays, from few Ω.m to few hundreds Ω.m for silty soils and from few hundreds Ω.m to several thousands Ω.m in sand, gravels and limestone. Fig. 4 shows the range of resistivity values of the main materials encountered in applied geophysics.

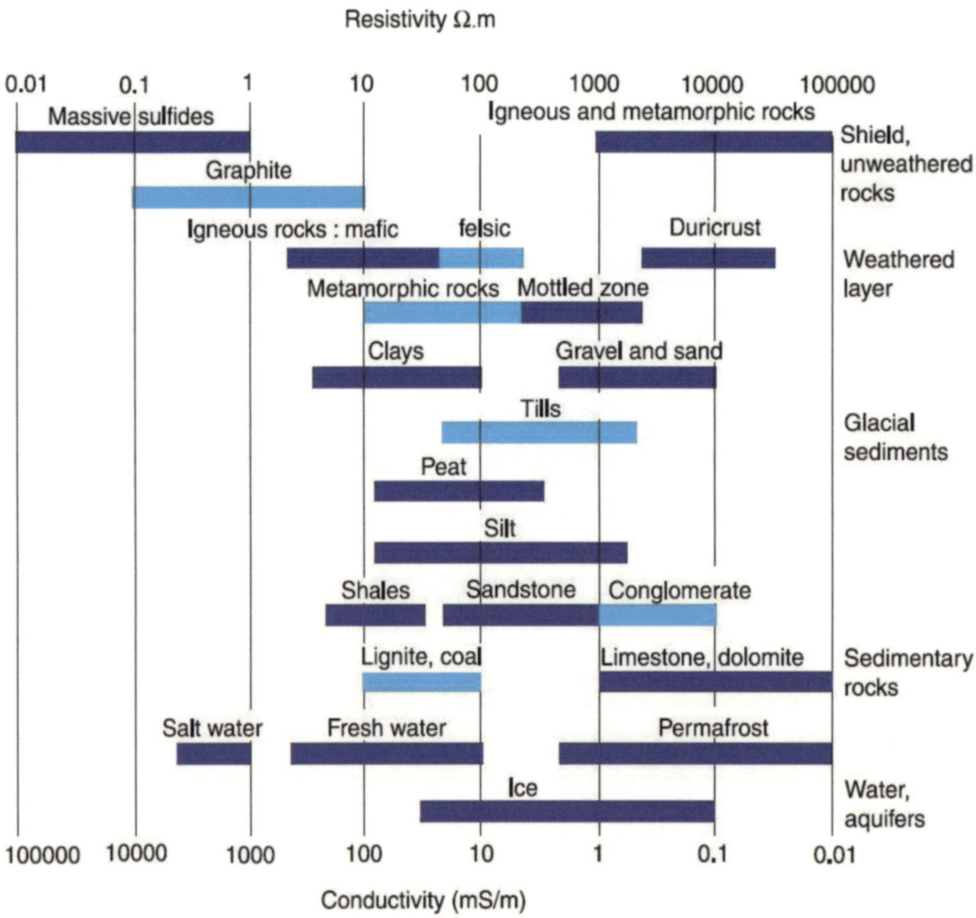

Fig. 4. Resistivity (and its inverse, conductivity) of the main earth materials (Palacky, 1991)

A rapid and cost effective technique is the electromagnetic method of **Slingram** (Mc Neill, 1980). It works in low frequency domain and measures the apparent conductivity of the dike body. The depth of penetration of such a method can reach 50 m, but common devices are designed to reach a classical depth of 10 m (mean height of dikes). A car can track the device if the dike crest has a pavement structure. The measurements are represented as a curve showing the variations of the conductivity (or resistivity) with regard to the distance. A significant variation compared to an average value is interpreted as a potential variation of dike body properties: it defines the area for a more detailed survey.

Nevertheless, these methods are highly sensitive to metallic environment, and their applications are still difficult in urban areas.

Airborne Electromagnetic Method (AEM) belongs to the Slingram family methods: the survey is performed from an airborne platform. It has been widely used for the levee in the region of New Orleans (Dunbar, 2003). Other high output methods were also carried out in the French context of the Loire Rivers. The **Radio MagnetoTelluric method,** identical to the Very Low Frequency in resistivity mode, but uses higher frequencies, has been designed for first zoning with good performances in the re-localization of repaired breaches. Today, this method has been discarded by most geophysicists mainly because its technical aspect is getting older, and because of the poor quality of incoming waves.

Another popular technique is the **Spectral Analysis of Surface Waves (SASW)** initially developed for marine seismic exploration (Gimble sensors). It is implemented for studying the contact surface between the dike body and its substratum. It evaluates the shear modulus. This method was described and implemented on the Loire levees (Samyn et al., 2009) for the detection of sinkhole in karstic substratum in the region of Orléans, France.

Ground Penetrating Radar is sometimes used on particular studies - on paved dikes for instance, or on very resistive dike bodies. This electromagnetic method is based on the radiation of electromagnetic waves in time domain and the reception of waves reflecting on dielectric contrasts encountered in the soil. Most of dike bodies absorb this kind of waves, and the provided information is often useless for a diagnosis.

2.2.3 Local zoning with geophysical methods

The local zoning consists in carrying out directly a more precise geophysical method. It is for instance an internal map (or tomography) of the dike body and of the top of its substratum. The best-suited method for this approach is the **Electrical Resistivity Tomography (ERT).** This approach takes more time and is more expensive than the others. However, it provides more accurate data entailing a better detection of potentially weak areas and therefore a better understanding of the structure stability.

In case of heading conditions (for dams or earthen embankments), the ERT can be applied for mapping internal structure and can be implemented for a time lapse monitoring. But here, the most important is to detect seepages and/or leakages, through or under the dike body. For that purpose **the self potential** is more appropriate. It can be implemented directly on the top of the structure, or could be carried out on river, along the dike. This method also leads to the estimation of the seepage flow through the structure (Bolève et al.,

2007). In that case, ERT provides additional measurements for processing the data. Some temperature probes could also be buried in the dike and the temperature variation could be correlated to the presence of water in the dike body (Radzicki & Bonelli, 2010).

2.3 Geotechnical testing

Geotechnical testing are generally carried out after the first investigations (prior knowledge of historical building and materials, localisation of heterogeneous areas). The final interpretation of geophysical measurements is only relevant when coupled with geotechnical testing. People interpreting the measurements have to decide to extrapolate - or not – the local tests to the rest of the dike.

Geotechnical testing locally provide physical parameters of the dike body that are required for a good diagnosis. A detailed methodology is given in (Lino et al., 2000)

Penetrometric tests are generally the first geotechnical method used to provide information about the soil density (derived from the measured dynamic resistance (in MPa) with regard to depth) and the layer thickness in the dike body. It consists in hammering a conical tip in soil with some characteristics depending on the penetrometric device. The depth of penetration can easily reach 10 m.

Permeability testing (e.g. the Lefranc test) consists in drilling a borehole, injecting and pumping water in an open-ended cavity, called a lantern, at the bottom of the borehole. It measures the variations of hydraulic head and its flow rate and gives the permeability around the lantern. Some devices evaluate both the soil density and the permeability.

Shear tests with phicometer provide the shear strength and the friction angle of soil. It consists in a probe – metal expansion shells - fitted with horizontal annular teeth inserted into the borehole. The shells move only laterally so that the teeth dig the soil. The method needs a good drilling quality with no lining – not the case in highly heterogeneous soils – and is not suited for soft soils.

A local investigation can be carried out with a mechanical shovel, digging a pit in the dike body or at its toe. It provides the distribution of materials.

Mechanical drilling basically provides the advance speed in borehole, and the location of interface layers. In case of destructive drilling, materials are breaking up and transported to the surface (cuttings) using a circulating fluid or an helicoidal cutting tool (auger). If percussion or rotopercussion conducts drilling (for cohesive and rocky soils), the analysis of cuttings can be difficult, but more information is provided by registered parameters like advance speed, tool pressure, circulation fluid pressure... The auger is applied mostly for loose and poorly cohesive soils and allows to take some material samples for lab-test analysis (water content, Atterberg limit, …). In case of core drilling - non-destructive testing – soil samples are extracted directly from borehole without modifying physical properties of soils. Then the samples can be packed and sent for lab testing. Core drilling is local, more expensive and more time consuming than destructive drilling, but provides very useful information for assessing dike properties.

All these methods require a free access to vehicle in the measuring location (crest and/or toe of the dike).

3. The airborne LiDAR as an efficient tool for topographical survey and detection of surface anomalies on dikes

3.1 Backgound on LiDAR systems

Airborne laser scanning (also called ALS) or LiDAR (Light Detection And Ranging) is an active remote sensing technique that provides georeferenced distance measurements between an airborne platform and the surface. It measures the time-of flight of a short laser pulse once reflected on the Earth surface. Strips of several kilometres, with a high overlapping ratio, provide the surveyed topography. The attitude of the airborne platform is acquired by both a GPS and an inertial measurement system. Distance measurements are then transformed into georeferenced 3D points. A detailed description of the processing chain can be found in (Mallet & Bretar, 2008) and (Shan & Toth, 2009).

The height accuracy (resp. horizontal accuracy), at the top end process, is less than 0.05 m (resp. about 0.40 m) or less depending on the flying conditions as well as on the surveyed topography.

Moreover, such active systems, called multiple echo LiDAR, allow detecting several return signals for a single laser shot. It is particularly relevant in case of vegetation areas since a single LiDAR pulse allows acquiring not only the canopy, but also points inside the vegetation layer and on the ground underneath.

In recent years this technique has been applied over natural landscapes to extract terrain elevation (Kraus & Pfeifer, 1998; Bretar & Chehata, 2010) or to classify land cover (Antonarakis et al., 2008; Yoon et al., 2008, Bretar et al, 2009).

In the particular case of dike monitoring, we need a high flexibility in the flight planning in terms of altitude (100-300 m) and heading, and also a high accuracy because dikes are civil engineering structures with a relative low height (less than 7 m) and with a lot of small surface singularities. As a result, it is advised to use a corridor mapping system like FLI-MAP (Fast Laser Imaging and Mapping Airborne Platform) developed by Fugro-Geoid (Gomes Pereira & Wicherson, 1999).

Embedded in an helicopter, FLI-MAP can provide†, over a 105 m wide corridor at a fly height of 150 m, a point density of 80 pts/m2, with an absolute height accuracy (Z) of 0.03 m. The Pulse Frequency Rate (PRF) of the latest version can reach 250 kHz with a field of view of 60 ° in the cross track direction. The survey is done following three scan plans in the flight direction (vertical for 50% of the points, front 7 ° and rear 7 ° for 2 x 25%), which reduce the effects of shadows.

The trajectory of the helicopter is recorded by two dual frequency GPS and an inertial measurement unit. A digital camera in nadiral position, synchronised with the LiDAR system, records the surveyed landscape and is used both to build a mosaic of georeferenced images and to colorize in real time the 3D point cloud so that a user should have a better understanding of the scene (Fig. 5). The system also includes two frontal and oblique cameras (photo and video). These data are particularly popular for dike managers who use

† Example based on a recent application of the FLI-MAP technique on the Loire levees near Orléans, in the context of the FloodProBE European research project.

the images for later processing and for marketing/communication actions towards the public or financial sponsors.

3.2 Surveying a dike with LiDAR data

A LiDAR system is able to acquire data on a dike structure of up to 80 km per day, which makes the use of this technique valuable in case of emergency situations (after a major flood, for example). Provided that it exceeds a length of up to 60 - 80 km (corresponding to a day of helicopter), the costs are competitive with regard to conventional field topographic techniques (in the order 2000 euros / km) and provide additional valuable products like precious information on dike slopes and crest or their near environment (river banks, etc.). The high-resolution digital images allow to measure with accuracy visible objects. Figure 6 shows the identification of a pump line through the base of a dike.

Fig. 5. Colored 3D point cloud over a dike (source FloodProBE - FUGRO)

Moreover, in case of vegetation, LiDAR data makes possible to study invisible structures from images. Fig. 7 illustrates the way the erosion of riverbanks under vegetation can be quantitatively analysed with laser profiles.

The field visit (Fig. 8) confirmed this erosion process. The possibility of studying the vegetation is also of high importance: the development of woody vegetation near or onto the dike is a major risk factor (Mériaux et al., 2006).

Surface singularities are often signs of disorder or suspected disorder in the dike itself: for example a subsidence or a sinkhole on a ridge may result from internal erosion or karst collapse. Such singularities, once pre-identified on the images are, of course, to be confirmed by field visits, but the contribution of high resolution LiDAR data is to improve the completeness of these visual clues (Clement & Mériaux, 2007). Geophysical survey or geotechnical testing will then characterize possible extension of surface singularities in the dike body or in the foundation.

Fig. 6. 0.02 m resolution digital image acquired during the LiDAR survey. It shows a pump line through the base of a dike (source Cemagref -FUGRO)

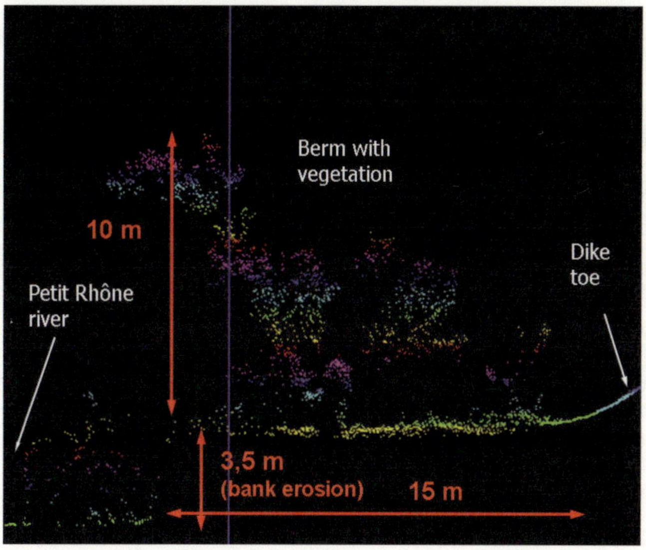

Fig. 7. Laser profile of a bank under the vegetation (source Cemagref-FUGRO)

Finally, high resolution topographic data contribute to build specific geomechanical model of the dike that, after incorporating data provided by geophysical and geotechnical surveys, are integrated in the calculations of the structure stability. The quality of the geomechanical model also depends on the accurate location of in situ geophysical and geotechnical surveys so that one should interpret the results with relevance. In this regard, a decimeter resolution DTM acquired with a LiDAR system or derived topographic plans at 1: 100 is of high interest for people in charge of operating field measurements.

Fig. 8. Bank erosion. The red arrow represents the 3.5 m elevation gap seen on Fig. 7 (source Cemagref)

4. Investigation and monitoring of dikes and dams with Electrical Resistivity Tomography

4.1 Quest for complementarity

Internal erosion processes and overtopping phenomenon represent more than 90% of dike failure (Foster et al., 2000; Fell & Fry, 2007). This section focuses on internal erosion processes which are more complex, and above all, should be detected by geophysical methods before the rupture of the earthwork.

Among them, the **DC-Electrical Resistivity Tomography (ERT)** is of particular interest (Johansson, 1997) for dike monitoring in heading condition. This technique is considered highly sensitive to the induced physical phenomenon such as changes in clay or water content, temperature and porosity. Fig. 9 presents the main interactions regarding the effects of internal erosion on electrical resistivity.

The main purpose of ERT campaigns is an insight of the subsurface via 1D, 2D, or 3D representations of the spatial and/or temporal variations of the electrical resistivity. One of the advantages of the method is its double resolution capacity:

- Low resolution imaging for high outputs with fast zoning techniques,
- High resolution imaging for selected short stretch

This double resolution capacity can be exploited in two ways:

- Instant survey for imaging the apparent resistivity distribution of the observed medium,
- Temporal monitoring to follow the evolution of the electrical resistivity of the earthwork.

For cost effectiveness purposes, in the case of dike survey, ERT is usually applied in a "classical" way (2D): a set of equidistant electrodes is aligned along the longitudinal direction on the dike crest, slope, or toe. Whereas the geo-electrical behaviour of dikes evolves in 3D, recorded and processed data are based on a two dimensional measurements and interpretation - 2D inversion software like Res2dinv® (Loke & Barker, 1996).

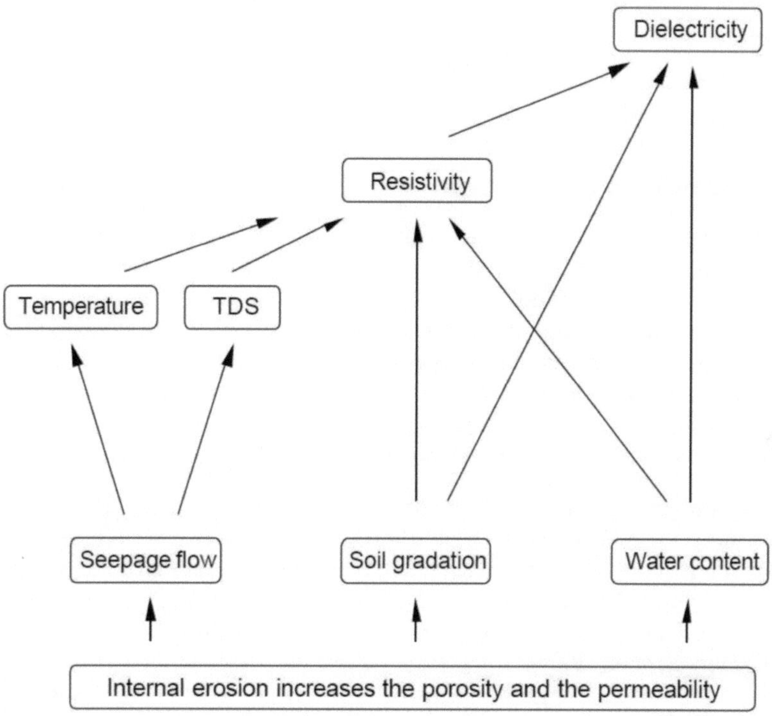

Fig. 9. Major influences of internal erosion processes on selected parameters (Johansson, 1997).

The principle of an inversion process is to find a model that best explains measurements obtained on the field plus other constraints. Consequently, in the case of a 2D inversion realized on a 3D medium the 2D inversion process inevitably leads to 3D artefacts. They are mainly due to the topographic effects, the siltation of the reservoir, the water reservoir effect, and the clay core effect.

However, external information collected from preliminary studies Fig. 1 can help to reduce these effects as presented in Table 1. Indeed, the location or depth of the anomaly can be given by a visual inspection or a morphodynamic study of the river to focus ERT acquisitions. The knowledge of the geomaterial of the dike and/or the foundation can be supply by a geological study to constrain the inversion. Then, the depth of the foundation or the thickness of a repaired breach can be available after historical investigations and can also constrain the inversion. Finally, the topography of the dike can be available (e.g. from LiDAR data or in situ measurements).

Here, we aim to illustrate pitfalls and misinterpretations of 2D-ERT inversion, before presenting methodological improvements without acquiring full 3D data.

	Explicit constrain on the model	Explicit constrain during the inversion
Topography	geometry	-
Geological study	substratum depth	material variation
Hydrogeological study	water table depth	roughness
visual inspection	focused measurement	-
Historical research	internal composition	smoothness

Table 1. External information gathered form preliminary studies.

4.2 Background on inversion artefacts

In case of ill-posed problem, the inversion process can become unstable and generate artefacts. Moreover, it is well known that the potential of inversion methods depends on the amount of information contained in the data (Tarantola, 1982). Therefore, the concepts of model resolution matrix or Region Of Investigation (Oldenburg & Li, 1999; Marescot et al., 2006) were introduced to assess the robustness of geophysical imaging methods. Neglecting 3D effects decreases the model resolution matrix or values of the region of investigation. This matrix represents a direct link between measurements and robustness of the result. Thus, this information can be used to enhance the method by:

- Allowing **image appraisal** (Stummer et al., 2004; Oldenburg & Li, 1999) by compensating the loss of spatial resolution and the topographic effects. This technique allows a quality control of the image and a better interpretation of the result;
- Creating optimized sequences of measurements to increase the quality of the information contained in the data (Tsourlos et al., 1999; Stummer et al., 2004; Sjödahl et al., 2006; Hennig et al., 2005);
- Finding an optimized design of electrode location to focus the survey and increase the reliability of the inversion result (Fargier et al., 2010).

4.3 Normalisation technique

The **normalisation technique** makes use of the original definition of apparent resistivity (Kunetz, 1966; Marescot et al., 2006). It indicates that the topography effects can be normalized and partly accounted during the inversion. This first technique is based on the definition of a general geometrical factor which is a generalisation of the conventional geometrical factor. This method requires an approximate knowledge of the topography (digital terrain model) as well as the resistivity of the media. As a result, this normalisation partly decreases 3D effects. However, in theory, the diffusion of an electrical field is a highly non-linear problem and the normalisation technique cannot completely take into account this non-linearity. This limitation can cause the formation of artefacts and can lead to misinterpretation.

4.4 2D$^+$ inversion strategy

Contrary to relevant works of Fox et al. (1980), Tong & Yang (1990) showed that previous normalisation techniques cannot completely take into account all non-linear effects. They showed also that the best way to remove non linear effects is to integrate explicitly the topography in the model. This principle is not only dedicated to the topography but can be extended to all finite media inside the observed domain.

Therefore, it is necessary to develop inversion methods capable of taking this non-linearity into account. To limit the financial cost of the acquisition and the computational cost of inversion, new inversion codes specifically dedicated to the dike and dam context have been developed (Fargier et al., 2011). The code InGEOTH-2D+ proposes a 2D inversion that integrates part of the full 3D geo-electrical behaviour of a dam (topography and water reservoir are included). The purpose of this code is twofold. The first purpose is to provide new discretization capabilities to better state the problem. The second purpose is to allow the inclusion of any explicit prior information that the geophysicist provides.

4.5 Results

To test the relevance of the presented techniques a measurement campaign has been carried out at the crest of a dam. An historical research, a topographic survey, a geological study, and a visual inspection were realized before the geo-electrical survey.

A dense Wenner-Schlumberger protocol was used because of its spatial resolution and robustness. Fig. 10 a) shows one electrical resistivity section obtained after inversion of the raw data without any external information. Fig. 10 b) represents the same section after normalization of the water reservoir effect and the topography. Fig. 10 c) shows the final result of the inversion obtained with InGEOHT - 2D+. Fig. 10 d) illustrates the inverse model used for the inversion shown in Fig 10 c). For all three results, and after four iterations, the convergence data criterion is less than 1%.

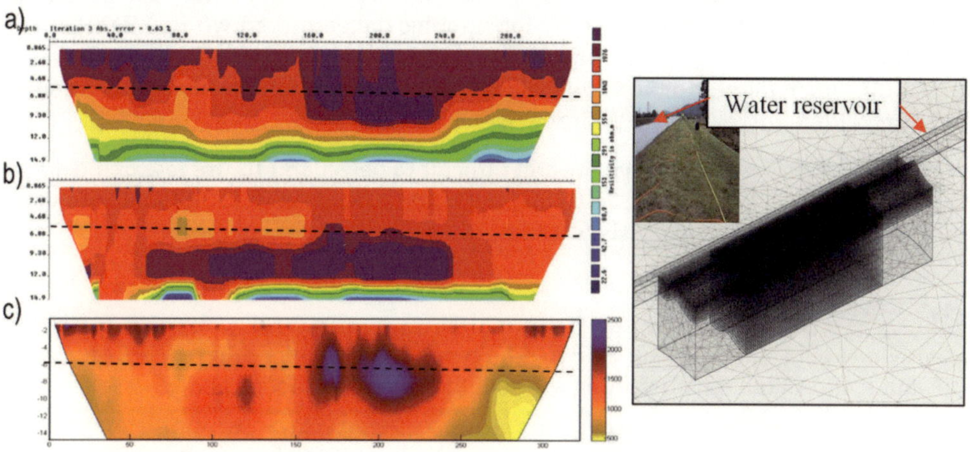

Fig. 10. Results of the the inversion process obtained a) without any correction procedure (Res2dinv®), b) with normalization of water reservoir effect and topography effect (Res2dinv®), c) with the InGEOHT - 2D+ inversion code. d) presents a view of the measurement campaign and The 2D+ inversion model used to inverse the result.

A first interpretation of the inverted section in Fig. 10 a) indicates that the medium is quite regular in the longitudinal direction and composed by two layers. The upper layer whose wall varies between 9 m and 12 m has a resistivity oscillating between 500 Ω.m and 2500 Ω.m. The resistivity of the lower layer decreases to 40 Ω.m. In Fig. 10 b) the electrical resistivity of the water reservoir was integrated in the inversion process (81 Ω.m). The effect of the normalization indicates that the influence of the lower layer decreases. This result suggests that the lower layer is in fact an artefact due to the presence of the water reservoir.

Figure 10.c), after applying InGEOHT - 2D+, shows that the effects of the water reservoir has been entirely removed (the lower conductive layer disappeared). The resistivity section is smoother than the two previous calculations except for three resistive anomalies between 9m and 12m depth. In conclusion, we think that this last result provide a better insight of the true behaviour of the dike and the detection of some suspicious zones will be further investigated by high resolution geophysical and geotechnical methods to validate this result.

5. Conclusion

In many countries, the regulation of hydraulic structures has recently been enhanced and is declined in classes depending on the issues in case of breakage. It stresses the need to know the level of safety of the structures, through in-depth diagnosis and analysis of risks, and to strengthen the surveillance. In France, these regulations now apply to 700 large dams, tens of thousands of small dams, and about 10 000 km of linear dikes.

We have presented in this chapter on going improvements in the global methodology of dike diagnosis in Europe. The stakeholders involved in the management of dikes must continue to integrate these improvements in their practices so that an efficient diagnosis should be drawn over time for a sustainable maintenance of the earthworks and dams.

Nowadays, the ERT method becomes the reference one for dike and dam geophysical investigations. Coupled with accurate 3D topographic data acquired with a LiDAR system, the 3D effects should be better integrated when interpreting the data. Those improvements will be all the more interesting for stakeholders (e.g. multi temporal analysis of long stretch dikes) as repetitive survey will be performed.

Phenomena like leakage or seepage are still difficult to detect and future research works on streaming potential (Bolève et al., 2007) and optic fibres (Khan et al., 2010) methods should supplement the available tools of stakeholders.

6. Acknowledgments

The authors would like to thank EDF-R&D, French Ministry of Ecology (DREAL Centre), SNCF and Fugro-Geoid for supporting a part of the works mentioned above.

This work has been partly supported by the European Community's Seventh.

Framework Programme through the grant to the budget of the FloodProBE project, and by EDF-R&D through the grant of the PAREOT project and the INTREPHYD project.

7. References

Antonarakis, A. S., Richards, K. S. & Brasington, J. (2008). Object-based land cover classification using airborne LiDAR, *Remote Sensing of Environment* Vol. 112: 2988–2998

Bolève, A., Revil, A., Janod, F., Mattiuzzo, J. L. & Jardani, A. (2007). A new formulation to compute self-potential signals associated with ground water flow, *Hydrology and Earth System Sciences* Vol. 11: 1–11

Bretar, F. & Chehata, N. (2010). Terrain modelling from lidar range data in natural landscapes : a predictive and bayesian framework, *IEEE Transactions on Geoscience and Remote Sensing* Vol. 48: 1568–1578.

Bretar, F., Chauve, A., Bailly, J. S., Mallet, C. & Jacome, A. (2009). Terrain surfaces and 3D landcover classification from small footprint full-waveform lidar data : Application to badlands, *Hydrology and Earth System Sciences* Vol. 13(No. 8): 1531–1544.

Clément, A & Mériaux, P. (2007). Apports de la télédétection LiDAR héliportée pour la reconnaissance des digues françaises : premiers résultats, *Revue Française de Photogrammétrie et Télédétection* Vol. 186: 87–91

Gomes Pereira, L.M. & Wicherson, R.J. (1999). Suitability of laser data for deriving geographical information : a case study in the context of management of fluvial zones, *Journal of Photogrammetry and Remote Sensing* Vol. 54: 105–114.

Dion, R. (1961). *Histoire des levées de la Loire*, Flammarion (ed.), Paris.

Dunbar, J. B., Ballard, R. F., Murphy, W. L., McGill, T. E., Llopis, J. L., Peyman-Dove, L. D. & Bishop, M. J. (2003). *Condition assessment of the U.S. International Boundary and Water Commission, Lower Rio Grande Valley levees, south Texas*. Vol I, Main Text and Appendices. ERDC TR-03-4. Vicksburg, MS: U.S. Army Engineer Research and Development Center.

Fargier, Y., Palma Lopes, S., François, D., Fauchard, C., & Côte, P. (2010). Optimization of 3D DC Electrical Resistivity Imaging (ERI) layouts for water infiltration monitoring, *Near Surface 2010 – 16th European Meeting of Environmental and Engineering Geophysics*, Zurich, Switzerland.

Fargier, Y., Palma Lopes, S., François, D., Fauchard, C., & Côte, P. (2011). DC-electrical resistivity tomography for earthen embankment dike survey : improvement strategies in between 2D and 3D approaches, *Annual Meeting of the European Working Group on Internal Erosion in embankment dams and their foundations (IEWG-ICOLD)*, Brno, Czech Republic

Fauchard, C. & Mériaux, P. (2007). *Geophysical and Geotechnical Methods for Diagnosing Flood Protection Dikes - Guide for Implementation and Interpretation*, Edition Quae, France, ISBN 978-2-7592-0035-1

Fell, R., and J.-J. Fry. (2007). *Internal erosion of dams and their foundations*. Taylor & Francis, London.

Foster, M., Fell, R., & Spannagle, M. (2000). *The statistics of embankment dam failures and accidents, Can. Geotech. J.* Vol. 37(No. 5): 1000–1024.

Fox, R., Hohmann, G., Killpack, T., & Rijo, L. (1980). Topographic effects in resistivity and induced polarization surveys, *Geophysics* Vol. 45: 75–93.

Hennig, T., Weller, A., & Canh, T. (2005). The effect of dike geometry on different resistivity configurations, *Journal of Applied Geophysics* Vol. 57: 278–292

Johansson, S. (1997). *Seepage monitoring in embankment dams*, PhD thesis, Royal Institute of Technology, Stockholm.

Khan, A., Vrabie, V., Mars, J., Girard, A., & D'Urso, G. (2010). Automatic monitoring system for singularities detection in dikes by DTS data measurement, *IEEE Trans. On Instrumentation and Measurement* Vol. 59(8): 2167–2175

Kraus, K. & Pfeifer, N. (1998). Determination of terrain models in wooded areas with airborne laser scanner data, *ISPRS Journal of Photogrammetry and Remote Sensing* Vol. 53: 193–203.

Kunetz, G. (1966). *Principles of direct current resistivity prospecting*, Geoexploration monographs 1.

Lino, M., Mériaux P., Royet P. (2000). *Méthodologie de diagnostic des digues appliquée aux levées de Loire moyenne*, Cemagref éditions, Paris.

Loke, M.H. & Barker, R.D. (1996). *Rapid least-squares inversion of apparent resistivity pseudosections using a quasi-Newton method, Geophysical Prospecting* Vol. 44: 211–226.

Mallet, C. & Bretar, F. (2008). Full-waveform topographic LiDAR: State-of-the-art. *ISPRS Journal of Photogrammetry and Remote Sensing*, Vol. 64: 1-16.

Marescot, L., Rigobert, S., Palma Lopes, S., Lagabrielle, R., & Chapellier, D. (2006). A general approach for DC apparent resistivity evaluation on arbitrarily shaped 3D structures, *Journal of Applied Geophysics* Vol. 60: 55–67.

McNeill, J. D. (1980). *Electromagnetic terrain conductivity measurement at low induction numbers, Technical Note TN-6*, Geonics, Ltd., Mississauga, Ontario, Canada.

Mériaux, P. & Royet, P. C. (2001). *Surveillance, Maintenance and Diagnosis of Flood Protection Dikes - A Practical Handbook for Owners and Operators*, ISBN 978-2-7592-0037-5, Cemagref Editions

Mériaux, P., Vennetier, M., Aigouy, S , Hoonakker, M. & Zylberblat, M. (2006). Diagnosis and management of plant growth on embankment dams and dykes, *ICOLD, 22nd Conference on Large Dams*, Barcelone, Spain, pp. 551–567

Oldenburg, D. W. & Li, Y. G. (1999). Estimating depth of investigation in dc resistivity and IP Surveys, *Geophysics* Vol. 64(No. 2): 403–416.

Palacky, G.J. (1991). Resistivity characteristics of geological targets, *in* M. N. Nabighian (ed.), *Electromagnetic Methods in Apllied Geophysics Vol. 1*, Soc. Explor. Geophys., pp. 53–129

Park, S. K. & Van, G.P. (1991). Inversion of pole-pole data for 3-D resistivity structure beneath arrays of electrodes, *Geophysics* Vol. 56: 951–960.

Radzicki, K., Bonelli, S. (2010). Thermal Seepage Monitoring in the Earth Dams with Impulse Response Function Analysis Model, *8th ICOLD European Club*, Innsbruck, Austria

Samyn, K., Bitri A., Blondel, A. & Lebert, F. (2009). *Détermination de la présence de karst sous les levées domaniales du bassin de la Loire et réduction du risque d'effondrement de la digue lié à la présence de ces conduits souterrains naturels*. BRGM/RP-57650-FR.

Shan, J. & Toth, C. K. (Eds) (2009). *Topographic Laser Ranging and Scanning. Principles and Processing*. CRC Press.

Sjödahl P., Torleif, D. & Zhou, B. (2006). 2.5D resistivity modeling of embankment dams to assess influence from geometry and material properties, *Geophysics*, Vol. 71(No. 3):, 107-114.

Stummer P., Maurer, H., & Green, A.G. (2004). Experimental design: electrical resistivity data sets that provide optimum subsurface information, *Geophysics*, Vol. 69(No. 1): 120-139.

Tarantola, A. & Valette, B (1982). Inverse Problems = Quest for Information. *Journal of Geophysics*, Vol. 50: 159-170.

Tong, L.-T. & Yang, C.-H. (1990). Incorporation of topography into two-dimensional resistivity inversion. *Geophysics,* Vol. 43: 1450-1469.

Tsourlos, P.I., Szymanski, J.E., & Tsokas, G.N. (1999). The effect of terrain topography on commonly used resistivity arrays, *Geophysics*, Vol. 64:, 1357-1363

Yoon, J. S., Shin, J. I. & Lee, K. S. (2008). Land cover characteristics of airborne LiDAR intensity data: a case study, IEEE *Geoscience and Remote Sensing Letters,* Vol. 9; 463-466.

15

Transforming Risk Assessment Tools from Paper to Electronic

Daniel Bergeron[1] and Kristiina Hyrkäs[1,2]
[1]*Maine Medical Center, Portland Maine;*
[2]*University of Southern Maine*
USA

1. Introduction

As the use of electronic health records (EHR) continues to grow, there is an increasing need for the development and translation of risk assessment tools for use in electronic media. The federal government of the United States is compensating hospitals that are meeting the usage guidelines of electronic health information systems (Kocher et al. 2010). Evidence suggests that use of electronic health records improves patient safety and outcomes (Radicki & Sittig, 2011). This incentive has dramatically increased interest in EHR and content that is easily translated to an electronic format.

Risk assessment is defined in this chapter as a process of evaluating a potential hazard, likelihood of suffering, or any adverse effects (Mosby, 2008). Assessment of a risk, and subsequent prevention is becoming standard practice in healthcare (Carayon et al. 2006). Hospitals in the United States are being held accountable for preventable iatrogenic events like; pressure ulcers, falls, catheter associated urinary tract infections, and central line associated infections (United States Federal Register, 2011). Use of risk assessment tools is thus necessary to prevent these and other hospital-acquired problems. Our hospital first converted a Fall and Injury Risk Assessment tool from paper to electronic (Chapman et al. 2011). Since that time we have been translating many of our risk assessment tools into the electronic medium. The purpose of this chapter is to describe the processes necessary to transform and implement a risk assessment tool in an electronic environment.

2. The process of transforming a risk assessment tool

The translation of risk assessment tools from paper to electronic is a multi-layered, complex process that aims at maintaining their integrity, scientifically tested properties and feasibility in clinical settings, thus, many factors need to be considered. Planning for this type of change requires an interdisciplinary team of nurses and Information Technology (IT) workers (see Wenzel 2002). The goal of this team is to determine the requirements of a successful transformation and implementation. The steps in preparation for this conversion include the following phases: (1.) Workflow Analysis, (2.) Design and Building an electronic version of the tool, (3.) Testing and Signoff, (4.) Training of staff/users of the tool, (5.) ' Go-Live' and Support, and (6.) Reporting (Courtney et al. 2005). Using electronic applications to deliver these risk assessment tools provide new opportunities for improving data integrity, decision support, and patient safety.

2.1 Workflow analysis

When a change in practice, such as use of an electronic risk assessment, is initiated, a workflow analysis is necessary to ensure that the design of the tool and requirements of the work/practice meet the needs of the clinician. Workflow analysis can be conducted in several different ways and examples are available in the current literature (Karsh, 2009). Most of the time the analysis begins with documentation of the current state of workflow for the process, in other words, a risk assessment conducted by a clinician. From the initial analysis, a 'future state' workflow is developed by the interdisciplinary team. The new workflow is created by combining the original workflow used in risk assessment with the functionality of the EHR system to create future state workflow.

The approaches that are described in the literature (Karsh, 2009) and commonly used in practice for workflow analysis emphasize different types of techniques for evaluating, documenting and re-designing work processes. These methods include: focus groups, interviews and observations.

Focus Groups. The use of focus groups is a helpful and quick method in determining if there is 'a standard' way of executing an assessment and related tasks (e.g. targeted interventions) in practice (AHRQ, 2011). The focus groups in this context mean meeting with several clinicians and staff who are familiar with the assessment tool and who can describe the steps used to complete the current risk assessment tool. During the focus group meetings, the 'moderator' collects information from clinicians and staff regarding the work flow process and attempts to capture the essential and concrete steps to obtaining the assessment. These may include, for example, looking for the patient's chart, assembling associated tools for measurement, bringing the assessment tool to the patient's bedside, and many more. The focus group method is helpful for looking at clinicians' and staff's group behaviors and especially the levels of consensus and agreement regarding the use of the assessment instrument and the steps of the workflow. The 'moderator' captures and summarizes the information most commonly on the flip-chart pages, sticky notes or a note-pad for further use (AHRQ, 2011). The information gained from focus groups provides a broad overview of the workflow with a variety of nuances from clinicians and staff, but it is important to point out that if very detailed information is required, one-on-one interviews allow a deeper exploration of workflow. (AHRQ, 2011)

The workflow analysis can reveal very important and unexpected issues. For example, in a recent conversion of a fall risk assessment tool at our hospital, we found out that the predominant workflow often led to misleading risk assessment scores and ultimately incorrectly targeted interventions. The workflow analysis revealed that the main reason for this inaccuracy was, in fact, caused by 'user variability'. Although the assessment tool being converted from a paper document to an electronic version of the instrument had already been found to be superior in both its specificity and sensitivity (Chapman et al. 2011), we detected that there were still inaccurate assessments. The original assessment tool was a paper document and all tabulations were calculated manually by the nurse. If there were any calculation errors, this led to inaccurate portrayal of patient's fall risk level (i.e. no risk, moderate and high risk). When the risk assessment tool was converted, we made a special effort during this process to make sure that the tabulations were completed automatically in the electronic version of the tool by the health information system. This new function reduced mathematical errors to zero and advanced the proper identification of the level of risk.

Interviews. Literature shows (AHRQ, 2011) that a workflow is not just composed of concrete tasks or a set of sequences, but that it also includes multiple complex mental and cognitive processes (see also Lopez et al. 2010). A workflow analysis based on interviewing can yield more important knowledge of these types of information processing activities. For example, during a risk assessment, nurses look for, and record certain observations and detailed characteristics regarding the assessments. The mental and cognitive processes that occur during this practice may include comparison of the current observation to previous-ones, or to certain standards or reference values. During paper-based risk assessment, mental and cognitive processes (e.g. reasoning) are often necessary and required to reach a conclusion of a certain amount or degree of risk. Analysis of these processes and the content of the related information are important considerations when designing an electronic assessment tool, because these help in identifying processes that may change the representation of the assessment in an electronic medium. During the interview, it is important to ask why a clinician takes some actions over others. Reasoning, calculations and critical thinking are examples of the mental and cognitive processes that are embedded in many risk assessments. A moderator/interviewer can collect the information from the interviews for the work-flow analysis by using a tape recorder or by taking hand-written notes.

Observations. Unlike the previous methods, observation can unveil discrete work processes that may not be readily shared during interviews and focus groups. It is possible that a combination of work and information flows are so complex and multi-layered that these are just simply too difficult to describe by a clinician. An observer focuses on watching the practice and the specific activities of the workflow as they are being performed (AHRQ, 2011). As nurses perform a risk assessment, the observer needs to note, most of all, the required activities (i.e. obtaining appropriate tools for measurement), their sequence (i.e. in the Fall Risk Tool we use, risk for falling precedes assignment of injury risk), and other related tasks (AHRQ, 2011).

Considerations of the future work-flow are important, but sometimes difficult to anticipate beforehand. In one instance, we found that we did not design a future workflow for Fall Injury Risk Assessment that included recommendations for interventions. When revisiting the workflow and using the observation method, we noted that once the assessment was completed at the bedside, the nurse then needed to recommend interventions to prevent falls. In addition, nurses needed to perform recommended interventions and document their completion. An integral part of the process for preventing falls was missed. The result of the translation process was indeed an electronic assessment tool, however the observations revealed that the nurse needed to go to a paper document and note the interventions there. This activity made the workflow inefficient and led to dissatisfaction with the tool. Subsequently, the electronic assessment tool was redesigned to allow nurses to select appropriate interventions and to document their completion.

2.2 Design & build of an electronic risk assessment tool

An analysis of system functionality is essential after the workflow's current state has been documented. Many organizations are using a clinical informaticist for this stage of the process (Hersh, 2009). Clinical Informaticists (informaticians) have a clinical background in addition to training in health informatics. Saba & McCormick (2005) has defined clinical informatics as a specialty that integrates clinical science/practice, computer science, and information science. It is also possible that a clinician with knowledge of information technology would be involved to work with clinicians and system analysts. The goal is to have a clinician and clinical informaticist explore, describe and/or demonstrate how the IT system accommodates data entry and utilization of the other functionalities in the design of the system. Once clinicians have an understanding of the possibilities for data entry and data manipulation, they are able to propose a design of an assessment tool that leverages the system's functionality to optimize the tool's feasibility. In the recent example, scoring for the pressure ulcer risk (Braden Pressure Ulcer Risk Assessment) was designed using dropdown menus and the fields were formed based on the sections on the assessment tool (see Clarke 2007). The defined terms in each field are assigned a value, so that their selection when combined with other fields yielded an overall level of risk. The values of each parameter were necessary to perform calculations of risk. In the Braden scale the sum of these parameters indicated a certain level of risk. The electronic tool requires that *all* fields to be completed before an assignment of risk is generated. This electronic tool has greatly improved the accuracy of the risk assessments (see also Gunningberg et al. 2009) and ultimately the targeted interventions by reducing the mathematical errors made on the paper document. (See Figures 1, 2, 3).

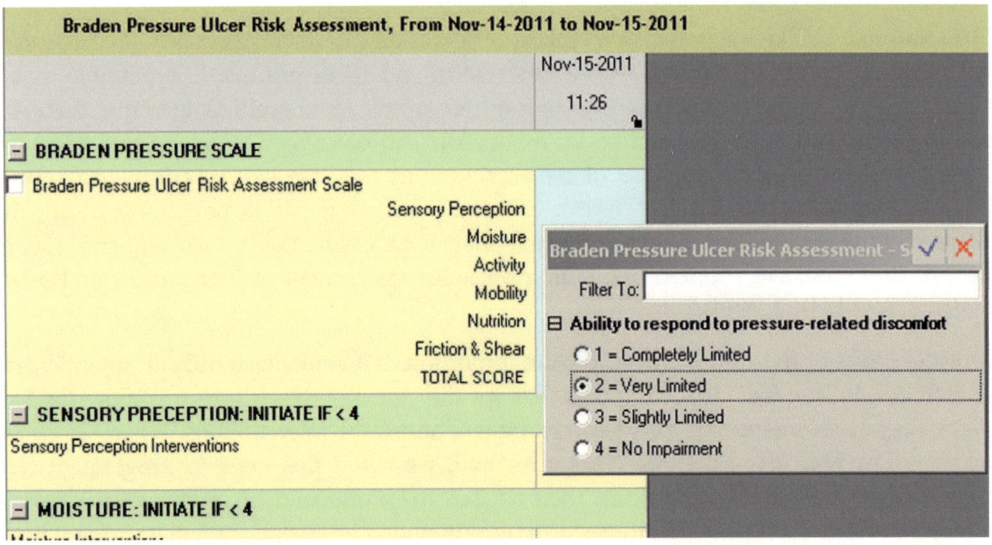

Fig. 1. Dropdown lists for risk assessment items. In this example, the Sensory Perception item presents four selections describing the patient's ability to sense discomfort from pressure. Selecting the option that most closely corresponds with the patient assessment yields a score that will populate the flow-sheet.

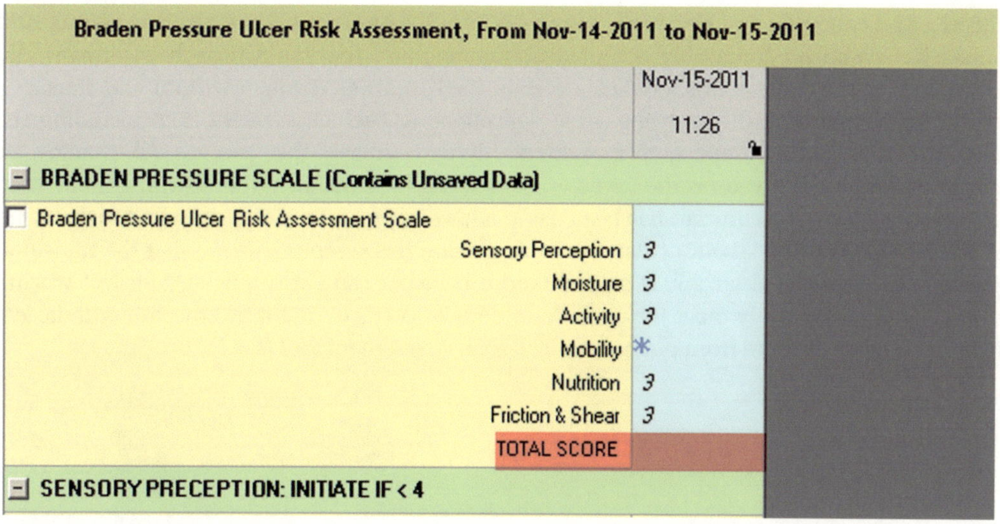

Fig. 2. In this example, one of the assessment parameters is not entered. The tool was designed to NOT provide a total score unless all of the assessments are documented. This ensures the instruments' integrity for identifying a pressure ulcer risk.

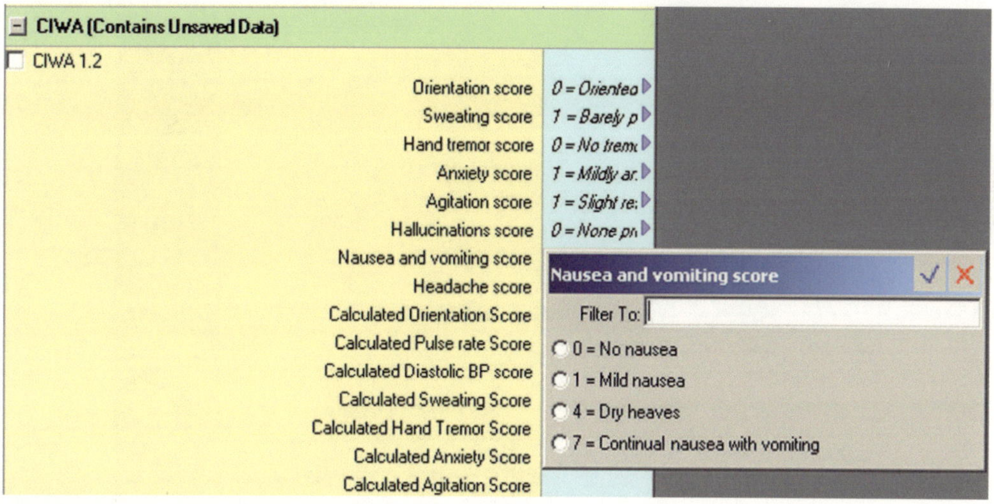

Fig. 3. Electronic assessment tool for CIWA (Clinical Institute Withdrawal Assessment for Alcohol). Each numerical value is associated with a description of the assessment option.

After the workflow analysis has been completed, the next step is to build the tool in the electronic environment. Careful consideration of the risk assessment tool's layout design and precise review of the tool's documentation functions are critical. This process is moderately easy for those risk assessment tools that have been developed using scientific methods (e.g.

theories and concepts) and statistical testing, because the items in these tools have been clearly stated (i.e. based on commonly accepted definitions) and the scales have been numerically expressed. These same characteristics are important in an electronic medium and thus it is crucial to transform and attach the same definitions to each assessment item including the criterion. It is highly important to underscore that during this process no changes or modifications are allowed on the items or the scoring systems once the instrument has been statistically tested. If a clinician has been using a hard copy of the risk assessment instrument, it is advisable that the layout of the electronic version has resemblance, so that the transition phase from paper to electronic is smooth and a clinician can learn quickly to select scoring options that most closely meet the cited definitions or correctly identifies the risk criteria. We have provided examples from our current Fall Risk Assessment tool (see Figure 4).

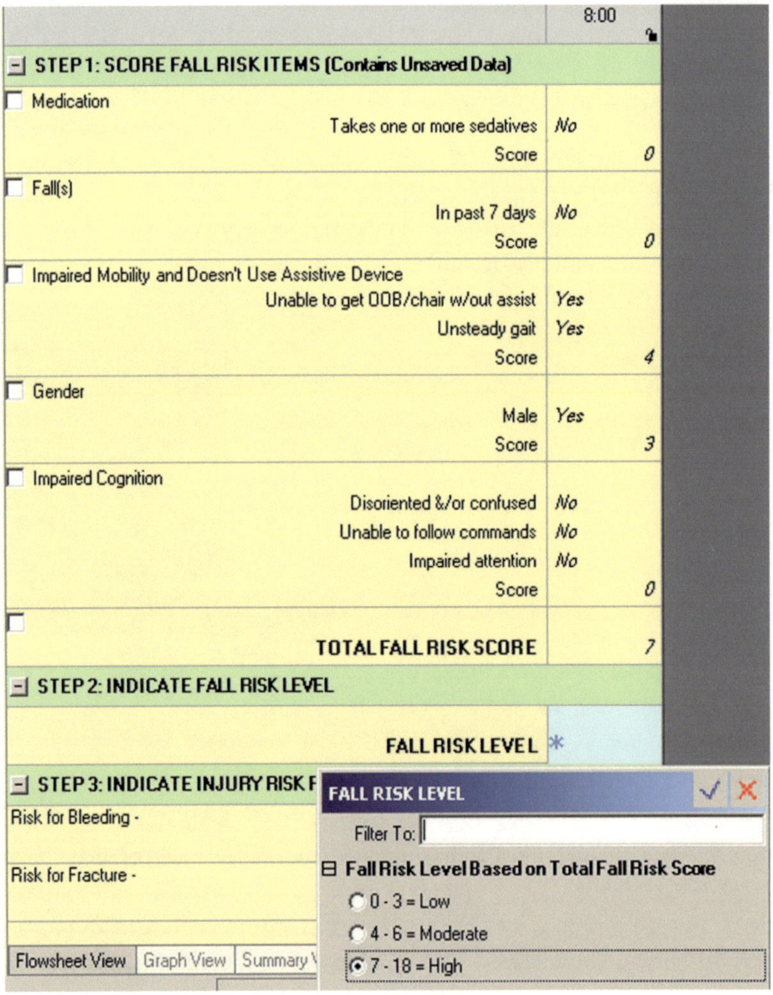

Fig. 4. Fall Risk Assessment Tool: as clinical questions get answered, numerical scores populate the electronic version of the instrument and yield total fall risk score.

It is important to note that some risk assessment tools may require sequencing of data element entry in order to maintain the accuracy and validity of the assessment. In fact, many tools require that data elements are gathered in a particular sequence for tabulation or there can be several decision points that must be sequenced appropriately to optimize the accuracy of the risk score for a clinician's decision making and to support practice. For example, when we were creating the electronic version of the Fall Risk Injury Tool, the required steps were to determine that the fall risk was first in the sequence of the assessment. Once the risk for falling was identified, the nurse could then proceed to assess the risk for injury. When both assessments were completed, an overall, total fall and injury risk score was possible to assign for a patient. In this example, the fall risk assessment score is a dependency for determining the overall risk of injury.

Once an electronic version of a risk assessment tool and its scoring scale are completed, the scores indicate a certain level of risk and, depending on the instrument in use, there may be corresponding actions or interventions necessary to minimize or eliminate a risk, or to prevent specific iatrogenic conditions. An important question to consider at this point is how the user/clinician will be alerted to a condition, a risk level or score, because this function is also developed in the tool's electronic format. The most commonly used options that are developed and built in an electronic risk assessment tool may provide either alerts or advisories to clinicians. It is also possible to develop functions in an electronic risk assessment tool that suggest actions, but since the practices may vary quite a lot between organizations, the development of this function requires that the actions are carefully analyzed and instituted by the clinicians. The alerts can be active (i.e. automatically activated by the IT system and triggered by the predetermined risk levels) or passive (i.e. informing a clinician regarding the pre-determined risk levels) in nature. It is very common that built-in alerts, regardless of their nature, also provide clinicians with care recommendations (Berner, 2009). The electronic risk assessment tool can be developed so that the care recommendations can either be added by the clinician, or if evidence-based recommendations are available, these predetermined interventions can be auto-populated by the information system. As with most risk assessments tools, the electronic versions provide some level of decision support for bedside clinicians. It is important to note that today clinical decision support is a common feature in most clinical information systems. What this means is that based on the availability of data in the EHR (electronic health record), the system can be designed to trigger alerts to numerous conditions to provide evidence-based care recommendations. For example, the presence of allergies to certain medications, conditions like pregnancy or lactation are possible to include in the alerts integrated with the best practice recommendations. Depending on the IT system's functionality, highly advanced alerts and decision support can be designed and created based on complex utilization of the data from the EHR.

Many IT systems have a feature that allows the user to review the previous assessment, or portions of it, in relation or as a part of the current assessment. This function goes by many names, with "copy forward" being the most common. The function is useful when designing and building an electronic risk assessment tool that requires data collection as a series of assessments at brief intervals. The benefit of this function is that it allows the clinician to copy the previous assessment forward and to edit the parameters that may have changed. Most of all, this function assists with clinical efficiency and tasks requiring fast paced and short time-interval execution. However, monitoring for improper use of this particular function is recommended, because busy clinicians may continue using the original assessment without editing changed parameters in order to save time. Unlike the 'copy' and 'paste' function, the copy forward function leaves an electronic audit trail which is commonly used to monitor the use of this function. (Mangalmurti et al, 2010).

2.3 Access and security

Unlike a paper record, the use and review of clinical information in electronic systems can be protected by controlling users' access, availability, or activity to portions of the electronic health record. However, privacy of patient information balanced with the clinicians' scopes of practice are important to take into consideration when creating appropriate levels of security in electronic health records. Access to the system alone, always requires a password. Once within the system, the type and amount of information that can be seen, entered or manipulated, is determined by a clinician's role in patient care. Most health care organizations use the 'minimum necessary' standard for access. As the phrase implies, access is granted for the least amount of information needed to perform in a clinician's role. For example, the access to enter or edit data related to a risk assessment today is usually limited to the Registered Nurse. A patient's health assessment is generally regarded as one of the most important tasks and core domain of nursing, therefore, most risk assessments are performed by Registered Nurses. The access to view and use the electronic risk assessment tool may be limited to licensed clinicians who are members of the care team. Depending on the defined details of the designed security in the clinical information system, some portions of the risk assessment tool may allow access to unlicensed caregivers. It is important to consider, whether there are health assessment data elements that are within the scope of practice of an unlicensed caregiver and how those fields may be available for data entry with appropriate security settings. The issues of access and security are important to consider early on, so that the electronic risk assessment tools allow for contribution by all members of the care team.

The following example demonstrates the care team's role in the risk assessment, its execution and documentation. In the conversion of the Braden Scale and Fall Risk Injury assessments tools from paper to electronic, the security configuration on the electronic version of the tool allowed only RNs to document the risk assessments. Once the assessment has been completed, the RN would select interventions to prevent either of these outcomes (i.e. pressure ulcers; falls and injuries). However, the interventions (e.g. skin care, repositioning; activation of bed alarms, hourly rounding) are often performed by an unlicensed staff member, therefore, providing limited access and ability to document the intervention is an important part of the care process. In both electronic risk assessment tool examples, the right to enter an assessment was only granted to Registered Nurses, and the portion of the tool with intervention selections was available for all care team members to document. (See: Figure 5) If the clinical information system does not have the functionality to restrict access to unlicensed personnel, another option is a hospital policy that clearly defines and prescribes the rights of each team member. However, this approach does require periodic audits to ensure appropriate use.

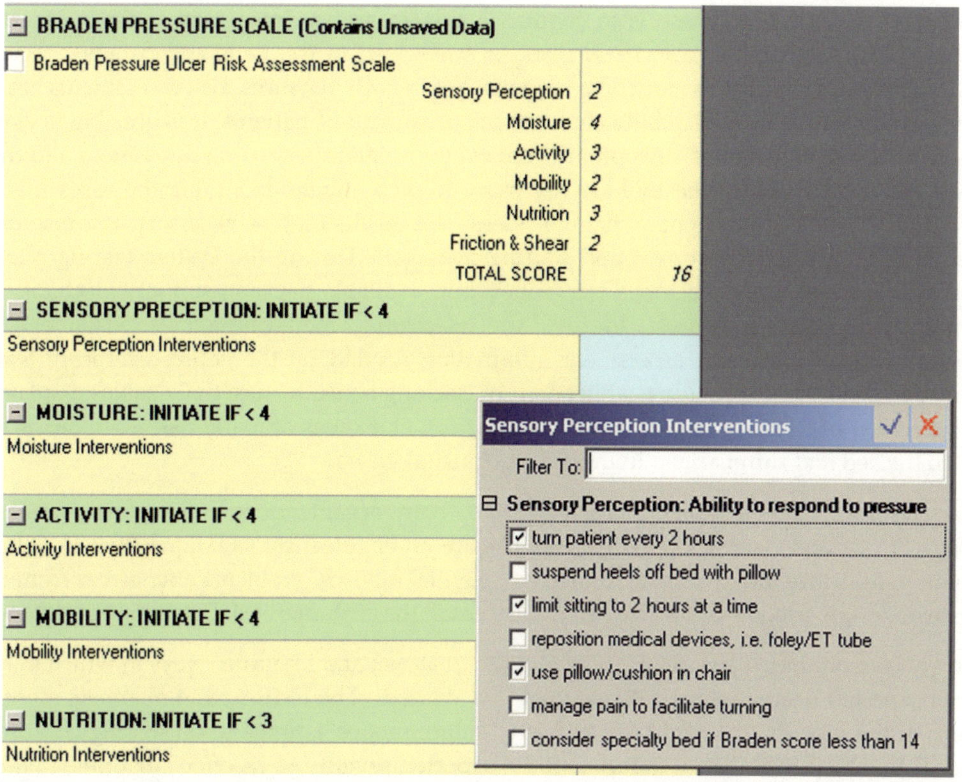

Fig. 5. In this figure, the top portion of the flow-sheet contains the assessment performed by the Registered Nurse. The interventions below the assessment allow both RN and unlicensed staff to document interventions.

2.4 Testing & signoff

Several testing phases are necessary to ensure accuracy of the electronic version of the risk assessment tool and its impact on decision-making after its configuration is available in the EHR test system. The electronic tool first undergoes **unit testing** (i.e. testing of the different sections and scoring system of the tool) to ensure that the developed assessment instrument and documentation perform as desired. During this phase the IT system analysts and staff nurses perform the testing, preferably using a test system and 'non-clinical' environment. The focus of the team is to enter data into each of the fields in the electronic instrument, and follow the same steps and approach as they would have when using a paper tool. As the team members continue to add data to the tools' fields, the clinicians are also asked to evaluate the function of the tool and determine whether it performs accurate risk-score calculations and yields the correct score based on the entered observations (cf. Weiner et al. 2007). This testing phase is very important, because the goal is that the electronic risk assessment tool accurately assigns the appropriate level of risk to inform the clinician, as it would in a paper format.

After the unit testing has been completed and the team is satisfied with the tools functions and accuracy, the next phase is to conduct **a system and/or integrated testing**. Different types of risk assessments are, in fact, only a small part of the EHR and patient's record, whether these are paper or electronic versions. Yet in both instances, risk assessments are an integral part of the overall quality care and safe treatment of patients. It is possible to claim that all the risk assessment tools provide critical information regarding assessment and thus most elements/fields of the tool are necessary to disseminate, including the assessment's outcome. It is important to note that the assessment results may be necessary as a resource-planning report for department and staffing managers. During **the system testing phase**, the goal is to evaluate the tool's integration into available decision-support mechanisms, reports and on-line views. Ideally, the electronic medium allows instantaneous dissemination of any risk assessment information needed by the entire healthcare team. Today many systems provide a whiteboard or tracking board feature that enables display of an overview of a unit or department (See Figure 6). Outcomes of many risk assessments are often tracked and summarized in these departmental views.

Once testing of the electronic tool is complete, many organizations will perform a Failure Mode Effects and Analysis (FMEA). FMEAs are an engineering technique used to define, identify, prioritize and eliminate known or potential failures, problems and errors from the system, design, process or service before they reach the customer. (McDermott et al, 1996)

An FMEA examines a process associated with great risk and identifies ways in which injury or unexpected death among patients might be averted. The FMEA team members identify steps in the process that could fail (termed "failure modes"), agree upon scoring criteria to determine the probability of failure and the expected severity of its effect, determine which failure modes could be eliminated or controlled by an intervention, and determine how the process might be changed, before a patient is injured (Kenney-Weeks et al., 2004).

2.5 Training

Patient health risk assessments are part of standard care and best practices and many clinicians are accustomed using different types of risk assessment tools in their daily patient care. If the electronic risk assessment tool is one that is already familiar to clinical staff, training need only focus on the use of the tool in a new, electronic environment. However, if a tool and its content (e.g. scoring scale, criteria for scoring, calculations) are new to the clinicians, this will require focused and carefully planned in-service education prior to the computer training. It is important that the clinicians comprehend the tool's use, how to interpret the results and most of all to feel confident and comfortable using the tool for a risk assessment. It is critical to identify any non-computer related content that needs to be included in the training and understood before using the tool in patient care. The content of the training regarding the use of the tool in an electronic environment is taught by the computer training staff, however it is recommended in the literature (Brettle, 2003) that any practice related information should be taught by a Registered Nurse at a separate time. The learning process is quicker and the outcomes (e.g. accuracy of the use of the tool, compliance) are better when the PC instruction is limited to use of the information system and the electronic tools. For example, in many IT systems the risk assessment tool must be selected and opened prior to use. Users will thus need to know how to access the tool and then how to enter the risk assessment scores (i.e. data elements of the assessment). Depending on the system's design, some fields of the tool can be automatically completed based on information (i.e. known data elements) from the patient's record. Such information as the patient's age, and gender are common data elements that are already existing elsewhere in the patient record. These can be pulled into the respective fields of a risk assessment tool. However, the IT system may require manual entry of all information (i.e. data elements) regarding the patient and the risk assessment. Similarly with the instrument's assessment scores and calculations, some systems are able to automatically present the risk level (i.e. outcome) or 'score' without user action. Conversely, there are systems that require a user 'action' after the risk assessment scores have been entered in order to obtain the overall risk level of a patient. These are just some examples of the important objectives necessary to consider and understand when developing the training curriculum.

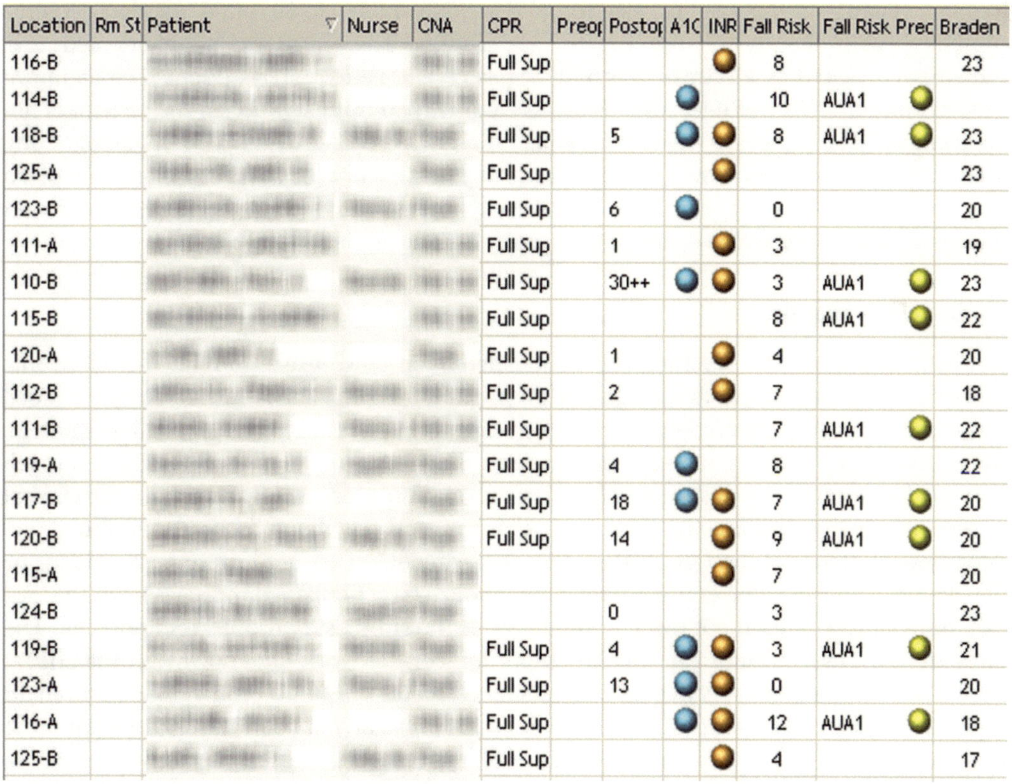

Location	Rm St	Patient	Nurse	CNA	CPR	Preo	Posto	A1C	INR	Fall Risk	Fall Risk Prec	Braden
116-B		[illegible]		[illegible]	Full Sup				●	8		23
114-B		[illegible]		[illegible]	Full Sup			●		10	AUA1 ●	
118-B		[illegible]	[illegible]	[illegible]	Full Sup		5	●	●	8	AUA1 ●	23
125-A		[illegible]		[illegible]	Full Sup				●			23
123-B		[illegible]	[illegible]	[illegible]	Full Sup		6	●		0		20
111-A		[illegible]		[illegible]	Full Sup		1		●	3		19
110-B		[illegible]	[illegible]	[illegible]	Full Sup		30++	●	●	3	AUA1 ●	23
115-B		[illegible]		[illegible]	Full Sup					8	AUA1 ●	22
120-A		[illegible]		[illegible]	Full Sup		1		●	4		20
112-B		[illegible]	[illegible]	[illegible]	Full Sup		2		●	7		18
111-B		[illegible]	[illegible]	[illegible]	Full Sup					7	AUA1 ●	22
119-A		[illegible]	[illegible]	[illegible]	Full Sup		4	●		8		22
117-B		[illegible]		[illegible]	Full Sup		18	●	●	7	AUA1 ●	20
120-B		[illegible]	[illegible]	[illegible]	Full Sup		14		●	9	AUA1 ●	20
115-A		[illegible]		[illegible]					●	7		20
124-B		[illegible]	[illegible]	[illegible]			0			3		23
119-B		[illegible]	[illegible]	[illegible]	Full Sup		4	●	●	3	AUA1 ●	21
123-A		[illegible]	[illegible]	[illegible]	Full Sup		13	●	●	0		20
116-A		[illegible]		[illegible]	Full Sup			●	●	12	AUA1 ●	18
125-B		[illegible]	[illegible]	[illegible]	Full Sup				●	4		17

Fig. 6. In this example, a nursing department displays key patient information for all members of the clinical staff. Three columns display the most recent Fall Risk and Braden scores. It also highlights all of the patients on Fall Risk precautions.

The timing of the training is critical in relation to the availability of the tool for its users. Depending on the assessment tool and its complexity (e.g. scoring scale, assessment criteria, number of assessment items, calculations), the staffs training period should be as close to the 'go-live' date as possible. The literature shows (Brettle, 2003) that training for any computer application is best learned with continued reinforced use of the application. Training too early may affect retention of the learned skills and training material. Today, many organizations maintain two copies of the developed IT system for training purposes so that one of these can be used as a practice environment. This identical, safe training environment allows users to practice the use of a new risk assessment tool, its functions and documentation and gain a learning experience that is possible to transfer directly to the clinical setting. Continuous evaluation of the training methods is important, because the goal is to allow nurses to learn, gain and maintain competency in the use of new electronic tools.

The literature shows (Brettle, 2003) that there are numerous teaching methods and approaches to training that should be considered. Most health care organizations have a computer training staff who can assist with curriculum development. The classes and training regarding the use of electronic risk assessment tools often employ lectures combined with 'hands-on' use of the application. These can be supplemented and sometimes replaced by electronic learning or (also known e-learning) options. The literature shows (Brettle, 2003) that e-learning is an efficient way to train staff when the content is not too complex. The benefit of this option is that it allows the user to learn the material at any time, day, or location. However, the downside of using e-learn for computer classes, is that a user does not have an opportunity to interact with an instructor/computer training staff and thus may be unable to receive clarification, on questions related to the tool.

2.6 Go-live & support

As with any change, planning the implementation and execution of the electronic assessment tool are critical to the success. To prepare the clinicians, units and organizations for a change in documentation, a strategy needs to be carefully developed. The strategy should obviously include, at least the date and time for 'go-live', as well as the approach that will be used during this phase of the transformation process. Depending on the clinical IT system being used, some 'go-lives' can be incremental (i.e. the tool is introduced to a few departments first, and then gradually other departments of an organization are added) or the 'go-live' can be for the entire organization at once (i.e. 'Big Bang' approach). For the 'go-live' of most risk assessment tools, a 'Big Bang' is the simplest and quickest approach. The tools are used by clinicians for only small portions of the work day, however, it is critical from a patient safety perspective that the risk assessments are conducted similarly within an organization and that the transition phase does not cause any gaps in the documentation of these assessments. It is also more economical to make the electronic version available organization-wide, all at once because this approach reduces the time period of support required by the users.

When an IT system change is introduced, the availability of on-site support needs to be re-visited and provided. Depending on the size of the organization, the number of users, and the complexity of the electronic risk assessment tool, an estimate regarding extra IT support staff needs to be planned beforehand, because the 'go-live' phase may require from 3 to 30 persons or more. The timing of planning of support staff is vital, because the support team members have to receive extensive training on the computerized assessment tool. They need to be familiar with the intricacies of the tool, its use and most of all be able to assist staff/clinicians who are using the tool for the first time. Our experience has demonstrated that one IT support team member can cover 1 or 2 departments at once in these types of 'go-live' processes.

2.7 Reporting

Once the risk assessment tool is being used by clinicians/staff throughout the organization, the progress of its use can be easily measured. Most clinical IT system databases are developed with the intent of retrieving data about care processes. The processes of Risk Assessment are important to monitor especially when a tool has been transformed from paper to an electronic version. The reports regarding the risk assessment tools can be very helpful, for example, for department (nursing) managers. The focus of the reports (i.e. data elements) for monitoring needs to be determined by the department leadership in collaboration with the report writers. Depending on the focus of monitoring, reports may include aggregate data, allowing managers to assess the resources necessary to mitigate overall risk in the department. Reports may also focus on compliance of use of the risk assessment tool. If the risk assessment is associated with a regular frequency (i.e. daily, every shift), then reporting allows (nurse) managers to check and audit the assessments done by all staff. This helps to identify staff/clinicians who are still reluctant to try the changed process. Publishing the reports of utilization rates of departments and staff members serves as motivation to use the tool regularly. The trends shown by the reports can also be used as discussion points and reveal barriers to regular use of the tool.

Aggregated reports can provide researchers in clinical settings support to the tool's validity, reliability and feasibility, but also can demonstrate the tool's effectiveness and impact on patient outcomes. For example, comparison of departmental utilization rates of a Fall Injury Risk Tools to the actual number of falls in those same departments may illuminate opportunities for improvement. It is possible to claim that, unlike paper- based tools, electronic risk assessment tools provide vital advantages in reporting: having information readily available allows for more timely data analysis and trending, which are extremely valuable when striving to meet certain patient safety and care goals.

3. Conclusions

This chapter introduces the ongoing need to make risk assessment tools available in an electronic medium. Advances in patient care and the clinicians' need for full availability of patient information, require the conversion of risk assessment tools from paper to electronic. This chapter outlines a process for achieving this conversion. The process described is one organization's method for conversion, although most organizations use a similar process for preparing to move any clinical documentation to an electronic environment. The tools that have been converted are those that this organization has chosen to use to assess health risks on hospitalized patients.

Perhaps the most critical phase of this process is the workflow analysis. Successful conversion to electronic documentation requires a good understanding of the processes of risk assessment, both in physical and cognitive terms. A complete understanding of any workflow provides a solid foundation for the design and configuration of tools in a computerized format. In our organization this process has been successfully repeated in the conversion of several risk assessment tools. Only in one instance, noted earlier in the chapter, the users found the converted tool less efficient, when in fact the tool's security settings made it less efficient. The initial workflow analysis did not include the associated selection of targeted interventions. In fact, a critical part of the complete workflow for risk analysis and risk prevention was missed. Revisiting the review of workflow confirmed the flaw in the tool's electronic design and build and access was provided to unlicensed workers assisting the Registered Nurse.

Although these conversions represent a change in how work is performed, there are numerous benefits to the patient, clinician and organization's leadership. For the patient, the outcome of the risk assessment is increased safety, because the electronic information is available to all members of the care team. This ensures continuous, high awareness among team members regarding different types and levels of risk and allows interdisciplinary contributions to prevention, when appropriate. For example, it is beneficial for all members of the care team to know that a patient is at risk for falling. Any team member can prevent falls, when assisting patients to chairs, bathrooms, or merely ensuring that the nurse call device is within the patient's reach. For the bedside nurse, this represents an interdisciplinary collaborative approach to preventing falls. Nurses are no longer the sole clinician responsible for assessing different types of risks and keeping patients safe. For the organization's leaders, electronic tools provide an opportunity to perform comprehensive, on-time monitoring and improved oversight of risk assessment and management. Electronic tools allow for more efficient auditing for compliance of policies regarding risk assessment, particularly if these are mandated by regulatory or accrediting organizations. Reports can also indicate the organization's progress toward its safety goals and decreasing risks for hospitalized patients. Many risk assessments are related to nursing sensitive indicators, and thus the data from the electronic IT system can provide a snapshot of an organization's performance in meeting the nursing quality measures.

Moving forward, it is likely that many risk assessment tools will be initially developed in an electronic medium. Conversion may become unnecessary, although attention to nursing workflow and practice will still need to be a part of the process. Development of risk assessment tools in an electronic medium will allow for more efficient testing of the tool, eliminating manual data review, and enabling greater efficiency of scientific research, and expedient statistical analysis of the tool's sensitivity, specificity, reliability and validity in different patient populations.

4. Acknowledgment

The authors would like to acknowledge the Analyst and Training Teams of the Information Services Department at Maine Medical Center in Portland, Maine, USA. These teams have made the transition of clinical assessment tools and instruments to an electronic health record. Our clinical community is indebted to their skills and expertise. It is their ability to work closely with our clinicians that has made these transitions occur easily.

5. References

AHRQ (2011), Workflow Assessment for Health IT Toolkit. November 14, 2011, Available from: http://healthit.ahrq.gov

Berner, ES. Clinical decision support systems: State of the Art. AHRQ Publication No. 09-0069-EF. Rockville, Maryland: Agency for Healthcare Research and Quality. June 2009.

Brettle, A. (2003) Information skills training: a systematic review of the literature. *Health Information and Library Skills Journal,* Vol. 20, No. s1. (June, 2003) pp. 3-9

Carayon, P. Schoofs Hundt, A. Karsh, B-T. Gurses, A. Alvarado, C. Smith, M. Flatley-Brennan, P. (2006). Safety by Design: Work system design for patient safety: the SEIPS model. *Quality and Safety in Healthcare,* Vol. 15 (2006) (Suppl 1) i50-i58.

Chapman J., Bachand D., Hyrkäs K. (2011) Testing the sensitivity, specificity and feasibility of four falls risk assessment tools in a clinical setting. *Journal of Nursing Management* Vol 1, No 1, 133-142. ISSN 0966-0429

Clarke M. (2007) *Evidence-Based Nursing: Monographs. Pressure Ulcers.* Mosby's Nursing Consult. Date of access: September 16, 2011. Available from: http://www.nursingconsult.com.library

Courtney KL, Demiris G, Alexander GL. (2005) Information technology: changing nursing processes at the point-of-care. *Nursing Administration Quarterly.*;Vol. 29, No. 4 (2005 Oct-Dec) pp.315-22.

Gunningherg, L. Fogelberg-Dahm, M. Ehrenberg, A. (2009) Wound Care and Pressure Ulcers: Improved quality and comprehensiveness in nursing documentation of pressure ulcers after implementing an electronic health record in hospital care. *Journal of Clinical Nursing,* Vol. 18, (2009), pp. 1557-1564.

Hersh, W. (2009) A Stimulus to Define Informatics and Health Information Technology. *BMC Medical Informatics and Decision Making,* Vol. 9, No. 24. doi: 10.1186/1472-6947-9-24.

Karsh, B-T. (2009) Clinical Practice Improvement and Redesign: How Change in Workflow Can Be Supported by Clinical Decision Support. Department of Industrial and Systems Engineering University of Wisconsin—Madison AHRQ Publication No. 09-0054-EF June 2009

Kenney-Weeks S. Bijkersma F. Hubbeartt E. Murphy B. Anderson, M. (2004) Failure Mode and Effects Analysis: a search for ways of preventing patients from falling. *American Journal of Nursing* 2004; 104(4)"72a-72d.

Kocher, R. Emanuel, E. DeParle, N-A. (2010) The Affordable Care Act and Future of Clinical Medicine: The Opportunities and Challenges. *Annals of Internal Medicine,* Vol. 153, Nos. 8, (October 19, 2010) pp. 536-539.

Lopez K., Gerling G., Cary M. & Lanak M (2010) Cognitive work analysis to evaluate the problem of patient falls in an impatient setting. *Journal of the American Medical Informatics Association* 17:312-321. ISSN 1527-974X

Mangalmurti, S. Murtagh, L. Mello, M. (2010) Medical Malpractice Liability in the Age of Electronic Health Records. *The New England Journal of Medicine,* Vol. 363, No. 21, (November 18, 2010), pp. 2060-2067.

McDermott RE, Mikulak RJ, Beauregard MR. *The Basics of FMEA*. Portland, Oregon, USA Productivity, Inc; 1996

Mosby's Dental Dictionary, 2nd edition. © 2008 Elsevier, Inc. All rights reserved.

Radicki, R. Sittig, D. (2011) Application of Electronic Health Records to the Joint Commission's 2011 Patient Safety Goals. *JAMA,* Vol. 306, No. 1, (January 2011), pp.92-93.

Saba V, McCormick KA. *Essentials of Nursing Informatics,* 4th ed. McGraw-Hill Companies, New York, NY, 2005

United States Federal Register. (2011) Medicaid Program; Payment Adjustment for Provider-Preventable Conditions Including Health Care-Acquired Conditions. *Department of Health and Human Services, Centers for Medicare and Medicaid,* Vol. 176, No. 108, (Monday, June 6, 2011), pp. 32816-32838.

Weiner, J. Kfuri, T. Chan, K. Fowles, J. (2007). "e-Iatrogenesis": The Most Critical Unintended Consequence of CPOE and other HIT. *Journal of the American Medical Informatics Association,* Vol. 14, No. 3, (June 2007), pp. 387-388.

Wenzel G. (2002) Creating an Interactive Interdisciplinary Electronic Assessment. *CIN, Computers, Informatics, Nursing*, Vol 20, No 6, 251-260. ISBN/ISSN: 15382931

16

A Bayesian Approach for Calibrating Risk Assessment Models

Michael S. Williams[1], Eric D. Ebel[1],
Jennifer A. Hoeting[2] and James L. Withee[3]
[1]*Food Safety and Inspection Service,*
United States Department of Agriculture, Fort Collins, Colorado
[2]*Department of Statistics, Colorado State University, Fort Collins, Colorado*
[3]*GigaYeast Inc., Belmont, California*
USA

1. Introduction

Monte Carlo simulation is a commonly used tool for constructing foodborne pathogen risk assessment models. Monte Carlo simulation enables an analyst to construct a probabilistic model of almost any desired complexity. It requires relatively little mathematical rigor and the models can be presented in an intuitive manner. It has some drawbacks, however. For example, Monte Carlo simulation requires that each parameter, as well as its uncertainty, be quantitatively described.

The models are typically used to make a projection of possible outcomes. In food-safety risk assessment applications, we typically construct a model to predict the number of human illnesses in the population. These calculations are based on the prevalence of contaminated production units and their microbial load. These are tracked through food production, consumer handing and consumption. The final step is converting predicted contamination into a human health impact via a dose-response model.

Foodborne illness is often the result of an acute microbial pathogen exposure. More than 75 countries have implemented surveillance systems to monitor occurrences of these illnesses (Allos et al., 2004; de Jong B & K., 2006; Herikstad et al., 2002). These surveillance systems do not capture every case of foodborne illness, so scaling factors are developed to estimate the total number of illnesses for the pathogen of interest (Ebel et al., 2012; Scallan et al., 2011). Additional scaling factors can be developed to extend these estimates to a specific product-pathogen pair (Hald et al., 2004).

A conundrum for risk assessors occurs when the illness estimates from a Monte Carlo-based risk assessment model do not match the estimates based on the surveillance data. When these two estimates do not match the risk assessment model must be calibrated. A common approach to calibrating the model is to adjust the parameters of the dose-response function to match predicted to observed illnesses. Alternative calibration approaches are to replace components of the model, such as changing the models used to calculate pathogen attenuation during cooking. The concern with these calibration approaches is their lack of objectivity and rigor.

Bayesian methods offer an alternative approach to the problem of calibration. Various Bayesian methods have been used in previous risk assessment applications (Albert et al., 2008; Hald et al., 2004; Parsonsa et al., 2005). The Bayesian models are similar in their structure to a Monte Carlo model. They do, however, offer the advantage that any data available at a stochastic node can be incorporated into the model in the form of a prior distribution. These methods allow the user to incorporate the data for observed human illnesses. The model then produces a Bayesian revision of the system's parameter estimates. A consequence of conditioning inferences on the human illness data is that the parameter distributions for each node of the model are shifted so that the predictions more closely match the observed illness data. These adjusted distributions can be thought of as posterior distributions, though some Bayesian methods use the terms pre- and post-model distributions (Givens, 1993; Raftery et al., 1995). The direction and degree to which each of the prior distributions are shifted is to a large extent determined by the relative degrees of uncertainty in the prior distributions (i.e., the parameters in the model that are highly uncertain will experience the largest degree of adjustment in the process of calibrating the model).

This chapter will focus on introducing Bayesian methods for use in food-safety risk assessment. The use of Bayesian methods requires first establishing a simple probabilistic model. Once a model is established, a number of different Bayesian techniques can be used for drawing inferences. We will introduce a relatively simple resampling algorithm that can be used to calibrate a food-safety risk assessment model. A number of examples of the application of this framework can be found in the literature (Williams & Ebel, 2012; Williams, Ebel & Hoeting, 2011; Williams, Ebel & Vose, 2011a) so we will present a rather unusual example where the probabilistic model and Bayesian resampling method are used to study the laboratory test sensitivity of a test for *Escherichia coli* O157:H7 in ground beef samples.

2. Probabilistic models for risk assessment

The probabilistic model we consider assumes that interest lies in modeling a count of events during a specified time period. For food-safety applications, this count will usually be the number of illnesses observed ($I_{observed}$) by a surveillance system during a single year.

We assume the count of sporadic illnesses detected by a surveillance system is reasonably modeled as Poisson random variable. Our model assumes that each food serving of the commodity of interest ($N_{servings}$) has a probability of causing illness of $P(ill)$. The product $N_{servings} \times P(ill)$ describes the rate parameter for a Poisson distribution that describes the total number of illnesses for the product-pathogen pair of interest.

Two factors relate the total number of illnesses for a single product-pathogen pair to the number of illness observed by the surveillance system. The first factor describes the proportion of illnesses, α, attributed to the product of interest. This attribution factor modifies a rate parameter for a specific product to describe the illness rate for the product-pathogen pairing of interest. The second factor describes the proportion of illnesses ρ that are reported by the surveillance system. These factors modify the product-pathogen rate parameter to describe the number of observed illnesses whose etiology is the pathogen. This leads to the basic model for observed illnesses being

$$I_{observed} \sim Poisson(\frac{\rho}{\alpha} N_{servings} P(ill)). \quad (1)$$

Other factors may be needed to relate the number of observed illnesses to the total number of illnesses. For example, the surveillance system may only cover a fraction of the population or a pathogen may be specific to a single product (e.g., BSE cases are associated with beef consumption so $\alpha = 1$). Thus, the inclusion of the adjustment terms will be specific to each surveillance system and product-pathogen pair. For this reason, the adjustment factor(s) will not be included for the remainder of the general model development and we note that $N_{servings}P(ill)$ is the rate parameter describing the illness rate in the population, denoted λ_{ill}. An extensive development of these scaling factors can be found in Williams & Ebel (2012) as well as Ebel et al. (2012).

In many applications, the objective of the risk assessment is to predict the change in the number of human illnesses that would occur if the production process where improved. This improvement is expected to reduce $P(ill)$ and the resulting reduction in illnesses can be modeled as

$$I_{avoided} \sim Poisson(N_{servings}(P(ill) - P_{new}(ill))), \quad (2)$$

where $P_{new}(ill)$ is the reduced probability of illness.

3. Partitioning $P(ill)$

The $P(ill)$ term is Equation 1 is one of the typical outputs of a Monte Carlo risk assessment model. Efforts to reduce the complexity of a risk assessment model begin with expanding this term and looking for biologically plausible situations where a simpler model is appropriate. Model simplifications of quantitative microbial risk assessments often begin from first principles: microbial contamination begets food exposure begets illness. In this approach, the interest is in determining the unconditional probability of illness among all servings.

It is also possible to derive these simplifications by applying Bayes Formula. In this case, the question to be answered is "what is the probability that an illness occurred given exposure to a contaminated food". Using Bayesian language, the answer to this question is termed a posterior distribution or $P(ill|exp)$. This is a conditional probability statement. From Bayes Theorem, we have

$$P(ill|exp)P(exp) = P(exp|ill)P(ill) \quad (3)$$

$$P(ill|exp) = \frac{P(exp|ill)P(ill)}{P(exp)}, \quad (4)$$

where $P(exp) = P(exp|ill)P(ill) + P(exp|ill^C)P(ill^C)$.

That the probability of exposure must be this sum can be appreciated from a simple Venn diagram in Figure 1 (i.e., the fraction of exposure servings includes those with and without illness)

This diagram also illustrates the triviality of the conditional probability $P(exp|ill)$. Because all illnesses result from exposure to a contaminated serving, the conditional probability of exposure given that a serving causes illnesses is unity. This conclusion generates a simpler calculation of the posterior probability that we are interested in, namely

$$P(ill|exp) = \frac{P(ill)}{P(exp)}. \quad (5)$$

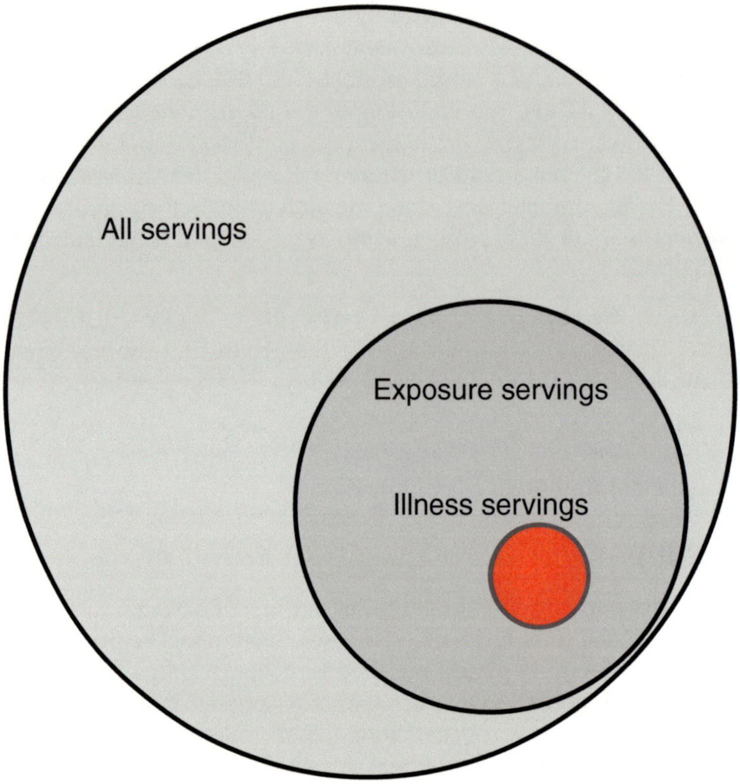

Fig. 1. Venn diagram describing the probability of exposure $P(exp)$ and the probability of illness for contaminated servings ($P(ill|exp)$).

A variation on this equation was employed by Bartholomew et al. (2005) to derive a linear risk model. In that example, the ratio of the estimated number of illnesses and number of exposures per annum derives a constant of proportionality that is ultimately used to project changes in illnesses from intentional changes in exposure. It should be noted that numbers of illnesses and exposures are simple transformations of $P(ill)$ and $P(exp)$ where each is multiplied by the number (or mass) of servings of a food consumed per year.

In microbial risk assessments, we usually have prior information about the number of illnesses per annum (i.e., $N_{servings} \times P(ill)$ where $N_{servings}$ is the number of exposure units), as well as prior information about the fraction of exposure units that are contaminated (i.e.,$P(exp)$). This evidence can be used to solve

$$P(ill|exp) = \frac{P(ill)}{P(exp)} \tag{6}$$

using Monte Carlo methods. The result is a posterior distribution of this conditional probability.

Contemplating this calculation, however, highlights a potential problem in using the posterior distribution to make risk projections that might result from changes in exposure. Because this posterior is derived as a ratio of two random variables (each describing uncertainty about a

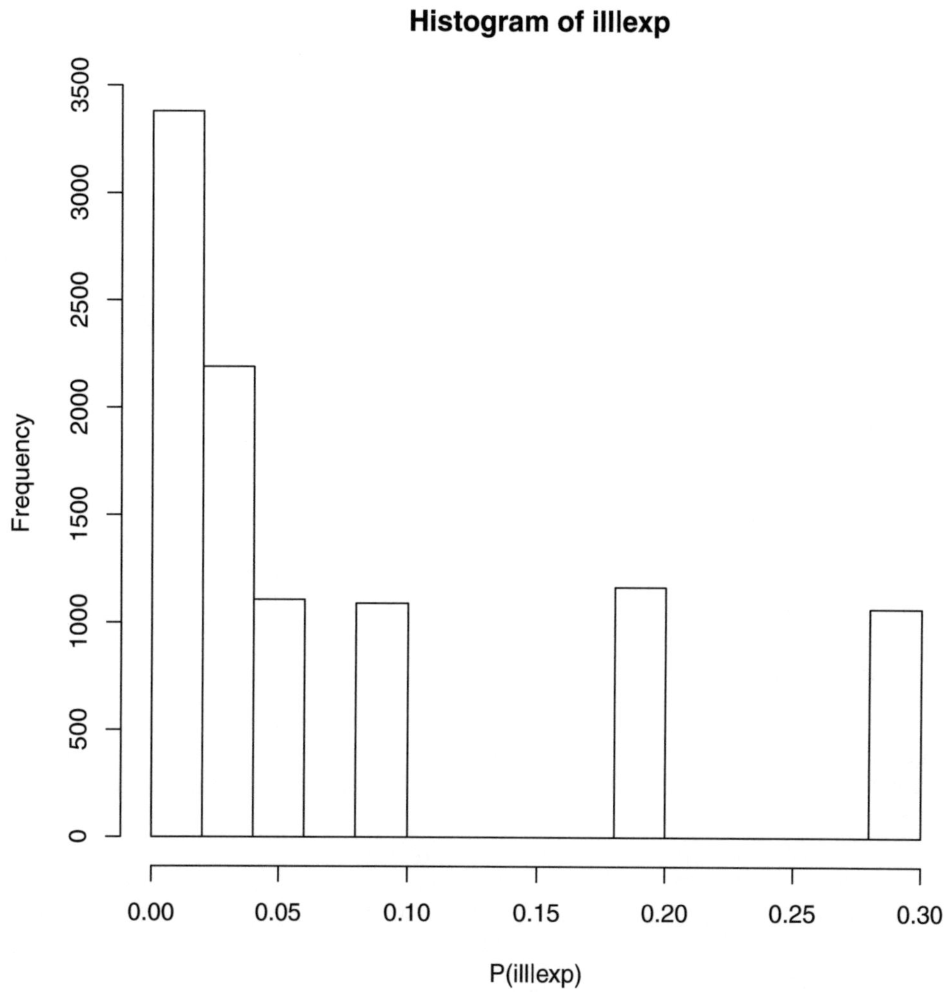

Fig. 2. Histogram for a simplistic Monte Carlo calculation of $P(ill|exp)$.

true parameter in nature), this distribution is necessarily informed by any covariance between these random variables. Furthermore, projections about future values of $P(ill)$ derived by $P_{new}(exp)P(ill|exp)$ need to account for the starting value of $P(exp)$.

A simple example may illustrate this point. Assume available prior evidence about numbers of illnesses (for a particular product-pathogen pair) is that 100, 200 or 300 cases occur per year with equal probability weights. Further, assume our understanding of exposures implies that 1%, 5% or 10% of 10,000 servings per year are contaminated with equal probability weights. A naive estimate of $P(ill|exp)$ would look like the histogram in Figure 2.

This ratio ranges from 0.01 (100 illnesses divided by 10,000 exposures) to 0.30 (300 illnesses divided by 1000 exposures). Furthermore, we can re-derive the uniform distribution of $P(ill)$ by simply multiplying the vector of $P(ill|exp)$ by the vector of $P(exp)$. But, if we consider these two independent random variables, we will generate a distribution for $P(ill)$ that is not at all what the prior $P(ill)$ looked like.

Instead, the appropriate distribution for $P(ill|exp)$ is a distribution particular to the value of $P(exp)$. For example, $P(ill|exp)$ for $P(exp) = 0.01$ is a discrete Uniform distributions of values $(100/1000, 200/1000, 300/1000)$.

If the analyst wants to predict the effect of a change in $P(exp)$, this dependence between $P(exp)$ and $P(ill|exp)$ should be borne in mind. Otherwise, incorrect representations of $P(ill)$ could result. Fortunately, the model simplifications developed in Williams, Ebel & Vose (2011a) avoid this trap because the $P(ill|exp)$ often cancels out of the equations and the change in illness occurrence can be estimated directly from changes in $P(exp)$. The term *prevalence-dependent* model is used to describe applications where this simplification is feasible.

4. Model simplification

A complete evaluation of the components of the model in Equation 1 can still be a complex task. Nevertheless, the factorization on $P(ill)$ into its exposure component ($P(exp)$) and hazard characterization component ($P(ill|exp)$) leads to situations where estimation of the number of illnesses can be greatly simplified. We outline two different models and describe methods for simplification.

4.1 Dose-dependent model

The first parameterization assumes that all servings have some level of contamination, where D describes the average number of pathogens in each serving. Note that when D describes an average concentration, it is possible for these concentration values to be much less than 1 unit per serving. Common examples are the description of pathogen levels in water. It may also be reasonable to model average concentrations for liquid and ground food products where no natural units exist. An exposure event from a particular food type will involve the ingestion of a random number of pathogenic organisms, where the distribution of organisms is described by the probability density $f(D)$. The lognormal distribution is a common and convenient choice (Limpert et al., 2001), so $f(D) \sim Lognormal(\mu_D, \sigma_D^2)$. The probability that a random person will become ill, given a microbial dose of size D, is $P(ill|D)$. Averaging across all possible doses yields the probability of a person becoming ill given exposure to the pathogen. When D describes an average dose, the probability of illness given exposures described by a continuous dose distribution is

$$P(ill) = \int_0^\infty P(ill|D) f(D) dD, \tag{7}$$

where $P(ill|D)$ is the dose-response function. The exponential and beta-Poisson dose-response functions are appropriate for continuous dose distributions. The term dose-dependent model will be used to denote this model.

The difficulty with this model is that data describing the dose at the point of consumption are not available. Instead, virtually all risk assessment models rely on a measurement of contamination, X, derived from data collected at a more convenient location in the farm-to-table continuum, such as during production or at retail. A typical risk assessment must rely on models of post-production activities to transforms measurements of microbial contamination following production (X) into measurements of human health risk at the point of consumption (D).

If it is assumed $X \sim Lognormal(\mu_X, \sigma_X^2)$ and the focus of a risk assessment is to determine how changes in the production process would lead to a change the number of illnesses, then the distribution of D can be derived from a single component Δ, that describes the cumulative change in average microbial level between production and consumption (i.e., it combines the effects of storage time and temperature as well as cooking and other process). Assuming that the cumulative change is distributed as $\Delta \sim Lognormal(\mu_\Delta, \sigma_\Delta^2)$, the distribution for D is $Lognormal(\mu_X + \mu_\Delta, \sqrt{\sigma_X^2 + \sigma_\Delta^2})$.

The dose-dependent model can be simplified by treating Δ as a latent variable, with its parameters $(\mu_\Delta, \sigma_\Delta^2)$ estimated during calibration. Williams, Ebel & Vose (2011a) provide an example based on *Campylobacter* contamination in chicken.

Additional simplifications of the model are possible in situations where pathogen numbers are uniformly low at the point of consumption (Williams, Ebel & Vose, 2011b).

4.2 Prevalence-dependent model

For the model in Equation 1, the number of illnesses avoided by reducing the prevalence of contaminated servings is readily predicted via Equation 2. Reduced prevalence of contamination might occur via changes in import practices or improved animal husbandry practices that reduce the occurrence of a pathogen among farms, herds, flocks or sheds. These changes are expected to reduce the prevalence of contaminated carcasses, but in a number of situations it is still reasonable be assumed that $P(ill|exp)$ would remain essentially unchanged.

For example, suppose that a country, where a specific pathogen is endemic, will begin importing animal products from a country that is free from the disease. If the importation of uncontaminated carcasses is such that prevalence is reduced by $P_{new}(exp) = \delta P(exp)$ where uncertainty about change in prevalence might be characterized as $\delta \sim Beta(a, b,)$, and it is reasonable to assume that $P(ill|exp)$ will remain unchanged, then the human health benefit is modeled as:

$$I_{avoided} \sim Poisson(N_{servings}(P(exp)P(ill|exp) - P_{new}(exp)P(ill|exp))) \quad (8)$$

$$\sim Poisson((1 - \frac{P_{new}(exp)}{P(exp)})\lambda_{ill}) \quad (9)$$

$$\sim Poisson((1 - \delta)\lambda_{ill}). \quad (10)$$

Note that this model relies only on the characterization of the number of illnesses (λ_{ill}) and the effect of the change in importation policy. Also note that measures of prevalence are not necessarily the prevalence of contaminated servings. Instead, one can argue that the prevalence of contaminated units at the point of data collection is proportional to the prevalence of contaminated servings. This constant of proportionality cancels out when $P(exp)$ and $P_{new}(exp)$ are measured at the same location in the farm-to-table continuum.

This formulation also obviates the need for modeling pathogen levels as well as eliminating the need to adjust for the difference between true and apparent prevalence. A linear relationship between contaminated carcass prevalence and human illnesses was demonstrated for chicken and campylobacteriosis in a previous risk assessment (Bartholomew et al., 2005).

5. Bayesian methods

Recent research interest has focused on the replacement of Monte Carlo models with a Bayesian approach that uses Markov chain Monte Carlo (MCMC) methods (Albert et al., 2008; Hald et al., 2004; Parsonsa et al., 2005). Along with this proposed approach comes the inevitable suggestion that the models be built using MCMC packages such as WinBUGS (Lunn et al., 2009; 2000; Vose, 2008). These packages often rely on the Gibbs sampler, or similar algorithms, to obtain a set of random samples from the posterior probability distribution of the risk assessment model. While it is possible to use software packages such as WinBUGS (Williams, Ebel & Hoeting, 2011), personal experience suggests that convergence is difficult to achieve given the high degree of uncertainty in the parameters of even a highly simplified food-safety risk assessment model. The underlying problem can be understood by examining the mechanics of an MCMC algorithm.

Sampling and numerical search algorithms generally follow two approaches. MCMC algorithms generate a new realization of the model parameters at each iteration (following a burn-in period). When the model converges, each iteration is an element or observation from the posterior distribution.

An alternative approach to MCMC algorithms are algorithms that first generate a large number of candidate values for each parameter using Monte Carlo simulation. Bayesian logic combines the new evidence, denoted by E, with the Monte Carlo parameter estimates to select or reweight a subset of the Monte Carlo generated parameters. In this application, the new evidence will be the illness count from a public health surveillance system (i.e., $E = I_{observed}$).

The algorithm we employ is the sampling importance resampling (SIR) approach proposed by Rubin (1987). This method generates an unequal probability sample where the sample weights are determined by the degree of agreement between the prior information and the new sampling evidence.

To demonstrate, let θ represent a vector of model inputs. Examples of model inputs are parameters describing the contamination distribution and dose-response function. The input parameters are not fixed values, so uncertainty is represented by the distribution $p(\theta)$. This distribution is referred to as the prior or pre-model distribution for the inputs. Consider a process model that predicts, among other parameters, the rate parameter describing the number of human illnesses (Equation 1). Let this model be denoted by $M()$. For a randomly sampled θ from $p(\theta)$, the observed output from the process model is $M(\theta)$. The $M(\theta)$ value will be compared to the observed number of illnesses from public health surveillance, which is denoted by $I_{observed}$.

The algorithm for implementing the SIR is:

1. Draw N samples $(\theta_1, \theta_2, ...\theta_N)$ from the prior distribution $p(\theta)$.
2. For each θ_i, use the model to determine $M(\theta_i)$.
3. Determine a weight for each $M(\theta_i)$. This weight describes the agreement between the model prediction and observe number of illnesses. For this application, the weight is $w_i = P(I_{observed}|M(\theta))$.

4. Draw an unequal probability with-replacement sample of size $m << N$ from $(\theta_1, \theta_2, ...\theta_N)$ using sample weights w_i.

As $N/m \to \infty$ the SIR algorithm produces an exact sample from the posterior distribution. Previous studies have found that values for N/m ranging from 20 to 40 are often sufficient (Rubin, 1987), but appropriate values must be considered on a case-by-case basis.

To illustrate the SIR algorithm, consider the problem of estimating the prevalence θ of a disease in a herd of animals when new sampling evidence is combined with prior information. Suppose the new evidence is a sample from the herd of size $n = 20$ of which $s = 4$ samples are positive. Suppose the prior evidence on the prevalence in the herd can be summarized by a beta distribution of the form $\theta \sim Beta(a = 1, b = 6)$. In this example the model, $M(\theta)$, uses the prior information on prevalence, and the number of additional test results, to predict the number of infected animals. Using the model $s \sim Binomial(n, \theta)$ and Bayes formula, it is known that the distribution the resulting posterior distribution $p(\theta|s, n) \sim Beta(s + a, n - s + b) = Beta(5, 22)$. The following *R* code (R Development Core Team, 2011) demonstrates the SIR algorithm and illustrates the equivalence of the SIR solution and the known posterior distribution.

```
# example of the beta-binomial by SIR

N=200000

# parameters for the beta prior
a=1
b=6

# evidence. 4-successes 20-trials
s=4
n=20

# draw a large sample from the prior (the "Sampling" step)
p.prior=rbeta(N,a,b)

# the "Importance" weight of each sample is determine by the likelihood
w=dbinom(s,n,p.prior)

# the "Resampling" step draws a weighted sample from the prior
m=round(N/40)
index=sample(seq(1,N),size=m,replace=T,prob=w)
p.posterior=p.prior[index]
```

6. Example

Previously published examples of the proposed probabilistic framework and Bayesian method (Williams & Ebel, 2012; Williams, Ebel & Vose, 2011a) focus on the prediction of changes in human illness in a farm-to-table model. The following example is a departure from the farm-to-table model and it is presented to highlight similarities and the utility of the proposed framework in applications where the data represent a surveillance system.

Surveillance sensitivity, in the context of testing for pathogens in food, is the probability that the pathogen is detected given that it exists in a sampled unit. Surveillance sensitivity is a function of test sensitivity in the sense that not only does a contaminated unit need to be sampled, but the results of testing must properly classify the unit as containing the pathogen of interest. When the sensitivity of a test is *Se*, and the number of units in the population is large in relation to the number of samples collected (n), and $p(S^+)$ is the proportion of the units that are contaminated. Then the surveillance system sensitivity is typically given by

$$P(\text{detecting one or more positives}) = 1 - (1 - Se \times p(S^+))^n. \quad (11)$$

Note that the role of test sensitivity is to modify the true prevalence term $p(S^+))$ to provide the apparent prevalence.

The concern with the standard approaches used in defining *Se* and equation 11 is that test sensitivity can decrease as the level of the pathogen drops, especially in cases where the average concentration is less than 1 cfu/tested unit (e.g., a test that uses 10 g of a 100 g food unit with only 1 organism will have an average concentration of 0.1 *cfu/g*).

In the testing for pathogens that occur at low levels, test sensitivity can be improved by employing enrichment techniques, increasing incubation time, and increasing the volume of material sampled. All of these methods increase the number of pathogens in the medium to be tested. The bonding and potential encapsulation of a microbe within fatty tissues, insufficient time for cells to leave a quiescent state during incubation, the possibility of cells entering a viable but nonculturable state (Oliver, 2005; Oliver et al., 2005), and the small volume of material tested all can lead to reductions in test sensitivity.

In this study, a situation is examined in which the prevalence of positive samples doubled over a one-year period. During this time, a minor modification was made to the enrichment technique used for testing. While the laboratory had performed testing to determine the equivalence of the new and old methodology, insufficient evidence exists to determine if the observed increase in prevalence is due to the change of the enrichment methodology or whether the change was due to an actual change in pathogen prevalence.

This study is predicated on the assumption that the observed increase in *E.coli* O157:H7 positive samples is the result of a change in enrichment media, rather than an actual increase in contamination. The analyses presented assess what the change in test sensitivity would be if this assumption where true. This initial analysis is used to specify both the required sample and the concentration of *E.coli* O157:H7 used to spike validation samples.

7. Data description

The Food Safety and Inspection Service of the United States Department of Agriculture has been collecting ground beef samples from all slaughter and grinding facilities producing ground beef products since the beginning of the year 2000. Nevertheless, the enumeration of positive samples was only begun in January 2007. This more limited dataset was used in this analysis. Each facility that produces ground beef for distribution is sampled on multiple occasions every year and the sample unit consists of approximately 900 *g* of ground beef collected at the end of production from approximately 5,000 kg lots.

The data used in this study spans the time period between January of 2007 and December 2009. The annual number of sampled collected was 9401, 10760, and 10774 for each of the three years. The number of positive samples observed in each year was 26,43 and 35, with the change in the enrichment media occurring in January of 2008.

The recorded data values are the result of two tests. The first being a qualitative test that provides a positive or negative result for each sample. The sample for this qualitative PCR test that consists of a 325 *g* subsample of the original 900 *g* sample that is incubated for 24 hours using an enrichment broth. The size of the sample implies a detection limit of 0.003 *cfu/g*.

The second test is only performed on the qualitatively positive samples. This test provides an estimate of the number of organisms per gram, derived from an additional 33.3 *g* sample of the remaining portion of the original 900 *g*. The estimate is derived using the most probable number (MPN) method (Harrigan, 1998) using three dilutions and three tubes per dilution. The dilutions used for MPN testing were 10, 1, and 0.1 *g*. Thus, the 33.3 *g* sample is divided into 3 10-*g*, 3 1-*g* and 3 0.1-*g* subsamples, incubated in a test tube, and tested for *E. coli* O157:H7 using the same PCR technique.

The two tests have different levels of detection and there are three possible outcomes for any sample (i.e., -/-,+/-, +/(estimated number or organisms per g)). Samples that are only positive on the first test will be referred to as qualitatively positive, while samples that were positive on both tests will be referred to as quantitatively positive.

The Food Safety Inspection Service (FSIS, 1998) provides additional descriptions of the testing methodology since the inception of the program.

8. Methods

8.1 Estimating the distribution for the average concentration of *E.coli* O157:H7

Ground beef at the point of production comprise no natural units, so it can be viewed as a homogeneously mixed product that mimics a viscous liquid. As such, the estimation of the levels of pathogens in ground beef follows a similar methodology as used for describing the distribution of contaminates in other fluids, such as water. The grinding process inevitably introduces some microscopic voids in the medium, so the unit of measurement is typically in grams rather than milliliters.

The sampling data are censored in the sense that the true number of organisms is not observed for all samples. This occurs because the test has a level of detection (*LOD*) at which the probability of a positive is low even though the original sample contained one or more viable organisms (Helsel, 2005).

Contamination generally occurs at very low levels for the vast majority of ground beef production. Nevertheless, there are situations where high contamination levels can occur. The commonly used biologically plausible model assumes that the average concentration of contamination varies according to a Lognormal distribution (i.e.,$X \sim Lognormal(\mu, \sigma^2)$ (Haas et al., 1999).

The parameters of the contamination distribution can be estimated using maximum likelihood. To account for the different levels of detection, the total number of samples, N, can be broken down into $N = N_{neg} + N_{qual+} + N_{quan+}$. The two levels of detection are given by $LOD_{qual+} = 0.003$ and $LOD_{quan+} = 0.03$.

The likelihood is given by

$$L = \left[F(LOD_{qual+})\right]^{N_{neg}} \times \left[F(LOD_{quan+}) - F(LOD_{qual+})\right]^{N_{qual+}} \times \prod_{i=1}^{N_{quan+}} f(x_i), \tag{12}$$

where x_i is the estimated number of organisms from the MPN analysis for the ith quantitatively positive sample, and F and f represent the cumulative and probability density function of the Lognormal distribution of contamination levels. Applying the specific LOD values for this application yields;

$$L = \left[F(0.003)\right]^{N_{neg}} \times \left[F(0.03) - F(0.003)\right]^{N_{qual+}} \times \prod_{i=1}^{N_{quan+}} f(x_i). \tag{13}$$

A nonlinear optimization routine (*optim* in R) was used to estimate $\hat{\mu}$ and $\hat{\sigma}^2$. This routine also provides the variance-covariance matrix so that $var[\hat{\mu}], var[\hat{\sigma}^2]$ and $cov[\hat{\mu}, \hat{\sigma}^2]$ are available for further analyses. These estimates will be used as hyper-priors describing uncertainty in the estimated pathogen levels.

8.2 Defining test sensitivity as a function of pathogen level

The process under which a positive test occurs requires two sequential outcomes:

1. The orignal 325 g sample must contain one or more viable organisms.
2. The DNA in a cell must be successfully amplified by PCR for detection to occur.

Two factors can affect the probability of detection. The first is that material from the single cell undergoes successful amplification. The alternative is that the cell leaves its quiescent state during the enrichment phase and begins the process of exponential growth.

Suppose there are Z cells in the sample and for each cell there is a probability Q of detection. Given the probability of detection for an individual cell, and the number of cells, the probability of a positive test is defined as the compliment of non-detection for each of the Z organisms, therefore

$$P(T+|Q = q, Z = z) = 1 - (1-q)^z. \tag{14}$$

This probability assumes fixed values for Q and Z, though there is likely to be variation among organisms as well as variation in the number of organisms.

Suppose for each cell in the sample the probability, Q, of detection is distributed in the population as a Beta distribution with parameters α and β. Similarly, let Z represent the number of viable organisms per gram initially in the sample. Assuming that the organisms are uniformly distributed in the medium, Z can be modeled as a Poisson distribution. The rate parameter (X) for any given sample is determined by a draw from the Lognormal distribution describing *E. coli* O157:H7 levels previously described.

A two-step process is used to incorporate the variability in Q and Z. First, assume a fixed probability of detection for each organism. Applying the Theorem of Total Probability across the Poisson distributed organism counts, with rate parameter X, gives

$$\begin{aligned} P(T+|Q=q,X=x) &= \sum_{z=0}^{\infty} P(T+|Q=q,X=x,Z=z)P(X=x,Z=z) \\ &= \sum_{z=0}^{\infty}(1-(1-q)^z)\frac{e^{-x}x^{-z}}{z!} \\ &= 1-e^{-x}\sum_{z=0}^{\infty}\frac{((1-q)x)^{-z}}{z!} \\ &= 1-e^{-x}\times e^{(x-qx)} \\ &= 1-e^{-qx}. \end{aligned} \tag{15}$$

The probability of detection for each cell is likely to vary across the population due to factors such as genetic diversity and the level of damage the cell may have received during the production process due to the possible application of antimicrobials and steam to carcasses, as well as drying, freezing and storage times prior to sample collection and testing. Assuming that the detection probability is $Q \sim Beta(\alpha,\beta)$ and rate parameter X, the probability of detection is given by:

$$\begin{aligned} P(T+|X=x) &= \int_0^1 P(T+|Q=q,X=x)f(Q=q)dq \\ &= \int_0^1 (1-e^{-qx})\frac{\Gamma(\alpha+\beta)}{\Gamma(\alpha)\Gamma(\beta)}q^{\alpha-1}(1-q)^{\beta-1}dq. \end{aligned} \tag{16}$$

The integral in this expression describes a confluent hypergeometric function. Using the results of (Haas et al., 1999) the exact solution can be written as

$$P(T+|X=x) = \frac{\Gamma(\alpha+\beta)}{\Gamma(\alpha)}\sum_{i=0}^{\infty}\frac{\Gamma(\alpha+i)}{\Gamma(\alpha+\beta+i)}\frac{(-1)^{i-1}x^i}{i!}. \tag{17}$$

The ultimate goal is to determine the probability of a positive test across the distribution of possible contamination levels. If $f_X(x)$ is the Lognormal probability density function describing contamination levels across the population of ground beef, then

$$P(T+) = \int_0^{\infty} P(T+|X=x)f_X(x)dx, \tag{18}$$

where $P(T+|X=x)$ is as defined in either 16 or 17. Note that the series expansion for the confluent hypergeometric in Equation 17 includes an alternating power series of x^i, and the range of integration for x is $(0,\infty)$. This causes numerical problems for levels of x greater than approximately 50 cfus and precludes the use of the numerical approximations based on the series expansion.

A solution to the problem is to combine 18 and 16 as

$$P(T+) = \int_0^{\infty}\int_0^1 (1-e^{-qx})\frac{\Gamma(\alpha+\beta)}{\Gamma(\alpha)\Gamma(\beta)}q^{\alpha-1}(1-q)^{\beta-1}dqf_X(x)dqdx. \tag{19}$$

The evaluation of the double integral that results from combining 18 and 16 is numerically difficult and time consuming to reliably implement in a Monte Carlo simulation. One solution is to use computer hardware and software that accommodates parallel processing, such as the *Snowfall* package in *R* (Knaus et al., 2009; Rossini et al., 2007).

Another alternative comes from noting that the model in Equation 16 also appears in epidemiology and food-safety applications where it is functionally identical to a beta-Poisson dose-response function (Haas et al., 1999). In this setting the model is used to determine the probability of illness $(P(ill))$ with a Poisson distributed number of organisms with mean value X, with an individual dose-response

$$P(ill|D = d) = 1 - (1 - q)^d. \tag{20}$$

In this equation $P(ill|D = d)$ is the individual probability for illness, d the individual dose, and q an individual measure of susceptibility that is distributed in the population as a beta distribution with parameters α and β. Note that this expression is equivalent to the probability of illness given exposure $(P(ill|exp))$ defined previously.

An approximation to the exact solution in 18 has been derived for the beta-Poisson dose-response equation. The approximation is

$$P(T + |X = x) = 1 - \left(1 + \frac{x}{\beta}\right)^{-\alpha}. \tag{21}$$

Furumoto & Mickey (1967) show, via a Taylor's series expansion, that this approximation is sufficiently accurate for $\beta > \alpha$ and x reasonably small. Prior experience indicates that the number of organisms is low for this application, however, the relationship between β and α is not known *a priori*. This model produces a sigmoidal curve in log-space and the interpretation of the two parameters is that α controls the slope of the curve while β controls the location.

The ultimate goal is to determine the probability of a positive test across the distribution of possible contamination levels. If $f_X(x)$ is the Lognormal probability density function describing contamination levels across the population of ground beef, then

$$P(T+) = \int_0^\infty 1 - (1 + \frac{x}{\beta})^{-\alpha} f_X(x)dx. \tag{22}$$

Two different probabilities of a positive test are of interest. The first is as described in Equation 22, which is assumed to be the probability of a positive test prior to the change in enrichment methodology. The sensitivity of the test is greatly improved if the organism has sufficient time to undergo the process of replication. If, as postulated, the increase in the number of positives samples is a side-effect of an improvement in enrichment media, the probability of detection for each organism would be greater. The magnitude of this increase would be captured in the parameters that describe the probability of detection. Let the parameters α_{new} and β_{new} represent the postulated increase in test sensitivity associated with the new enrichment technique, then the probability of a positive test is

$$P_{new}(T+) = \int_0^\infty 1 - (1 + \frac{x}{\beta_{new}})^{-\alpha_{new}} f_X(x)dx \tag{23}$$

8.3 Description of surveillance data

The goal of the analysis is to quantify the parameters $\alpha, \beta, \alpha_{new}, \beta_{new}$. The differences in the resulting test sensitivity models can be used to inform additional analyses related to determining the necessary sample size to test for equivalence of the two enrichment media.

The enumerated values from the surveillance data are used to characterize the distribution of contamination in ground beef (i.e., $f_X(x)$). These surveillance data from 2000-2006 represent what is thought to be a time period of relative stability in the level of contamination and are characterized by annual rate of positives samples of approximately 1 positive for every 500 samples (0.2 %).

In the following three years there were $T_{2007} = 26$, $T_{2008} = 43$ and $T_{2009} = 35$ positive tests observed. The number of tests performed was $N_{2007} = 9,401$, $N_{2008} = 10,760$ and $N_{2009} = 10,774$. The 2007 data represent the last year during which no changes were made to the laboratory protocols. The 2008 and 2009 data represent the first years in which the new enrichment technique was implemented and there was an immediate increase in the number of positives samples observed, with the percent positive rate jumping from a long-term average of approximately 0.20 to 0.36%.

8.4 Bayesian model

Application of the SIR routine requires the definition of the model $(M(\theta_i)$ and the likelihood equation used to determine the sample weights (i.e., $w_i = P(E|M(\theta))$.

In this application, the model generates estimates of the number of positive tests during the 2007 and the combined time period 2008-2009. Define these two predictions as $M_{2007}(\theta_{2007}) = N_{2007}P(T+)$ and $M_{2008-9}(\theta_{2008-9}) = (N_{2008} + N_{2009})P_{new}(T+)$. The parameter vectors consist of $\theta_{2007} = (\mu, \sigma^2, \alpha, \beta)$ and $\theta_{2008-9} = (\mu, \sigma^2, \alpha_{new}, \beta_{new})$. Uncertainty in the μ, σ^2 parameters is modeled by a multivariate Normal distribution where the mean values and variance-covariance matrix were derived from the maximum likelihood solution of Equation 13. Uniform priors were used for the α and β parameters of the test sensitivity models. Equations 22 and 23 were used to calculate $P(T+)$ and $(P_{new}(T+)$, respectively.

There is no evidence of plant-level clustering amongst the positive tests collected across the year 2000-2009, so it is reasonable to assume that the number of positive tests can be modeled as a Poisson distribution, giving

$$T_{2007} \sim Poisson(N_{2007}P(T+)) \tag{24}$$

and

$$T_{2008-9} \sim Poisson((N_{2008} + N_{2009})P_{new}(T+)). \tag{25}$$

Each of these expressions serves as a likelihood function in a Bayesian model where uncertainty in $P(T+)$ and $P_{new}(T+)$ provide extra Poisson variability. The model was solved using the SIR algorithm where the weights w_i were the product of the two likelihoods, so

$$w_i = P(T_{2007}|N_{2007}P(T+)) \times P(T_{2008-9}|(N_{2008} + N_{2009})P_{new}(T+)). \tag{26}$$

The linkage between the two likelihoods is the common distribution explaining the level of contamination. The sample sizes used in the SIR algorithm, were $N = 1,000,000$ and $m = 25,000$.

9. Results

The test sensitivity models are essentially equivalent for extremely low concentrations and for concentrations substantially greater than the LOD, with the probability of detection being near zero at low levels and essentially one at higher concentrations (Figure 3). Any laboratory testing to determine the equivalence between the two enrichment methods could require an enormous sample size to detect a significant difference if the concentration were too high because essentially no difference exists between the two methods when samples contain more than four or five viable cells. However, there exists a range over which the models no longer overlap Figure. 3.

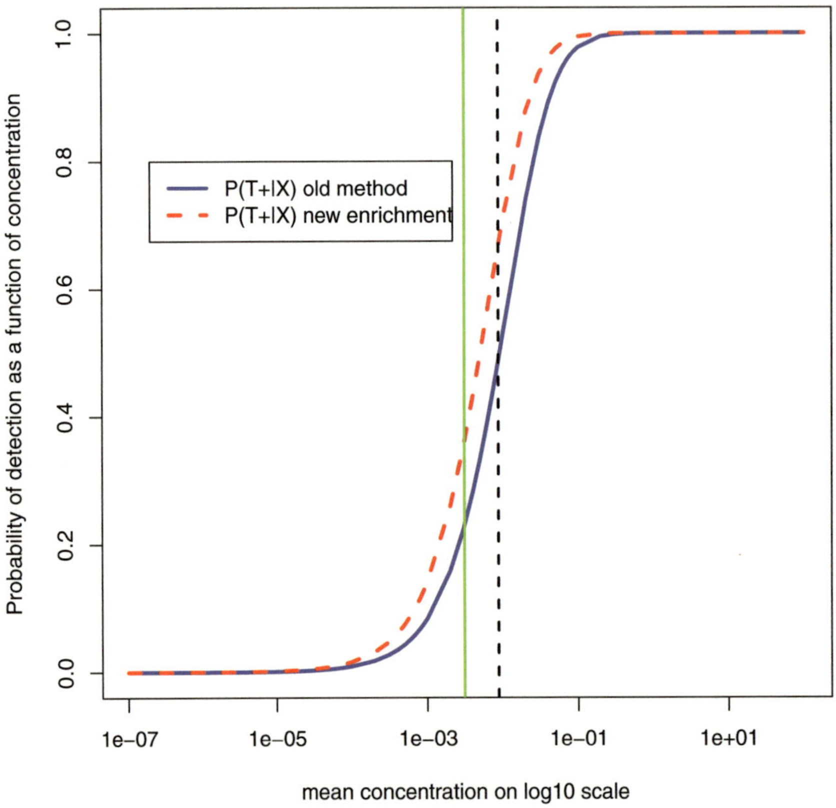

Fig. 3. Summary of the test sensitivity models. The discrepancies between the two models predominantly occur at concentrations much lower than 1 cfu per g. The *LOD* is given by the solid vertical line. The dashed vertical line indicates the concentration at which the maximum difference in sensitivity between the old and new enrichment methods occurs.

The vertical line in Figure 3 represent the mean concentration at the theoretical *LOD* of a single organism. If this is interpreted to mean that the test could truly identify 1 organism in a 325 *g* sample, than there will be one or more organisms contained in the sample with probability $P(T+) = 1 - e^{-1} = 0.63$. A test with less than perfect sensitivity will have a lower probability of detection. The probabilities of detection at this concentration for both tests are substantially less than this theoretical upper bound, with the probabilities of a positive test being 0.21 and 0.40.

The level of the target organism at which the maximum difference between the two tests occurs can be estimated from the two models and is presented graphically in Figure 4. This value could then be used to determine both the appropriate concentration for testing and the number of samples required to achieve a test with a specified power. From a practical standpoint this may not be a reasonable approach, given that for this application the concentrations are very low and it becomes very difficult to accurately dilute concentrations at these low levels.

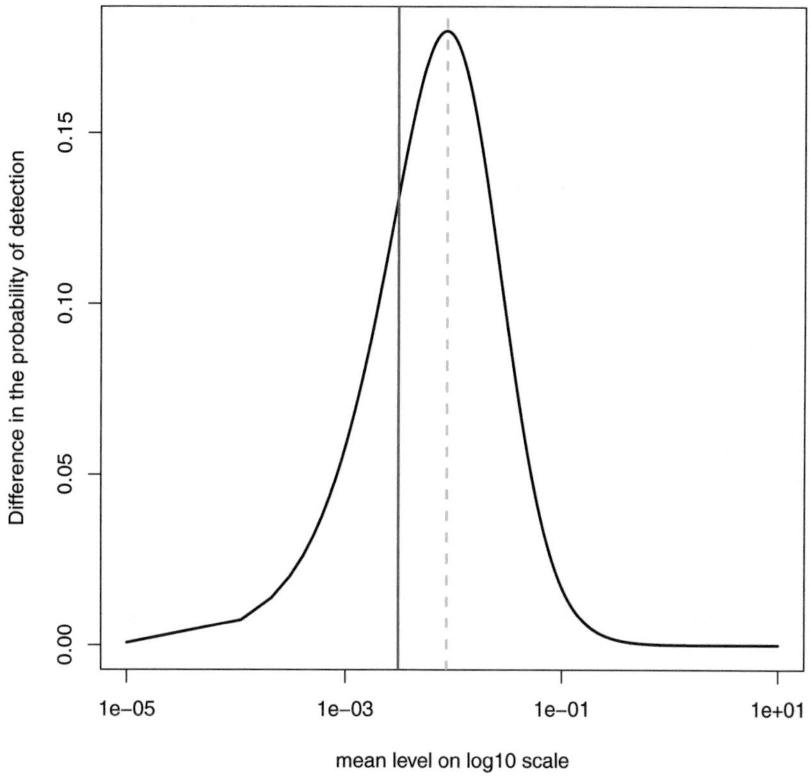

Fig. 4. Results of the model can be used to determine the average concentration at which the maximum discrepancies occurs. This concentration can be used in a validation test to determine if differences in the two enrichment media exist. The *LOD* is given by the solid vertical line. The dashed vertical line indicates the concentration at which the maximum difference in sensitively between the old and new enrichment methods occurs.

The solid vertical line on Figure 4 represents the difference in the probability of detection at the mean level of 1 organisms per 325 *g*, which is the *LOD* for the current sampling program. The maximum difference between the two test sensitivity models occurs for a concentration of roughly 1 organism per 114 *g* (dashed vertical line), at which point the probability of a positive test result is 0.45 and 0.67 for the old and new enrichment techniques, respectively.

In retrospect is seems fairly obvious that equivalency testing for enrichment methodologies should be performed at levels near the *LOD*. The reasoning for this claim is twofold:

1. Qualitative tests, such as PCR methods, are so sensitive that only a small number of cells are required for detection.
2. The purpose of enrichment techniques is to selectively instigate reproduction of the target organism. Given the exponential growth in the number of organisms once reproduction has begun, a positive test result is expected.

10. Conclusions

There are surprising number of applications where this simplified framework can be used to integrate surveillance data with data describing basic demographics of the population (e.g., prevalence and/or levels of contamination) and counts derived from a surveillance system. The advantage of using Bayesian methods with simplified risk assessment models is their ability to combine the available data, objectively calibrate the model to match the surveillance data, and to estimate latent variables in the model for which there are few data.

11. Acknowledgement

This research utilized the Colorado State University ISTeC Cray HPC System supported by NSF Grant CNS-0923386.

12. References

Albert, I., Grenier, E., Denis, J. & Rousseau, J. (2008). Quantitative risk assessment from farm to fork and beyond: A global bayesian approach concerning food-borne diseases, *Risk Analysis* 28: 557–571.

Allos, B., Moore, M., Griffin, P. & Tauxe, R. (2004). Surveillance for sporadic foodborne disease in the 21st century: The foodnet perspective, *Clinical Infectious Diseases* 38: 115–120.

Bartholomew, M., Vose, D., Tollefson, L. & Travis, C. (2005). A linear model for managing the risk of antimicrobial resistance originating in food animals, *Risk Analysis* 25(1): 99–108.

de Jong B & K., E. (2006). The comparative burden of salmonellosis in the european unionmember states, associated and candidate countries, *BMC Public Health* 6(4). URL: *www.biomedcentral.com/1471-2458/6/4*

Ebel, E., Williams, M. & Schlosser, W. (2012). Parametric distributions of under-diagnosis parameters used to estimate annual burden of illness for five foodborne pathogens, *Journal of Food Protection* 74: (in press).

FSIS (1998). Microbiology laboratory guidebook. 3rd ed, *Technical report*, Food Safety and Inspection Service, US Department of Agriculture. `http://www.fsis.usda.gov/Science/Microbiological _Lab_Guidebook/index.asp`; accessed October 23, 2011.

Furumoto, W. & Mickey, R. (1967). A mathematical model for the infectivity-dilution curve of tobacco mosaic virus: Theoretical considerations, *Virology* 32: 216–223.

Givens, G. H. (1993). *A Bayesian framework and importance of sampling methods for synthesizing multiple sources of evidence and uncertainty linked by a complex process model*, PhD thesis, University of Washington, Seattle, Washington.

Haas, C., Rose, J. & Gerba, C. (1999). *Quantitative Microbial Risk Assessment*, John Wiley & Sons, New York, New York.

Hald, T., Vose, D., Wegener, H. & Koupeev, T. (2004). A bayesian approach to quantify the contribution of animal-food sources to human salmonellosis, *Risk Analysis* 24: 255–269.

Harrigan, W. (1998). *Laboratory Methods in Food Microbiology*, Academic Press, San Diego, California.

Helsel, D. (2005). *Nondetects and Data Analysis: Statistics for Censored Environmental Data*, John Wiley & Sons., Hoboken, New Jersey.

Herikstad, H., Motarjemi, Y. & Tauxe, R. (2002). Salmonella surveillance: A global survey of public health serotyping, *Epidemiology and Infection* 129(1): 1–8.

Knaus, J., Porzelius, C., Binder, H. & Schwarzer, G. (2009). Easier parallel computing in r with snowfall and sfcluster, *The R Journal* 1: 54–59.

Limpert, E., Stahel, W. & Abbt, M. (2001). Log-normal distributions across the sciences: Keys and clues, *Bioscience* 51: 341–352.

Lunn, D., Spiegelhalter, D., Thomas, A. & Best, N. (2009). The bugs project: Evolution, critique and future directions (with discussion), *Statistics in Medicine* 28: 3049–3082.

Lunn, D., Thomas, A., Best, N. & Spiegelhalter, D. (2000). Winbugs - a bayesian modelling framework: Concepts, structure, and extensibility, *Statistics and Computing* 10(4): 325–337.

Oliver, J. (2005). The viable but nonculturable state in bacteria, *Journal of Microbiology* 43: 93–100.

Oliver, J., Dagher, M. & Linden (2005). Induction of *escherichia coli* and *salmonella typhimurium* into the viable but nonculturable state following chlorination of wastewater, *Journal of Water Health* 3: 249–257.

Parsonsa, D., Ortona, T., D'Souza, J., Mooreb, A., Jonesc, R. & Dodd, C. (2005). A comparison of three modelling approaches for quantitative risk assessment using the case study of *salmonella* spp. in poultry meat, *International Journal of Food Microbiology* 98: 35–51.

R Development Core Team (2011). *R: A Language and Environment for Statistical Computing*, R Foundation for Statistical Computing, Vienna, Austria. ISBN 3-900051-07-0. URL: *http://www.R-project.org/*

Raftery, A., Givens, G. & Zeh, J. (1995). Inference from a deterministic population dynamics model for bowhead whales, *Journal of the American Statistical Society* 90(430): 402–416.

Rossini, A., Tierney, L. & Li., N. (2007). Simple parallel statistical computing in r, *Journal of Computational and Graphical Statistics* 16(2): 399–420.

Rubin, D. (1987). The calculation of posterior distributions by data augmentation: Comment: A noniterative sampling/importance resampling alternative to the data augmentation algorithm for creating a few imputations when fractions of missing information are modest: The sir algorithm, *Journal of the American Statistical Society* 82(398): 543–546.

Scallan, E., Hoekstra, R., Angulo, F., Tauxe, R., Widdowson, M., Roy, S., Jones, J. & Griffin, P. (2011). Foodborne illness acquired in the united states major pathogens, *Emerging Infectious Diseases* 17(1): 7–15.

Vose, D. (2008). *Risk Analysis: A Quantitative Guide. 3rd edition*, John Wiley & Sons, West Sussex, England.

Williams, M. & Ebel, E. (2012). Estimating changes in public health following implementation of hazard analysis and critical control point in the united states broiler slaughter industry, *Foodborne Pathogens and Disease* 9: 59-67.

Williams, M., Ebel, E. & Hoeting, J. (2011). Bayesian analysis for food-safety risk assessment: Evaluation of dose-response functions within winbugs, *Journal of Statistical Software* 43: 1–14.

Williams, M., Ebel, E. & Vose, D. (2011a). Framework for microbial food-safety risk assessments amenable to bayesian modeling, *Risk Analysis* 31(4): 548–565.

Williams, M., Ebel, E. & Vose, D. (2011b). Methodology for determining the appropriateness of a linear dose-response function, *Risk Analysis* 31(3): 345–350.

17

Polynomial Response Surfaces for Probabilistic Risk Assessment and Risk Control via Robust Design

Sergey Oladyshkin and Wolfgang Nowak
SRC Simulation Technology, Institute of Hydraulic Engineering, University of Stuttgart
Germany

1. Introduction

Many engineering systems represent challenging classes of complex dynamic systems. Lacking information about their system properties leads to model uncertainties up to a level where quantification of uncertainties may become the dominant question in modeling, simulation and application tasks. Uncertainty quantification is the prerequisite for probabilistic risk assessment and related tasks. Current numerical simulation models are often too expensive for advanced application tasks that involve accurate uncertainty quantification, risk assessment and robust design. This Chapter will present recent approaches for these challenges based on polynomial response surface techniques, which reduce massively the initial complex model at surprising accuracy. The reduction is achieved via projections on orthonormal polynomial bases, which form a so-called response surface. This way, the model response to changes in uncertain parameters and design or control variables is represented by multi-variate polynomials for each output quantity of interest. This technique is known as polynomial chaos expansion (PCE) in the field of stochastic PDE solutions. The reduced model represented by the response surface is vastly faster than the original complex one, and thus provides a promising starting point for follow-up tasks: global sensitivity analysis, uncertainty quantification and probabilistic risk assessment and as well as robust design and control under uncertainty. Often, the fact that the response surface has known polynomial properties can further simplify these tasks. We will emphasize a more engineering-like language as compared to otherwise intense mathematical derivations found in the literature on PCE. Also we will make use of most recent developments in the theory of stochastic PDE solutions for engineering applications. The current Chapter provides tools based on PCE for global sensitivity analysis, uncertainty quantification and risk analysis as well as design under uncertainty (robust design).

2. Response surface via polynomial chaos expansion

In the present Chapter, we consider the response surface in a closed polynomial form. Obviously, a response surface can be constructed in different ways, e.g. it can be constructed directly on a dense Cartesian grid of input parameters at extremely high computational efforts. Likewise, conceptually straightforward numerical Monte Carlo (MC) simulation techniques are computationally demanding since the statistical accuracy of their predictions depend on the number of realizations used. In this Chapter we explore an alternative methodology which demands anly minimum number of model evaluations to construct a response surface (see details in Section 2.3).

Our alternative approach is through the polynomial chaos expansion (PCE) introduced by Wiener (1938). Generally, all PCE techniques can be viewed as an efficient approximation to full-blown stochastic modeling (e.g., exhaustive MC). The basic idea is to represent the response of a model to changes in variables through a response surface that is defined with the help of an orthonormal polynomial basis in the parameter space. In simple words, the dependence of model output on all relevant input parameters is approximated by a high-dimensional polynomial. This projection can be interpreted as an advanced approach to statistical regression. The PCE offers an efficient and accurate high-order way of including non-linear effects in stochastic analysis, see e.g. Fajraoui et al. (2011); Foo & Karniadakis (2010); Zhang & Lu (2004). One of the attractive features of PCE is the high-order approximation of error propagation (Ghanem & Spanos, 1991; 1990) as well as its computational speed when compared to MC (Oladyshkin, Class, Helmig & Nowak, 2011b).

The original PCE concept (Wiener, 1938) can be used only for Gaussian distributed input parameters. In recent years, the classical PCE technique was extended to the generalized polynomial chaos (gPC) (Wan & Karniadakis, 2006) which accommodates for the use of an increased, yet limited number of statistical distributions (Askey & Wilson, 1985). The PCE methods discussed above assume an exact knowledge of the probability density functions of all input parameters and they are optimal only when applied to a finite number of certain parametric probability distributions. Unfortunately, information about the distribution of data or input parameters is very limited in many realistic applications, especially in environmental engineering and sciences. Data that characterize model parameters often indicate a variety of statistical distribution shapes (e.g., bounded, skewed, multi-modal, discontinuous, etc). Also, empirical parameter distributions derived from raw data sets do in general not follow analytically known distribution shapes. For such reasons, application tasks demand further adaptation of the chaos expansion technique to a larger spectrum of distributions.

To accommodate for a wide range of data distributions, a recent generalization of PCE is the arbitrary polynomial chaos (aPC). There are only very few studies that have used aPC before, and they can only be found in mathematical stochastics (Ghanem & Doostan, 2006; Soize & Ghanem, 2004) and aerospace engineering (Witteveen & Bijl, 2006; Witteveen et al., 2007). These studies focused on proofs of existence, constructing the basis by Gram-Schmidt orthogonalization and on related, quite theoretical issues. A notable exception is the very recent study of Li et al. (2011) in the field of petroleum engineering. That study did not discuss, however, the aPC in the light of data availability which plays a critical role in applications such as those presented by Oladyshkin, Class, Helmig & Nowak (2011a); Oladyshkin & Nowak (2011).

Compared to earlier PCE techniques, the aPC adapts to arbitrary probability distribution shapes of input parameters and, in addition, can even work with unknown distribution shapes when only a few statistical moments can be inferred from limited data or from expert elicitation. The arbitrary distributions for the framework can be either discrete, continuous, or discretized continuous. They can be specified either analytically (as probability density/cumulative distribution functions), numerically as histogram or as a raw data sets.

The aPC approach provides improved convergence in comparison to classical PCE techniques, when applied to input distributions that fall outside the range of classical PCE (Oladyshkin & Nowak, 2011). The necessity to adapt to arbitrary distribution is discussed in more details in Section 4 of the Chapter.

PCE techniques can mainly be sub-divided into intrusive (Ghanem & Spanos, 1993; Matthies & Keese., 2005; Xiu & Karniadakis, 2003) and non-intrusive (Isukapalli et al., 1998; Keese & Matthies, 2003; Li & Zhang, 2007; Oladyshkin, Class, Helmig & Nowak, 2011b) approaches, i.e., methods that require or do not require modifications in the system of governing equations and corresponding changes in simulation codes. Intrusive approaches require symbolic manipulations of the governing equations and can sometimes provide semi-analytical solutions for stochastic analyses of simple problems. The most well-known method from this group is the stochastic Galerkin technique. However, the necessary symbolic manipulations may become very complex and analytically cumbersome, and cannot easily be implemented in commercial codes. For this reason, non-intrusive approaches like sparse quadrature and the probabilistic collocation method (PCM: see Li & Zhang (2007); Oladyshkin, Class, Helmig & Nowak (2011b)) have lately been receiving a quickly increasing attention. In a simple sense, PCM can be interpreted as a smart (mathematically optimal) interpolation rule of model output between different parameter sets. The polynomial interpolation may be interpreted as a response surface of the model. It is based on a minimal and optimally chosen set of model evaluations, each with a defined set of model parameters (called collocation points). The challenge here is to find a compromise between computational effort and a reasonable approximation of the physical processes.

2.1 Definitions and polynomial chaos expansion

In this Section, we will introduce the aPC-based framework. aPC is for arbitrary Polynomial Chaos which is a most recent generalization of PCE methods (see Section 2.2). Let $\boldsymbol{\omega} = \{\omega_1,...,\omega_N\}$ represent the vector of N input parameters for some model $\Omega = f(\boldsymbol{\omega})$. The model $\Omega(\omega)$ may be an explicit or implicit expression (e.g., a partial or ordinary differential equation or a coupled system). To perform sensitivity analysis, uncertainty quantification and risk assessment, we wish to investigate the influence of all parameters $\boldsymbol{\omega}$ on the model output Ω. The model output can be space and time dependent, i.e. $\mathbf{x} = (x_1,x_2,x_3)$.

In the following, we will approximate the model response by a truncated polynomial expansion for each point in space $\mathbf{x}$ and time t. According to polynomial chaos theory (Wiener, 1938), the model output Ω can be approximated by polynomials $\Psi_j(\boldsymbol{\omega})$ as follows:

$$\Omega(\mathbf{x}, t; \boldsymbol{\omega}) \approx \sum_{j=0}^{M} c_j(\mathbf{x}, t)\Psi_j(\boldsymbol{\omega}), \tag{1}$$

where the number M of polynomials depends on the total number of analyzed input parameters (N) and the order d of the polynomial representation, according to the combinatoric formula $M = (N+d)!/(N!d!) - 1$. The coefficients c_j in Eq. (1) quantify the dependence of the model output Ω on the input parameters $\boldsymbol{\omega}$ for each desired point in space $\mathbf{x}$ and time t. The symbol Ψ_j is a simplified notation of the multi-variate orthonormal polynomial basis for $\boldsymbol{\omega}$ including all cross-terms between different parameters, as explained below. Let us mention that, in the current state of science for polynomial chaos expansion, the random variables have to be statistically independent or may be linearly correlated, which can be removed by adequate linear transformation. Construction of a polynomial basis for statistically dependent random variables beyond linear dependence is a very important issue for future research.

2.2 Data-driven orthonormal basis

The support interval, weighting function or probability distribution for $\boldsymbol{\omega}$ is determined from available information (modeler's experience, expert opinion, general prior information or field data) and reflects the uncertainty or expected range of variation of input parameters. This implies that the polynomial basis and related projections should adapt to the acquired information, such that they approximate the model best where the probability density of the parameters is highest. In order to construct such a data-driven polynomial basis that considers all the available information about the input parameters $\boldsymbol{\omega}$, let us define the set of polynomials $\{P_j^{(0)}, \ldots, P_j^{(d)}\}$ of degree d for the parameters ω_j as an orthonormal basis in the parameter space. The polynomial $P_j^{(k)}(\omega_j)$ of degree k in an individual parameter ω_j can be written as a simple linear combination of the different powers i of ω_j:

$$P_j^{(k)}(\omega_j) = \sum_{i=0}^{k} p_{i,j}^{(k)} \omega_j^i, \quad k = 0 \ldots d, \quad j = 0 \ldots N. \tag{2}$$

Here $p_{i,j}^{(k)}$ are the coefficients within the polynomial $P_j^{(k)}(\omega_j)$. Assuming that the input parameters within $\boldsymbol{\omega}$ are independent (Ghanem & Spanos, 1991), the multi-dimensional basis can be constructed as a simple product of the corresponding univariate polynomials:

$$\Psi_k(\boldsymbol{\omega}) = \prod_{j=1}^{N} P_j^{(\alpha_j^k)}(\omega_j), \quad \sum_{j=1}^{N} \alpha_j^k \leq M, \quad k = 1 \ldots N, \tag{3}$$

where α_j^k is a multivariate index that contains the combinatoric information how to enumerate all possible products of individual univariate basis functions. In other words, the index α can be seen as $M \times N$ matrix, which contains the corresponding degree (e.g. 0, 1, 2, etc.) for parameter number j in expansion term k. The multivariate basis allows to represent the reaction of a model Ω to several (N) parameters ω_j ($j = 0 \ldots N$), as an N-dimensional polynomial response surface, defined by the expansion in Eq. (1).

We will show now, how to construct the data-driven orthogonal polynomial basis for each individual component ω_j from the vector $\boldsymbol{\omega}$. The main idea of the data-driven approach, see Oladyshkin, Class, Helmig & Nowak (2011a); Oladyshkin & Nowak (2011), consists in constructing the coefficients $p_{i,j}^{(k)}$ for Eq. 2 in such a way that the polynomials in Eq. (2) form a basis that is orthonormal in precisely the given input distributions of model parameters. It does so without posing any restrictions to the statistical distribution shapes or weighting functions that available data, expert opinion or modeler experience may assume.

According to Oladyshkin & Nowak (2011), an orthogonal polynomial basis up to order d can be constructively defined for any arbitrary probability measure, given that ω_j has finite statistical moments (e.g., mean, variance, skewness, etc) up to order $2d - 1$. The unknown polynomial coefficients $p_{i,j}^{(k)}$ can be defined (among other available options for construction:

Abramowitz & Stegun (1965); Stieltjes (1884); Witteveen & Bijl (2006)) from the following matrix equation:

$$\begin{bmatrix} \mu_{0,j} & \mu_{1,j} & \cdots & \mu_{k,j} \\ \mu_{1,j} & \mu_{2,j} & \cdots & \mu_{k+1,j} \\ \cdots & \cdots & \cdots & \cdots \\ \mu_{k-1,j} & \mu_{k,j} & \cdots & \mu_{2k-1,j} \\ 0 & 0 & \ldots & 1 \end{bmatrix} \begin{bmatrix} p_{0,j}^{(k)} \\ p_{1,j}^{(k)} \\ \cdots \\ p_{k-1,j}^{(k)} \\ p_{k,j}^{(k)} \end{bmatrix} = \begin{bmatrix} 0 \\ 0 \\ \cdots \\ 0 \\ 1 \end{bmatrix}. \tag{4}$$

Here, $\mu_{i,j}$ are the ith non-central (raw) statistical moments for random variable ω_j. It becomes evident from Eq. (4) that statistical moments are the only required form of information on the input distributions. This property carries over to all other types of Taylor or polynomial chaos expansions.

The above orthogonal polynomial basis can be used directly for analysis. However, a normalized basis has further useful properties. For example, the mean and variance of $\Omega(\boldsymbol{\omega})$ according to the expansion (1) is given by simple analytical relations (see Eq. (15) in Section 4 of this Chapter), by virtue of the orthonormality property. This follows from the general properties of Fourier expansions, which encompass all expansions in orthonormal bases. The squared coefficients of such expansions are called the spectrum (compare with the Fourier transformation). The sum of squared coefficients is the total energy or variance. Due to orthonormality, most terms cancel out in subsequent steps. For example, if we consider a stochastic process in the probability space (Λ, A, Γ) with space of events Λ, σ-algebra A and probability measure Γ (see e.g. Grigoriu (2002)) and if $\widehat{P}_j^{(k)}(\omega_j)$ is an orthonormal basis, then by definition of orthogonality:

$$\int_{\omega_j \in \Lambda} \widehat{P}_j^{(k)}(\omega_j) \widehat{P}_j^{(l)}(\omega_j) d\Gamma(\xi) = \delta_{kl}. \tag{5}$$

The orthonormal polynomial basis can be obtained as:

$$\widehat{P}_j^{(k)}(\omega_j) = \frac{P_j^{(k)}}{\left\| P_j^{(k)} \right\|}, \quad \left\| P_j^{(k)} \right\|^2 = \int_{\omega_j \in \Lambda} \left[P_j^{(k)}(\omega) \right]^2 d\Gamma(\omega_j), \tag{6}$$

where $\|P_j^k\|$ is the normalizing constant of the polynomial P_j^k for space of events Λ (where $\omega_j \in \Lambda$) with probability measure Γ.

Because the square of a polynomial of order k yields a polynomial of order $2k$, normalization according to Eq. (6) requires the statistical moments of ω up to order $2d$.

When the input data set is small, the sample moments are only uncertain estimates of real moments. Hence, a direct application of the method presented becomes less robust. In that case, it would be useful to apply some standard methods to achieve robustness in the estimation of moments, such as bootstrapping (e.g. Efron (1987)). Moreover, especially in such cases, expert opinion can be very useful. In the presented approach, an expert will have total freedom of data interpretation (not restricted to the selection among standard PDFs) and can provide much more sophisticated information (e.g. lower and higher moments, etc.). Such an expert opinion (in a most general sense) will be incorporated directly without any additional transformation or additional subjectivity when translating it to the stochastic numerical framework.

2.3 Non-intrusive determination of the coefficients

The remaining task is to evaluate the coefficients c_j in Eq. (1). Therefore, we will use the non-intrusive probabilistic collocation method (PCM) (Li & Zhang, 2007; Oladyshkin, Class, Helmig & Nowak, 2011b). The collocation formulation does not require any knowledge of the initial model structure, i.e., of Ω. It only requires knowledge on how to obtain the model output for a given set of input parameters, which allows to treat the model Ω like a "black-box". The distinctive feature of non-intrusive approaches is that any simulation model can be considered a "black-box", i.e. commercial software can be used without any modifications required. The idea of PCM is to evaluate the model exactly M times, which allows to directly fit the polynomial representation of Ω, see Eq. (1), with its M unknown coefficients c_j to the obtained M model results. The M model evaluations are performed with M different parameter sets $\{\omega_1^{(i)},...,\omega_N^{(i)}\}$, $i = 1,..,M$, called collocation points.

This leads to the following system (Villadsen & Michelsen, 1978) of linear equations:

$$\mathbf{M}_{\Psi}(\boldsymbol{\omega})\mathbf{V}_c(\mathbf{x},t) = \mathbf{V}_{\Omega}(\mathbf{x},t;\boldsymbol{\omega}) \tag{7}$$

where $\mathbf{V}_c$ is the $M \times 1$ vector of unknown coefficients c_j in expansion (1), the $M \times 1$ vector $\mathbf{V}_{\Omega}$ contains the model output for each collocation point, and the $M \times M$ matrix $\mathbf{M}_{\Psi}$ contains the polynomials evaluated at the collocation points:

$$\begin{aligned}
\mathbf{M}_{\Psi} &= \left\{\Psi_i(\omega_1^{(i)},...,\omega_N^{(i)})\right\}, \quad i = 1\ldots M, j = 1\ldots M; \\
\mathbf{V}_{\Omega} &= \left\{\Omega_i(x,y,z,t,\omega_1^{(i)},...,\omega_N^{(i)})\right\}, \quad i = 1\ldots M; \\
\mathbf{V}_c &= \{c_i(x,y,z,t)\}, \quad i = 1\ldots M.
\end{aligned} \tag{8}$$

The vectors $\mathbf{V}_c$ and $\mathbf{V}_{\Omega}$ are space- and time-dependent, whereas the matrix $\mathbf{M}_{\Psi}$ does not depend on space and time and can be generated once for the given expansion degree and parameter number.

The solution $\mathbf{V}_c$ of the system (7) depends on the selection of collocation points. According to Villadsen & Michelsen (1978), the optimal choice of collocation points corresponds to the roots of the polynomial of one degree higher $(d+1)$ than the order used in the chaos expansion (d). Once the orthonormal polynomial basis is constructed using data (or assumptions on data), the collocation points become as well data-driven and optimally distributed in the space of input parameters. This strategy is based on the theory of Gaussian integration (e.g., Abramowitz & Stegun (1965)), and allows exact numerical integrations of order d given $d+1$ values of the function to be integrated.

The data-driven polynomial basis (see Section 2.2) defines the positions of the collocation points specific to the distribution of input parameters at hand and, thus, indicates what are the optimal parameter sets for model evaluation using all available information about the input parameters.

For multi-parameter analysis, the number of available points from the original optimal integration rule is $(d+1)^N$, which is larger than the necessary number M of collocation

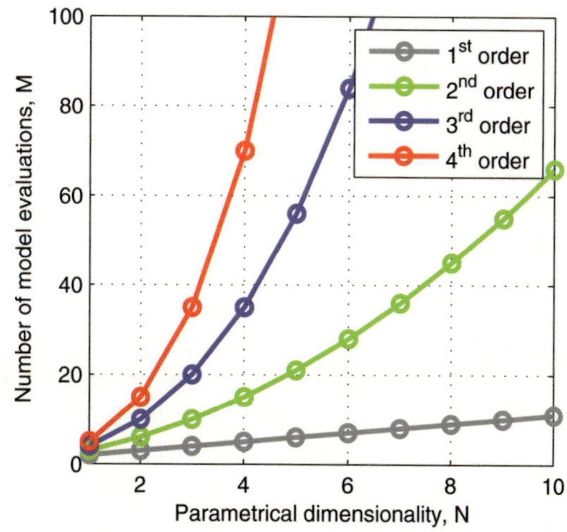

Fig. 1. Computational costs: number of model evaluations for different orders

points. The full tensor grid can be used for low-order (1st, 2nd) analysis of few parameters, but for higher-order analysis of many parameters the tensor grid suffers from the curse of dimensionality ($(d+1)^N$ points). In that case, a smart choice of a sparse subset of the tensor grid becomes necessary. For this reason, the collocation approach became more popular in the last years. Probabilistic collocation (Li & Zhang, 2007; Oladyshkin, Class, Helmig & Nowak, 2011b) chooses the collocation points from the full tensor grid according to their probability weight, i.e. their importance as specified by the available probability distribution of ω. This simply means to select the collocation points from the most probable regions of the input parameters' distribution (see Oladyshkin, Class, Helmig & Nowak (2011b)) and the modeler can extract a lot of information in the main range of the parameter distribution.

From the practical point of view, the computational costs of the data-driven chaos expansion combined with non-intrusive collocation approach are proportional to the number of terms in the chaos expansion multiplied by the time of a single model evaluation. The number of model evaluations (number of expansion terms M) for the construction of response surface via collocation approach depends on crossing order of expansion and presented in Figure 1.

2.4 Efficient convergence of data-driven expansion

The data-driven polynomial chaos expansion presented in Section 2.2 provides a simple and efficient tool for analysis of stochastic systems. Let us illustrate the efficiency of analysis within the data-driven versus the conventional expansion. For that, we will consider the simple exponential decay differential equation which was already used in Xiu & Karniadakis (2003) for illustration of the Askey scheme:

$$\frac{dY(t)}{dt} = -\xi_{Ph} Y, \quad Y(0) = 1 \tag{9}$$

Let Y_{PC} be the solution obtained using the polynomial chaos expansion (1) for the problem defined in equation (9). We use a Monte Carlo simulation as reference solution and define the relative error between the polynomial chaos expansion solution Y_{PC} and the Monte Carlo solution Y_{MC} as $\epsilon = |Y_{PC} - Y_{MC}| \,/\, |Y_{MC}|$.

We will apply the chaos expansion to equation (9) with both intrusive, like the Galerkin method (Ghanem & Spanos, 1993; Matthies & Keese., 2005; Xiu & Karniadakis, 2003), and non-intrusive approaches , like the probabilistic collocation method (Isukapalli et al., 1998; Li & Zhang, 2007; Oladyshkin, Class, Helmig & Nowak, 2011b). Both approaches depend on the distribution of the random variable ξ. If the random variable ξ is not distributed in same space as the polynomial basis, an additional conversion is required, such as e.g. Gaussian anamorphosis (Wackernagel, 1998). For each method, we will construct two expansions: conventional PCE mapped to Gaussian and data-driven aPC. Figure 2 demonstrates the convergence of the mean value (at time $t = 1$) for a small selection of exemplary distributions of the model input ξ (e.g. Rayleigh, Weibull, Log-normal). A similar situation has been observed for the convergence of the variance. More details are presented in the study by Oladyshkin & Nowak (2011). Chaos expansion in the data-driven (optimal) polynomial basis without transformation shows at least an exponential convergence. Let us mentione, that the data-driven basis provides identical results for both intrusive and non-intrusive methods, because numerical integration is exact to the necessary degree when using the roots of the $d+1$ order polynomial from the optimal basis, and because no transformation from physical to normal space is necessary. Convergence with a non-optimal basis (here: Hermite) after transformation strongly depends on the nonlinearity of the required transformation.

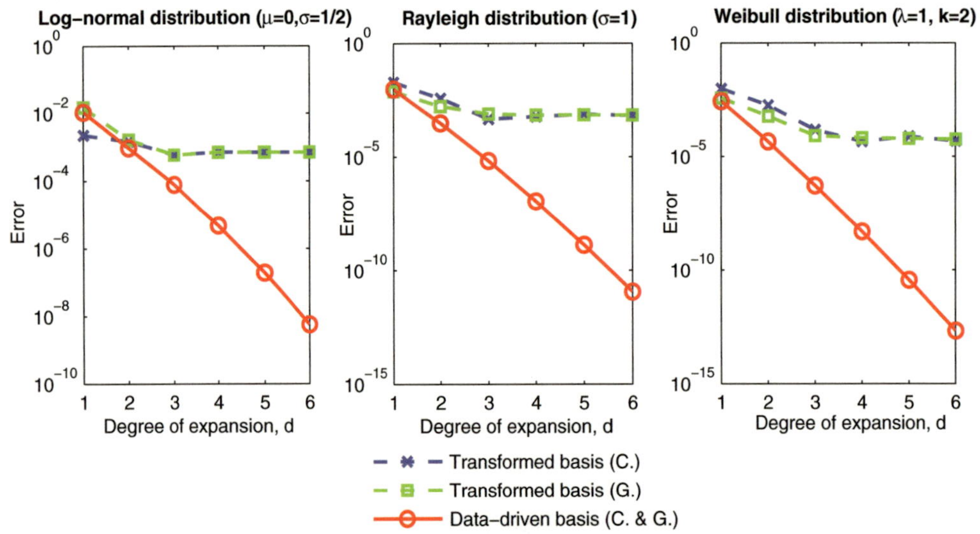

Fig. 2. Convergence estimationg the mean value based on optimal data-driven basis and non-optimal transformed basis using Galerkin projection (G.) and Collocation approach (C.)

2.5 Conclusions to section 2

We presented the data-driven approach (aPC) for construction of a response surface based on a global orthonormal polynomial basis for arbitrary distributions. The arbitrary distributions can be either discrete, continuous, or discretized continuous and can be specified either through a few statistical moments, analytically as PDF/CDF, numerically as a histogram, or theoretically through the even more general format of a probability measure. The aPC approach provides improved convergence in comparison to classical PCE techniques, when applied to input distributions that fall outside the range of classical PCE (Oladyshkin, Class, Helmig & Nowak, 2011a; Oladyshkin & Nowak, 2011).

The method only demands the existence of a finite number of moments, and does not require the exact knowledge or even existence of probability density functions. An interesting aspect is that only moments up to twice the order of expansion matter. Therefore, any PDFs fitted to input data will lead to the same statistical moments for model output, if the PDFs coincide in the required moments up to twice the order of expansion. For the same reason, if one still desires to fit a single PDF to the input data set, we recommend maximum entropy or minimum relative entropy methods applied such that the moments relevant for PCE are matched. This differs drastically from fitting low-parametric distributions to only lower moments of the available data, because this would modify the remaining relevant moments up to twice the order of expansion. More details on this issue will be provided in Section 4.

3. Global sensitivity analysis

Understanding the general role of parameters in models and the impact of varying model parameters on the response of prediction models is a relevant subject in various fields of science and engineering. Characterizing the impact of parameter variations is known as sensitivity analysis and can be subdivided into local and global analysis (Saltelli et al., 2008; Sobol, 1990; Sudret, 2008). In many cases of practical interest, we wish to perform a Global Sensitivity Analysis (GSA) in order to analyze a model as such or to investigate, quantify and rank the effects of parameter variation or parameter uncertainty on the overall model uncertainty. GSA can also be used to: (1) quantify the relative importance of each individual input parameter in the final prediction (Anderson & Burt, 1985; Sobol, 1990; Winter et al., 2006); (2) aid engineers to produce more robust designs; and finally (3) help decision makers to allocate financial resources towards better uncertainty reduction.

For example, the field of subsurface contaminant hydrology requires uncertainty estimates due to the ubiquitous lack of parameter knowledge caused by spatial heterogeneity of hydraulic properties in combination with incomplete characterization (Dagan, 1989; Rubin, 2003). For such reasons, we need to rely on probabilistic tools to predict contaminant levels and their overall health effects, and to quantify the corresponding uncertainties. Having efficient computational approaches to estimate uncertainty and to perform GSA in hydrogeological applications (and many other fields of science and engineering that feature uncertain dynamic or distributed systems) is desirable. It can inform modelers about the relevance of processes or parameters in the models they compile, and can inform engineers and decision makers about which parameters require most attention and where characterization efforts should be allocated such that prediction uncertainty can be minimized. Hence, there is an ever-increasing demand for having a GSA method that efficiently quantifies uncertainty and parameter relevance in complex and non-linear systems. An important recommendation to keep in mind is that GSA should be global not only in the sense of looking at the entire range of possible parameter variations. It should also be used to assess the importance of parameters on a global, final model output or post-processing result that is relevant to generate new insight, or relevant for final decisions. GSA should not merely be applied to model-internal quantities that are of secondary importance for the scientific or application task at hand (Saltelli et al., 2008).

In this Section, we tackle GSA based on the aPC technique, following the line of work on aPC by Oladyshkin, Class, Helmig & Nowak (2011a); Oladyshkin & Nowak (2011) and synthesizing the GSA approach by Oladyshkin, de Barros & Nowak (2011). Because the presented framework accounts for arbitrary bounds or weighting functions for input parameters, it provides a weighted global sensitivity. In some sense, the novel sensitivity indexes introduced by Oladyshkin, de Barros & Nowak (2011) can be perceived as a generalization of the Morris method (Morris, 1991) to weighted analysis. Up to presence the Morris method considers a uniform importance of input parameters within pre-defined intervals. We also generalize the Sobol indices (Sobol, 1990) for GSA to the aPC context, and provide a novel GSA measure which resembles a weighted square norm of sensitivities. Compared to Sobol indices, the new weighted sensitivity measure is absolute rather than relative. The advantage of an absolute index over a relative one is that it is a quantitative expression for the (averaged) derivative (slope), and hence keeps the original meaning of a sensitivity as known from linear, local analysis.

Performing GSA requires to evaluate the model at many points in the space of the input parameters. The correct choice of such points within the parameter space is important for adequate and efficient assessment of sensitivity. The aPC approach explicitly offers a method for optimal choice of these points, based on the generalized mathematical theory of Gaussian integration (e.g., Abramowitz & Stegun (1965)), see Section 2.

Foglia et al. (2007) pointed out that, according to their experience, one can get 70% of the insight from 2% of the model runs when using local sensitivity analysis methods versus global methods. The big advantage of aPC-based GSA (or more generally: GSA based on any PCE technique), is that one can obtain global sensitivity information at computational costs that are hardly larger than those for local analysis. The reason is the following: Local methods use infinitesimally small spacing between parameter sets for model evaluation to get numerical derivatives evaluated at a single point. The aPC based-method places the parameter sets for model evaluation at an optimized spacing in parameter space. This can be interpreted as fitting secants (or polynomials for non-linear analysis) to the model response. These secants (polynomials) approximate the model over the entire parameter space in a weighted least-square sense (compare with the best unbiased ensemble linearization approach described by Nowak (2009)). This is more beneficial to computing a tangent or local second derivatives (compare FORM, SORM methods, e.g., Jang et al. (1994)) that approximate the model well just around one point in the parameter space.

This Section provides an alternative procedure to perform GSA that is computationally efficient and highly flexible. In particular, due to aPC, the GSA introduced here and in Oladyshkin, de Barros & Nowak (2011) can be interpreted as exploiting a smart (mathematically optimal) interpolation rule of model output between optimally chosen sets of input parameters, where the number of model evaluations is minimal. Compared to earlier works that related GSA to classical PCE the presented approach: (1) emphasizes a more engineering-like language as compared to otherwise intense mathematical derivations and is based on a clear 3-step procedure to perform sensitivity analysis, (2) provides easy-to-use semi-analytical expressions for frequent use in applications, (3) generalizes PCE-based GSA to arbitrary probability distributions of the investigated parameters, and moreover, (4) allows to align the complexity and order of analysis with the reliability and detail level of statistical information on the input parameters.

3.1 Sobol sensitivity indices

Classical PCE has already been used for sensitivity analysis in different fields of applications (Crestaux et al., 2009; Sudret, 2008). Within the data-driven PCE framework (de Barros, Oladyshkin & Nowak, 2011; Oladyhskin et al., 2011) used in this work, the so-called Sobol indices for sensitivity estimation (Sobol, 1990) can be computed analytically based on the PCE (see Crestaux et al. (2009); Plischke (2010); Sudret (2008)) using Eq. (10):

$$S_{i_1,...,i_s} = \frac{\sum_{j=1}^{M} \chi_j c_j^2}{\sum_{j=1}^{M} c_j^2}, \quad \chi_j = \begin{Bmatrix} 1, & if \quad \alpha_j^k > 0, & \forall j \in (i_1,...,i_s) \\ 0, & if \quad \alpha_j^k = 0, & \exists j \in (i_1,...,i_s) \end{Bmatrix},$$

where $S_{i_1,...,i_s}$ is the Sobol index that indicates what fraction of the of total variance of Ω can be traced back to the joint contributions of the parameters $\omega_{i_1},...,\omega_{i_s}$. The index selection operator χ_j indicates where the chosen parameters ω numbered as $i_1,...,i_s$ (i.e., $\omega_{i_1},...,\omega_{i_s}$) have simultaneous contributions within the overall expansion. In plain words, it enumerates all polynomial terms that contain the specified combination $i_1,...,i_s$ of model parameters.

A complementing metric for sensitivity analysis is the *Total Index* introduced by Homma & Saltelli (1996). It expresses the total contribution to the variance of the model output Ω due to the uncertainty of an individual parameter ω_j in all cross-combinations with other parameters:

$$S_j^T = \sum_{(i_1,...,i_s):j\in(i_1,...,i_s)} S_{i_1,...,i_s}, \tag{10}$$

where S_j^T simply sums up all Sobol indices in which the variable ω_j appears, both as univariate and joint influences.

3.2 Weighted sensitivity indices

The weighted sensitivity index introduced by Oladyshkin, de Barros & Nowak (2011) reflects the square slope $\partial\Omega/\partial\omega_j$, but averaged over the statistical distributions or weighing functions of $\omega_1,...,\omega_N$:

$$S_{\omega_j}^2 = \int_{\omega_1\in\Lambda} \cdots \int_{\omega_N\in\Lambda} \left[\frac{\partial\Omega(\boldsymbol{\omega})}{\partial\omega_j}\right]^2 d\Gamma(\omega_1)...d\Gamma(\omega_N) \tag{11}$$

So, the weighted global index for GSA using an arbitrary probability measure Γ can be used for reflection of the expected magnitude of parameter variation. Compared to Sobol indices (see Section 3.1) which analyze the spectral energy (variance) contribution of individual parameters to the model Ω, the new weighted measure looks at the spectral energy within the derivatives $\partial\Omega/\partial\omega_j$. The averaged sensitivity index S_{ω_j} can be explicitly expressed as:

$$S_{\omega_j}^2 = \sum_{k=0}^{M} c_k^2 \sum_{i=0}^{\alpha_j^k-1} \left[b_i^{(\alpha_j^k-1)}\right]^2 P_j^{(i)}(\omega_j). \tag{12}$$

where the re-collection coefficients $b_i^{(\alpha_j^k-1)}$ are defined as solution of the corresponding linear system:

$$\begin{bmatrix} b_0^{(\alpha_j^k-1)} \\ b_1^{(\alpha_j^k-1)} \\ \cdots \\ b_{\alpha_j^k-1}^{(\alpha_j^k-1)} \end{bmatrix} \begin{bmatrix} p_{0,j}^{(0)} & 0 & 0 & 0 \\ p_{0,j}^{(1)} & p_{1,j}^{(1)} & \cdots & 0 \\ \cdots & \cdots & \cdots & \cdots \\ p_{0,j}^{(\alpha_j^k-1)} & p_{1,j}^{(\alpha_j^k-1)} & \cdots & p_{\alpha_j^k-1,j}^{(\alpha_j^k-1)} \end{bmatrix}^T = \begin{bmatrix} p_{1,j}^{(\alpha_j^k)} \\ 2p_{2,j}^{(\alpha_j^k)} \\ \cdots \\ \alpha_j^k p_{\alpha_j^k,j}^{(\alpha_j^k)} \end{bmatrix}. \tag{13}$$

This index reflects the influence of a parameter ω_j onto the model output Ω in a similar fashion to the total index defined in Eq. (10). However, the new multivariate index in Eq. (11) does not rely on comparison among different parameters, i.e., it is an absolute measure. This is an advantage over the existing Sobol-based total index which is only a comparative and relative measure.

3.3 Three-step algorithm for data-adaptive global sensitivity analysis

This Section summarizes the computational algorithm for GSA based on aPC. The important feature in this computational algorithm is that it can be performed for arbitrary distributions of the input data ω. The entire algorithm for the desired degree of precision d and number of parameters N is based on the following 3 steps:

Step 1. Characterize the model parameters to be investigated: Compute the raw moments $\mu_{k,j}$ ($k = 1 \ldots 2d$) of the input data for each input parameter ω_j ($j = 1 \ldots N$). If a probability density function (PDF) is provided, then we can evaluate the actual theoretical moments *in lieu* of the raw data moments of a given data set. Alternatively, expert elicitation may serve to provide opinions on these moments, e.g., by guessing a distribution shape and using its moments.

Step 2. Approximate the model response surface by polynomials: For the specific moments $\mu_{k,j}$, compute the coefficients of the optimal polynomial basis in Eq. (3) using the system of linear equations Eq. (4) and the normalization in Eq. (6). In addition, compute the coefficients of the expansion using the system of linear equations, Eq. (7), to represent the model response.

Step 3. Compute the desired sensitivity information from the polynomials: With the original model reduced to a multi-variable polynomial, the sensitivity indices can be obtained analytically without any heavy additional computational efforts. This is achieved by using the relations provided in Eqs.(10) or (12) provided in the two upcoming Sections.

In summary, the algorithm described above has the following advantages:

1. It constructs an optimal orthonormal polynomial basis for any desired distribution of data, whereas previous PCE techniques were limited to a small number of statistical distributions. Since only statistical moments are relevant, the input distributions may be either discrete, continuous, or discretized continuous and can be specified either through some statistical moments, analytically as PDF, numerically as histograms, or theoretically through the even more general format of probability measures.
2. The algorithm performs an optimal projection of the physical model onto a polynomial basis with minimum computational effort. This reduces the original model to a set of polynomials with many useful properties that allow immensely fast evaluation and offer a list of analytical relations.

3. The presented algorithm can easily be implemented without any deep knowledge related to the theory of chaos expansion and projection techniques.
4. The model Ω does not have to be modified and no specific properties are required. Hence, it may by given in any arbitrary form for simple analytical solution up to a multi-scale multi-physical simulation software framework.

3.4 Illustration: sensitivity analysis for human health risk

3.4.1 Problem formulation

We will demonstrate the presented GSA methodology for a contaminant transport problem in a 3D heterogeneous aquifer and the resulting human health risk for an exposed population. The aquifer has a hydraulic conductivity tensor $\mathbf{K}(\mathbf{x})$ and constant effective porosity n_e. For illustration purposes, we consider flow to be incompressible, single-phased, at steady-state, free of boundary effects. The Random Space Functions for the log-conductivity $Y = \ln K$ is statistically characterized by its mean $\langle Y \rangle$, variance σ_Y^2, covariance model C_Y (assumed here exponential) and its integral scale $I_{Y,i}$, where $i = 1, 2$ and 3 for $\mathbf{x} = (x_1, x_2, x_3)$. We will consider $I_Y \equiv I_{Y,1} = I_{Y,2}$ and $I_{Y,v} \equiv I_{Y,3}$ with anisotropy ratio $f = I_3/I_Y$. Statistical stationarity for Y is assumed. Flow is uniform-in-the-average with mean velocity $\langle \mathbf{u}(\mathbf{x}) \rangle \equiv (U, 0, 0)$. In this test case, we will make use of the semi-closed expressions for one-particle displacement covariance function from Dagan (1988).

In most cases, decision makers are interested in quantifying adverse effects in human health due to contaminated groundwater exposure. For this case, the environmental performance metric of interest is human health risk, which depends not only on the hydrogeological parameters but also on the physiological and behavioural parameters of the exposed individual. For illustration purposes, we will consider the increased lifetime cancer risk model from the EPA (USEPA, December 1989;D), although many other risk models exist as discussed in the literature (de Barros, Ezzedine & Rubin, 2011; Maxwell & Kastenberg, 1999; Siirila et al., 2011). The increased lifetime cancer risk formulation for the groundwater ingestion pathway is given by:

$$r = a\overline{C}, \quad \overline{C} = \max \left[\frac{1}{ED} \sum_{t}^{t+ED} C(t) \right]_{t=0}^{\infty}, \tag{14}$$

where $\overline{C}$ is the maximum running average over the exposure duration ED (years) defined in Maxwell & Kastenberg (1999) and a is an uncertain health parameter (see Oladyshkin, de Barros & Nowak (2011) for details).

3.4.2 Importance ranking of modeling parameters

We will apply the presented framework introduced in Section 3.3 for the human health risk analysis. These sensitivity indices are useful to quantify the simultaneous influence of model parameters, especially when the number of parameters becomes large and visualization of a multivariate model response becomes unfeasible and confusing. We will analyze an uncertain contaminant spill location and uncertain parameters in the model for health risk. Let us consider the integral scale I_Y, variance σ_Y^2, longitudinal velocity U and position x_s, y_s, z_s of the source with the source volume V_0 and the slope of the risk model a as analyzed parameters ω_j in our analysis.

According to Step 1 of the algorithm presented in Section 3.3, we first characterize all input parameters. For illustrative purpose we will consider the following distributions of modelling parameters:

- For ω_1: $\omega_1 = 1 + 2\omega_1'$, with ω_1' following a beta distribution with $\alpha = 2$ and $\beta = 2$;
- For ω_2: Uniformly distributed within the interval $[0.1, 0.7]$;
- For ω_3: Log-normally distributed with $\mu = 3.6$ and $\sigma = 0.3$.
- For ω_4: $\omega_4 = 9 + 2\omega_4'$, where ω_4' is beta distributed with $\alpha = 2$ and $\beta = 2$;
- For ω_5 and ω_6: $\omega_{5,6} = -1 + 2\omega_{5,6}'$, where $\omega_{5,6}'$ is beta distributed with $\alpha = 2$ and $\beta = 2$.
- For ω_7: ω_7 is lognormal distributed with $\mu = 1.7$ and $\sigma = 0.14$.

These PDFs are used in Step 1 to calculate the corresponding raw moments of input data. According to Step 2 of the algorithm (Section 3.3), we compute the coefficients of the orthonormal polynomial basis using Eq. (4) and Eq. (6) for given raw moments and for the desired degree of expansion d. Also, we consider the response surface of the model according to Eq.(1) and Eq.(7). At the last phase of analysis, Step 3 of the algorithm (Section 3.3), the desired quantitative sensitivity information is extracted from the polynomial response surface.

All sensitivity indices can be constructed as it was shown above. For brevity, we will focus on the total and weighted indices only. Figure 3 illustrates the total (left plot) and weighted (right plot) sensitivity indices at different degrees of expansion. This figure shows the overall influence of all model parameters on the total human health effect (final prediction), which implicitly includes a time integral over the concentration history (see Eq. 14). For this particular case, we observe that the risk is more sensitive towards hydrogeological parameters than the health parameters. In the case study related to health risk (Figure 3), the new weighted measure converges faster in comparison to Total indices with increasing order of the expansion. Total indices show the main reaction of the model to the analyzed parameters, but as relative quantities do not provide faster stabilization in the ranking of analyses parameters in comparison to absolute weighted indices. The presented approach for GSA provides a relatively accurate approximation even for moderate orders (see also the convergence study for aPC in general in Oladyshkin & Nowak (2011)). Thus, global and weighted sensitivity analysis can already be performed at computational costs that are only slightly larger than those of local analysis.

3.5 Conclusions to section 3

We have presented an alternative method for global sensitivity analysis (GSA), which is based on the arbitrary polynomial chaos expansion (aPC). Compared to existing polynomial-based GSA methods, it can accommodate for all types of statistical distributions or weighting functions of the input parameters. This approach is denoted as *data-adaptive* because the aPC method can be applied even in situations where precise statistical information (e.g., known parametric distributions) for the input parameters is not available. If desired, the method can work directly with raw sampled data sets to represent the uncertainty and possible variational ranges of input data. The presented methods allow experts to choose freely of technical constraints the shapes of their statistical assumptions, and they allow to align the complexity and order of analysis with the reliability and detail level of statistical information on the input parameters.

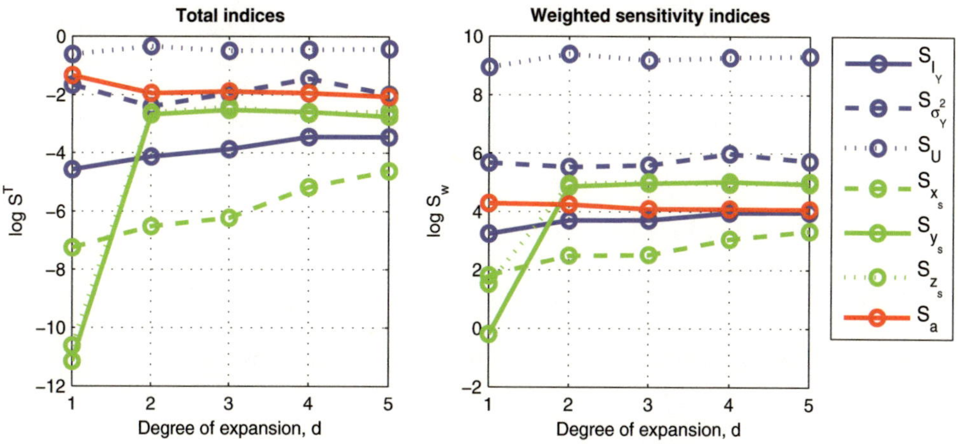

Fig. 3. Convergence of total (left plot) and weighted (right plot) sensitivity indices for the health risk prediction.

The proposed method incorporates the full range of possible simulation outcomes for the investigated model parameters as it approximates the model's full response surface by multivariate polynomials. The existing polynomial-based method to compute Sobol indices was extended to the much more general aPC framework. While Sobol's indices look at the contributions from individual parameters to the energy (variance) of the model, our weighted sensitivity indices look at the energy norm of derivatives of the model with respect to individual parameters. The resulting sensitivity measure shows better convergence properties than Sobol analysis. It can convey the information of a global sensitivity analysis at computational costs that are almost low as those of linear analysis.

4. Uncertainty quantification and probabilistic risk assessment

This Section presents a reasonably accurate method for uncertainty quantification and probabilistic risk-assessment at acceptable computational costs. The lack of information about distributed properties leads to model uncertainties up to a level where the quantification of uncertainties becomes the dominant question in application tasks and may override the influence of secondary physical processes. Often numerical simulation models are inadequate for stochastic simulation techniques based on brute-force Monte Carlo simulation and related approaches, because even single deterministic simulations may require parallel high-performance computing. In the current Section, we suggest and apply a massive stochastic model reduction technique based on non-intrusive polynomial expansion as defined in Section 2. As a second focus for this Section, we continue discussion of data availability and subjectivity raised in the Section 2.

Unfortunately, precise information on distribution shapes for all uncertain input parameters is very rare in realistic applications, such as underground reservoir simulations, groundwater modeling, etc. Applied research on real-world systems, like modeling CO_2 storage, groundwater flow, etc., often faces the problem of immensely limited information about the model parameters involved, e.g. reservoir permeability, porosity, etc. With only limited data available or even in total absence of data, not even probability density functions representing the lack of knowledge can be easily inferred in a justified manner. Moreover, even if some amount of data is available, the statistical distribution of the corresponding model parameters can be nontrivial, e.g. bounded, skewed, multi-modal or discontinuous. In any case, the attempt to construct probability density functions from samples of limited size or from sparse information introduces additional room for subjectivity into the analysis. Thus, applied tasks demand the direct handling of arbitrary distribution shapes and sparse data sets without additional assumptions.

The purpose of the current Section is to work with a highly parsimonic and yet purely data-driven description of uncertainty, applying the arbitrary polynomial chaos (aPC). It was shown in Section 2 that statistical moments are the only source of information that is propagated in all types of polynomial expansion-based stochastic approaches. Thus, exact probability density functions for uncertain input parameters do not have to be known and do not even have to exist. This avoids the necessity of assuming subjectively or speculating on the exact shapes of probability distributions. The new freedom in distribution shapes gained with aPC opens the path to accessing with PCE even those applications where data samples of limited size merely allow the inference of a few moments, and one would not be able to construct a probability density function without introducing subjective assumptions and hence dangerous sources of bias.

4.1 Evaluation of statistics

In the presented approach, the statistics of the model output are based directly on the model and the specified moments of input data, see Section 2. If a model output $\Omega(\omega)$ is expanded in the normalized polynomial basis Eq. (6), then characteristic statistical quantities of $\Omega(\omega)$ can be evaluated directly. For example, the mean and variance of $\Omega(\omega)$ are given by the following simple analytical relations:

$$\texttt{mean}(\Omega) = c_1, \quad \texttt{var}(\Omega) = \sum_{j=2}^{N} c_j^2, \tag{15}$$

where the latter is a result of Parseval's Theorem (e.g. Siebert (1986)).

Likewise, all moments of Ω up to the order of expansion can be obtained analytically, based only on expansion coefficients and the used moments of the input parameters. Therefore, all probability distributions of input parameters that share the same $2d$ moments will, in any polynomial chaos expansion of order d, lead to the same moments of model output of order d. For the same reason, PCE expansions of order d are unaffected by subjective choices concerning input parameter distributions that only affect moments beyond the order $2d$. This is the case, for example, for maximum entropy PDFs.

Cumulative distribution functions (CDF), probability density functions (PDF) and other arbitrary statistics of model output can be evaluated via Monte Carlo analysis of the polynomial response surface that results from the expansion Eq. (1). This is very fast, because the polynomial surface is much faster to evaluate than the original model equations. However, the latter approach will ask for full knowledge of the input probability density function involved, which we try to avoid. If Monte Carlo cannot be avoided, e.g. for computing exceedance probability in risk analysis, we suggest using maximum entropy PDFs, or alternatively, a sufficiently large data set (if available) that can be used directly as Monte-Carlo realizations of input parameters.

4.2 Problem of data availability

Unfortunately, information is not available for all sources of uncertainty. In such cases, additional assumptions become necessary. The classical way would be to introduce a theoretical (parametric) probability distribution, e.g. with an assumed mean and variance. For example, some expert could introduce a lognormal distribution for the leaky-well permeability (see details in Section 4.3.1) with parameters defined from the benchmark values (Class et al., 2009), see Figure 4.

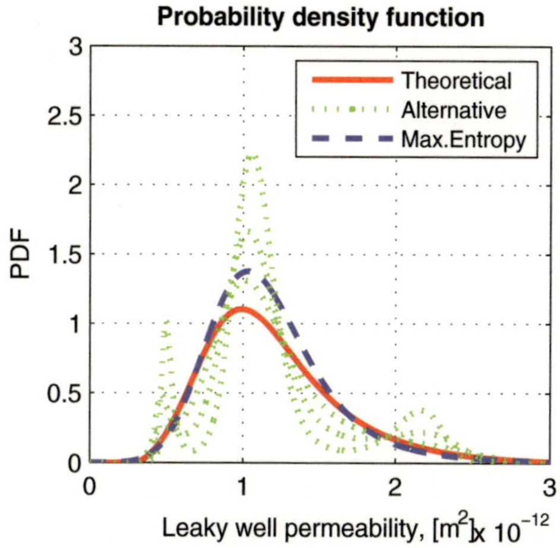

Fig. 4. Assumed stochastic distribution: theoretical PDF (red line), alternative PDFs (green lines) with the same first four moments and Maximum Entropy PDF with the same first four moments (blue line).

Establishing a full theoretical probability density function involves a strong assumption on all higher moments up to infinite order, and assumes implicit knowledge of the exact shape, e.g. also of the extreme-value tails. The data-driven approach strongly alleviates this situation, because it can handle a set of moments directly (e.g. the mean, variance, skewness, peakedness), without any further assumptions on higher-order moments and without having to introduce a PDF. In the current Section, we will use this freedom, and obtain only a small number of required moments directly from a large database or via experts, without asking for a full PDF.

For evaluating arbitrary complex output statistics without having a sufficiently large raw data set, we would recommend the Maximum Entropy PDF also shown in Figure 4 to draw a sufficiently large sample for Monte Carlo analysis of the polynomial. For example, a second-order expansion requires knowing the first four moments. Figure 4 illustrates possible distributions of stochastic variables, where the red line represents a theoretical PDF and the green lines represent a small collection of alternative PDFs with the same first four moments. The data-driven method does not require a choice between these alternatives, but only uses common information in the form of the required moments (here: up to order four). Existence of finite moments (up to a certain required order) is a necessary and sufficient condition for the proposed framework. Usually, this condition is easily fulfilled for a large spectrum of practical applications, especially for moderate degrees of expansion.

4.3 Illustration: risk assessment for carbon dioxide storage

4.3.1 Problem formulation

We consider the benchmark leakage problem of injected CO_2 into overlying formations through a leaky well defined by Class et al. (2009). In the current Section, we present an illustrative example for a homogenized system, i.e. we consider spatial heterogeneity only through zonation according to different geological media. However, the technique presented can be extended to many classes of heterogeneous systems, where spatially correlated heterogeneous parameter fields can be decomposed into their uncorrelated principal components using the KL-expansion (e.g. Li & Zhang (2007)), if heterogeneity does not span over too many scales. CO_2 is injected into a deep aquifer, spreads within the aquifer and, upon reaching a leaky abandoned well, rises to a shallower aquifer. The goal of the simulation is to quantify the leakage rate which depends on the pressure build-up in the aquifer due to injection and on the plume evolution. The leaky well is at the center of the domain and the injection well is 100m away. Both aquifers are 30m thick and the separating aquitard has a thickness of 100m. The leaky well is modeled as a porous medium with a higher permeability than the formation. The CO_2 leakage rate is defined in the benchmark study as the total CO_2 mass flux integrated over a control plane midway between the top and bottom aquifer, divided by the injection rate, in percent.

The benchmark problem assumes that fluid properties such as density and viscosity are constant, all processes are isothermal, CO_2 and brine are two separate and immiscible phases, mutual dissolution is neglected, the pressure conditions at the lateral boundaries are constant over time, the formation is isotropic rigid and chemically inert, and capillary pressure is negligible. Within the presented approach, the physical complexity of the problem can be increased to an arbitrary extent because of the non-intrusive black-box conception of the probabilistic collocation method (Isukapalli et al., 1998; Li & Zhang, 2007). All relevant parameters used for the simulation are given in the paper by Oladyshkin, Class, Helmig & Nowak (2011a). In the current study, the benchmark problem is simulated using DuMuX, a multi-scale multi-physics toolbox for the simulation of flow and transport processes in porous media (Flemisch et al., 2007).

To illustrate the proposed methodology, we will consider three parameters uncertain: reservoir absolute permeability, reservoir porosity and permeability of the leaky well. The distributions of absolute permeability and porosity were taken from the U.S. National Petroleum Council Public Database (which includes 1270 reservoirs), see also the work of Kopp et al. (2009). This choice reflects the situation of site screening, where site-specific data and data that allow heterogeneity within geological units to be resolved are not yet available. Instead, we use data sets that represent macroscopic properties of supposedly similar sites as prior knowledge.

Formally, the investigated uncertain parameters correspond to the components of the vector $\omega = \{\omega_1, \omega_2, \omega_3\}$ and represent the input parameters of Equation (1). The model output quantities Ω considered here are pressure and saturation values as a function of space and time, and the CO_2 leakage rate through the leaky well as a function only of time.

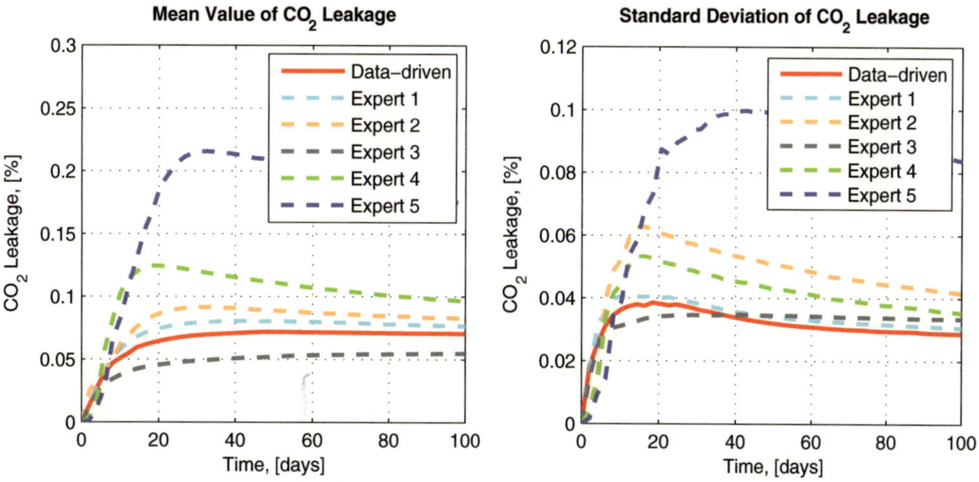

Fig. 5. Estimation of mean value (left plot) and standard deviation (right plot) of the CO_2 leakage rate: expert opinions (dashed lines) and data-driven approach (solid lines)

4.3.2 Data-driven uncertainty quantification

Similar to GSA in Section 3 the procedure for quantifying uncertainty in CO_2 storage can be divided into three main steps. The first step is to construct the polynomial basis according to the data considered in Section 4.3.1. The second step is to set up the chaos expansion and obtain the required coefficients c_i for expansion (1) using the non-intrusive probabilistic collocation method. In the third step, we evaluate mean and variance of output statistics according to Section 4.1, see Figure 5.

The three steps mentioned above are straightforward and exploit the input data directly, i.e. in a data-driven manner. Involving additional assumptions on input data is possible, but can be dangerous and will import more room for subjectivity inti the analysis. To illustrate the drastic impact of subjectivity that can be introduced into analysis, we performed the following simple experiment. The sample data (reservoir permeability and porosity) are distributed to five different and independent experts. The task of each expert consisted in constructing a theoretical probability density function for each parameter, which in their opinion would describe the statistics of the raw data best. As a common way of data description, all experts provided two-parametrical distributions for each input parameter (permeability and porosity). All experts proposed very different assumptions and techniques to match the permeability and porosity distributions. The responses of all experts were collected and used as input for modeling the benchmark problem defined in Sections 4.3.1. The results of this experiment are illustrated in Figure 5 using second-order polynomial chaos expansion. It shows the resulting mean value (left plot) and standard deviation (right plot) of the CO_2 leakage rate over time. Here, dashed lines correspond to the results based on subjective expert opinions, and solid lines correspond to the results from our purely data-driven approach. All test cases presented in Figure 5 were performed under the same conditions in all other aspects. The differences in predicting leakage rates based on the different experts are a consequence of their different interpretations and opinions on how to treat the data. This example clearly demonstrates that the room for subjectivity (when assigning theoretical stochastic distributions to real data and thus modifying their higher-order moments) can lead to significant differences in the predicted values.

The data-driven polynomial chaos expansion proposed here for stochastic analysis is based directly on the moments of sampled data without intermediate steps of data re-interpretation. This avoids the subjectivity usually introduced when choosing among a small limited number of theoretical distributions to represent a natural phenomenon, and so avoids the problem illustrated in Figure 5. Complex models often exhibit stronger non-linearity, which can amplify the impact of subjectivity especially for assumptions on extreme value tails and higher-order moments. Subjectivity within data interpretation can cause the same order of magnitude in uncertainty or error as numerical, conceptual and stochastic approximation error, see details in the paper by Oladyshkin, Class, Helmig & Nowak (2011a).

4.3.3 Efficient probabilistic risk assessment

The data-driven polynomial chaos expansion provides a simple but powerful tool for stochastic modeling and, in this case study, for the probabilistic risk assessment of CO_2 storage. The most integrative characteristic of the overall benchmark problem is the total leakage of CO_2. To compare a quantitative characteristic that is most important in probabilistic risk assessment, we computed the cumulative probability function of the CO_2 leakage rate after 30 days (Figure 6). The cumulative density function represents the probability that the CO_2 leakage is less than or equal to a particular value. In the current work, we also apply the conventional PCE approach for direct comparison with the data-driven approach. The principal difference lies in constructing the polynomial basis. The conventional approach is based on Hermite polynomials, which are optimal for Gaussian random variables. Because the random variables $\boldsymbol{\omega}$ are not distributed Gaussian in compliance with the Hermite polynomial basis, an additional conversion is required. A large number of methods are based on transformations of the model variables $\boldsymbol{\omega}_{Ph}$ from physical space to corresponding normal variables $\boldsymbol{\omega}_N$ via Gaussian anamorphosis (Wackernagel, 1998). This would implicitly define an exact PDF, which once again is an unjustifiably strong assumption. In contrast, the data-driven approach is based directly on the considered moments of the distributions or raw data of the uncertain input variables $\boldsymbol{\omega}$ in physical space. The left plot in Figure 6 corresponds to the conventional approach, i.e. involving Gaussian anamorphosis and Hermite polynomials. The right plot in Figure 6 corresponds to the data-driven approach without additional transformation. The convergence of both types of chaos expansions (dashed lines) was validated by traditional Monte Carlo simulations (solid lines) with 1270 realizations, where we sampled directly from the available data sets. We repeated the comparison study for different degrees of the chaos expansion, such as first order (4 evaluations), second order (10 evaluations), third order (20 evaluations) and fourth order (35 evaluations). Figure 6 demonstrates that the data-driven approach provides fast convergence. Even small degrees of the data-driven expansion (even the linear one in our specific case study) ensure adequate representation of the stochastic processes in the considered multiphase flow system. This fact can be very useful for fully resolved and complex real-world application challenges, where computational costs are very high even for a single model evaluation.

Convergence with the conventional approach (here: based on Hermite polynomials) strongly depends on the non-linearity of the required transformation from $\boldsymbol{\omega}_{Ph}$ to $\boldsymbol{\omega}_N$ (Oladyshkin & Nowak, 2011). There are two sources of slow convergence (or errors) for chaos expansions based on non-data-driven polynomial bases. First, the transformation from physical space to normal space introduces additional non-linear behavior into the overall problem, which will require a higher order of expansion to obtain comparable accuracy. Second, for non-intrusive methods that rely on numerical integration to obtain the coefficients c_i, the accuracy of numerical integration strongly depends on the choice of integration or collocation points. For example, in Gauss-Hermite integration (Abramowitz & Stegun, 1965), the polynomial basis of degree d defines the positions of integration points by the roots of the polynomial of degree $d+1$. These integration points are optimal only if the weighting function (here: the probability measure) and the polynomial basis are in direct correspondence. Any non-linear transformation from ω_{Ph} to ω_N destroys this direct relation. Thus, using a non-data-driven polynomial basis leads to a non-optimal placement of integration points, which causes a reduced accuracy of the integration (see Oladyshkin & Nowak (2011)). As a consequence, in comparison with the classical PCE, the same order of precision can be achieved with a smaller degree of expansion.

4.4 Conclusions to section 4

In this Section, we presented a data-driven approach for uncertainty quantification and risk assessment, based on polynomial chaos expansion (PCE). The data-driven approach provides a response surface based on a global orthonormal polynomial basis for arbitrary distributions. Thus, the data-driven approach provides freedom for modeling even physical systems with unknown probability distribution functions, when only data sets of very limited size or only little prior knowledge is available. It offers a new approach to defining parameter uncertainty in stochastic analysis that avoids the subjectivity of assigning theoretical probability distributions. In a small fictitious example, we asked independent experts to select and fit parametric distributions to two raw data sets. The example demonstrated that subjectivity in data interpretation can easily lead to prediction bias. Modeling is a chain of many tasks. Like any chain, it is only as strong as its weakest link. Modeling results indicate clearly that the statistical treatment of input data is part of the chain, and that the subjectivity in assuming theoretical curves can weaken that link immensely.

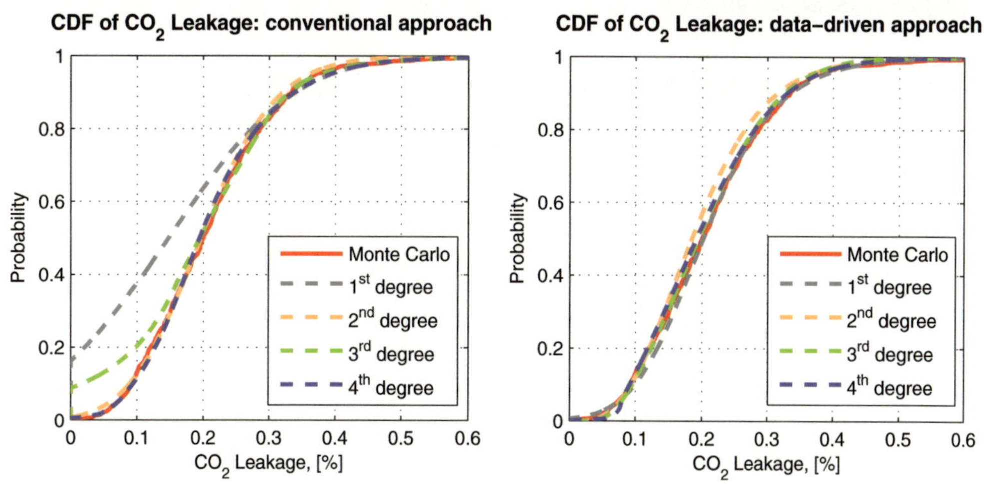

Fig. 6. Cumulative distribution function of CO_2 leakage rate after 30 days: conventional approach (left plot) and data-driven (right plot)

The data-driven stochastic approach was validated on the basis of Monte Carlo simulation using a common 3D benchmark problem. The proposed approach yields a significant computational speed-up compared with Monte Carlo, and provides faster convergence than conventional polynomial chaos expansions. Even for small degrees of expansion, the data-driven expansion can be very accurate, which can save a lot of computational power for probabilistic risk analysis.

Data-driven polynomial chaos expansion is based directly on raw data or other arbitrary sources of information without auxiliary assumptions. This increases the efficiency of chaos expansion and minimizes subjectivity, providing valuable support for risk-informed decision making as well as for robust design and control, allowing a better assessment and reduction of risk.

5. Robust design under uncertainty

The task of finding optimal system design while accounting for uncertainty leads to robust designs. In specific, risk can be controlled to stay below a given risk acceptance level, if robust design is performed under corresponding constraints. Obviously, the system performance and failure probability depends on uncertain parameters, but also depends on design variables. Evidently, the decision making will depend on the interplay between the response to design parameters, system uncertainty and, finally, the probability of failure or some statistical expectation of benefit. We propose to combine the model response to all design variables and uncertain parameters into a single approach based on an integrative (joint) response surface (obtained via PCE), which allows to reflect the non-linear dependence of the original model on all these parameters. This integrates the design task into the reduced stochastic model, allowing us to find robust designs with controlled risks at low computational costs. For example, designing system behaviors for maximum load at a guaranteed specific safety level can be achieved by quick Monte-Carlo methods on the resulting polynomials.

5.1 Integrative response surface

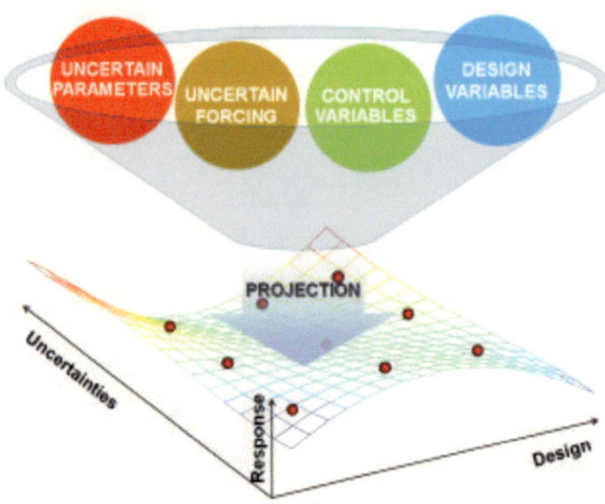

Fig. 7. Integrative response surface

Stochastic response surface methods (Isukapalli et al., 1998) deal with the characterization of uncertainties in systems describing the dependence of model output on the uncertain input parameters. Usually in many applied tasks, the model parameters can be classified in two classes: design or control variables that can be chosen by the operator of a system, and uncertain parameters that describe our (incomplete) knowledge of the system properties. On the other hand, the system's performance and failure probability will also depend on design parameters. Evidently, the decision-making for design parameters will depend on the interplay between the response to design parameters, system uncertainty and, finally, the probability of failure. In terms of system response, sensitivity and expansion, this distinction is artificial. Therefore, we drop the distinction between uncertain and design parameters. Instead, we project the model response to all design and uncertain parameters onto a single integrative model response surface. It is a multidimensional surface and contains the integral information about the system behavior under all possible conditions at all points in space and time. Thus, the notion of stochastic response surfaces introduced by Isukapalli et al. (1998) is expanded to integrative response surfaces forming an effective basis for robust design under uncertainty.

We investigate the influence of uncertain and design parameters $\boldsymbol{\omega}$ on the model output Ω (integrative response surface approach). This means that the considered model has a multivariable input $\boldsymbol{\omega}$ for the expansion Eq. (1):

$$\omega_1, ..., \omega_{N_U}, \omega_{N_U+1}, ... \omega_{N_U+N_D} \tag{16}$$

where N_U is the number of uncertain parameters and N_D is the number of design parameters. The total number of input parameters is $N = N_U + N_D$.

Of course design parameters do not have probabilistic distributions, but suitable weighting functions for such parameters can be described by user-defined feasibility functions that select the feasible range or preferences of the designing engineer concerning the values of design parameters. Feasibility functions provide a freedom for scenario analysis and can be used as an entry point for monetary punishment terms.

5.2 Illustration: robust design for carbon dioxide storage

5.2.1 Problem formulation

We again consider the problem of CO_2 leakage (Class et al., 2009) presented in Section 4.3. In the current Section we will consider the design task of finding an optimal injection regime. As in the Section 4.3 we consider three uncertain parameters: reservoir absolute permeability, reservoir porosity and permeability of the leaky well. For the current case study we also included two design parameters for describing the injection strategy: the CO_2 injection rate (fluctuating around 8.87 kg/s) and the size of the screening interval (up to 30m), see details in the paper by Oladyshkin, Class, Helmig & Nowak (2011b). The choice of the design parameters in this study is only exemplary and serves to demonstrate how engineering decision-making can be supported by the approach presented here. Both the injection rate and the screening interval directly affect the ratio of forces in the reservoir during the injection. The choice of feasibility functions is arbitrary, and modelers have full freedom to introduce feasibility functions and to weight them according to their personal experience or preferences. For example, we chose a beta distribution for the size of the screening interval with lower and upper bounds of z=0[m] and z=30[m], respectively, reflecting the physical bounds of the reservoir.

5.2.2 Design under uncertainty vs. conventional design

The greatest challenge to modeling consists of finding a healthy and reasonable compromise between an accurate system representation and computational efforts. Unfortunately, neglecting uncertainty in design tasks can be a strong simplification for modeling, and the consequences can be stronger than when neglecting several physical phenomena.

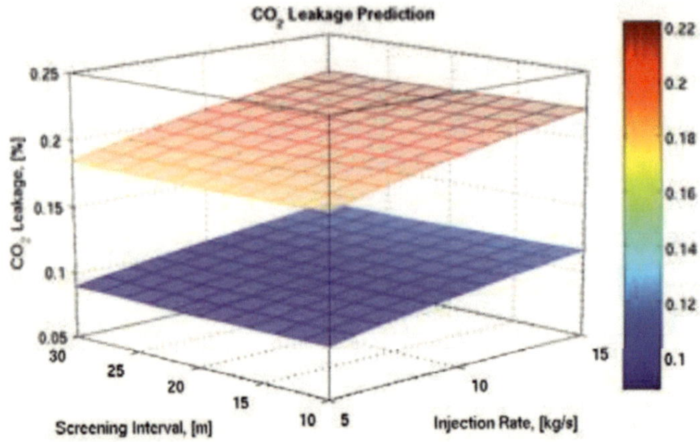

Fig. 8. Influence of design parameters on prediction of CO_2 leakage rate after 1000 days: top surface (robust design) - expected CO_2 leakage rate (average over uncertain parameters); bottom surface (conventional design) - CO_2 leakage rate evaluated deterministically with expected values of parameters

Figure 8 demonstrates how the injection rate and screening interval influence the leakage rate of CO_2. An important advantage of the integrative response surface approach is that parameter uncertainty is easily included in such predictions. The top surface in Figure 8 is the CO_2 leakage rate expected after 100 days as a function of the design parameters, averaged over the uncertain parameters (robust design). The bottom surface in Figure 8 is the CO_2 leakage rate using the expected values of the uncertain parameters, i.e. as in deterministic simulations (conventional design). It is easy to see that the impact can be extremely important for non-linear systems (here, a factor of about two), especially in long-term simulations. Instead of looking at mean values, robust design can also look at failure probability (see below). Thus, designs hat admit uncertainty are much more robust than design under deterministic assumptions.

5.2.3 Robust design of failure probability

The integrative response surface approach provides a constructive solution to the problem of robust design under uncertainty and provides valuable support for risk-informed decision making.

In a similar fashion as in the previous Section 5.2.2, the dependence of the leakage probability or any other statistical characteristics on design parameters can be evaluated, so that the

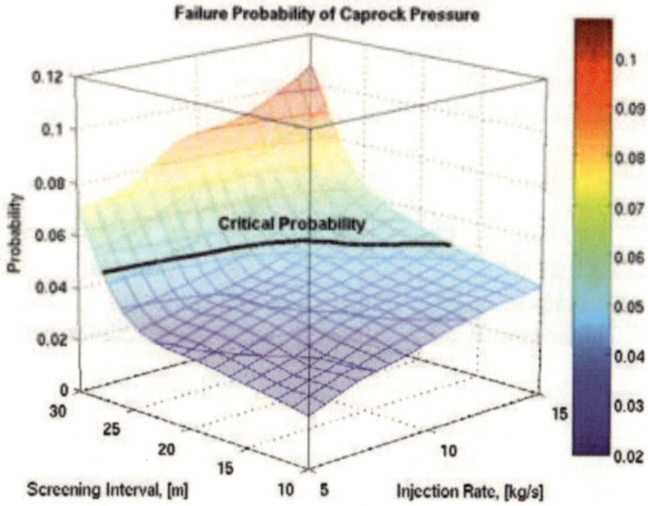

Fig. 9. Choice of design parameters based on caprock pressure after 1000 days: critical pressure 330 bar at a significance level of 5 %

injection regime can be chosen according to a maximum allowable failure probability. Figure 9 illustrates the choice of design parameters based on the caprock pressure after 1000 days. In this test case, a critical caprock pressure equal to 330 bar was chosen at a significance level of 5%, i.e. the maximal acceptable probability of failure is set to 0.05 (solid black line on surface). Figure 9 demonstrates acceptable strategies of injection where the caprock pressure does not exceed the limit of 330 bar, which corresponds to an injection-induced pressure build-up of about 40 bar.

5.3 Conclusions to section 5

This Section explores a massive stochastic model reduction via polynomial chaos expansion for robust design. This approach offers fast evaluation for statistical quantities and their dependence on design or control parameters. In particular, we map the response of a model to changes in all the design parameters and uncertain parameters onto one single integrative response surface. Based on this integrative concept, the design task explicitly includes uncertainty, which leads to robust designs with minimum failure probability.

We demonstrated that neglecting parametric uncertainty in design can be a strong simplification for modeling. Due to the non-linearity of processes, including uncertainty can lead to a systematic and significant shift of the predicted values, affecting both risk estimates and the design of injection scenarios. Thus, design under uncertainty, that openly admits uncertainty and seeks for roust solutions is much reliable in comparison to conventional design.

6. Acknowledgement

The authors acknowledge the German Research Foundation (DFG) for its financial support of the project within the Cluster of Excellence in Simulation Technology (EXC 310/1) at the University of Stuttgart.

7. References

Abramowitz, M. & Stegun, I., A. (1965). *Handbook of Mathematical Functions with Formulas, Graphs, and Mathematical Tables*, New York: Dover.

Anderson, M. & Burt, T. P. (1985). Modelling strategies, *in Anderson, M.G. and Burt, T.P. (eds.), Hydrological Forecasting* pp. 1–13.

Askey, R. & Wilson, J. (1985). *Some basic hypergeometric polynomials that generalize Jacobi polynomials*, Memoirs of the American Mathematical Society, AMS, Providence.

Class, H., Ebigbo, A., Helmig, R., Dahle, H., Nordbotten, J. N., Celia, M. A., Audigane, P., Darcis, M., Ennis-King, J., Fan, Y., Flemisch, B., Gasda, S., Jin, M., Krug, S., Labregere, D., Naderi, A., Pawar, R. J., Sbai, A., Sunil, G. T., Trenty, L. & Wei, L. (2009). Abenchmark-study on problems related to co_2 storage in geologic formations, *Computational Geosciences* 13: 451Ű–467.

Crestaux, T., Le Maitre, O. & Martinez, J.-M. (2009). Polynomial chaos expansion for sensitivity analysis, *Reliability Engineering and System Safety* 94(7): 1161–1172.

Dagan, G. (1988). Time-dependent macrodispersion for solute transport in anisotropic heterogeneous aquifers, *Water Resour. Res.* 24(9): 1491–1500.

Dagan, G. (1989). *Flow and Transport in Porous Formations*, Springer Verlag, Berlin.

de Barros, F. P. J., Ezzedine, S. & Rubin, Y. (2011). Impact of hydrogeological data on measures of uncertainty, site characterization and environmental performance metrics, *Advances in Water Resources* .

de Barros, F. P. J., Oladyshkin, S. & Nowak, W. (2011). An integrative data-adaptive approach for global sensitivity analysis: application to subsurface flow and transport, *Geophys. Res. Abstr.*, Vol. 13, EGU General Assembly 2011.

Efron, B. (1987). *The Jackknife, the Bootstrap, and Other Resampling Plans*, Society for Industrial Mathematics.

Fajraoui, N., Ramasomanana, F., Younes, A., Mara, T. A.and Ackerer, P. & Guadagnini, A. (2011). Use of global sensitivity analysis and polynomial chaos expansion for interpretation of non-reactive transport experiments in laboratory-scale porous media, *Water Resour. Res.* (doi:10.1029/2010WR009639).

Flemisch, B., J., F., Helmig, R., Niessner, J. & Wohlmuth, B. (2007). Dumux: a multi-scale multi-physics toolbox for flow and transport processes in porous media, *In A. Ibrahimbegovic and F. Dias, editors* ECCO3MAS Thematic Conference on Multi-scale Computational Methods for Solids and Fluids, Cachan, France, November 28–30, 2007.

Foglia, L., Mehl, S., W., Hill, M., C., Perona, P. & Burlando, P. (2007). Testing alternative ground water models using cross validation and other methods, *Ground Wate* 45(5): 627–641.

Foo, J. & Karniadakis, G. (2010). Multi-element probabilistic collocation method in high dimensions, *Journal of Computational Physics* 229(5): 1536–1557.

Ghanem, R. & Doostan, A. (2006). On the construction and analysis of stochastic models: Characterization and propagation of the errors associated with limited data, *Journal of Computational Physics* 217: 63–Ű81.

Ghanem, R. G. & Spanos, P. D. (1991). *Stochastic Finite Elements: A Spectral Approach*, Springer-Verlag, New York.

Ghanem, R. & Spanos, P. (1993). A stochastic galerkin expansion for nonlinear random vibration analysis, *Probabilistic Engineering Mechanics* 8: 255–264.

Ghanem, R. & Spanos, P. D. (1990). Polynomial chaos in stochastic finite elements, *Journal of Applied Mechanics* 57: 197–202.

Grigoriu, M. (2002). *Stochastic Calculus: Applications in Science and Engineering*, Birkhauser, Boston.

Homma, T. & Saltelli, A. (1996). Importance measures in global sensitivity analysis of nonlinear models, *Reliability Engineering and System Safety* 52(1): 1–17.

Isukapalli, S., S., Roy, A. & Georgopoulos, P., G. (1998). Stochastic response surface methods (srsms) for uncertainty propagation: Application to environmental and biological systems, *Risk Analysis* 18(3): 351–363.

Jang, Y., S., Sitar, N. & Kiureghian, A., D. (1994). Reliability analysis of contaminant transport in saturated porous media., *Water Resources Research* 30(8): 2435–2448.

Keese, A. & Matthies, H. G. (2003). Sparse quadrature as an alternative to mc for stochastic finite element techniques, *Proc. Appl. Math. Mech.* 3: 493Ű–494.

Kopp, A., Class, H. & Helmig, H. (2009). Investigations on co_2 storage capacity in saline aquifers - part 1: Dimensional analysis of flow processes and reservoir characteristics, *Int. J. of Greenhouse Gas Control* 3: 263–276.

Li, H., Sarma, P. & Zhang, D. (2011). A comparative study of the probabilistic collocation and experimental design methods for petroleum reservoir uncertainty quantification, *SPE Journal* pp. SPE–140738–PA–P.

Li, H. & Zhang, D. (2007). Probabilistic collocation method for flow in porous media: Comparisons with other stochastic methods, *Water Resources Research* 43: 44–48.

Matthies, H., G. & Keese., A. (2005). Galerkin methods for linear and nonlinear elliptic stochastic partial differential equations, *Comp. Meth. Appl. Mech. Engrg.* 194: 1295–1331.

Maxwell, R. M. & Kastenberg, W. E. (1999). Stochastic environmental risk analysis: An integrated methodology for predicting cancer risk from contaminated groundwater., *Stochastic Environmental Research and Risk Assessment* (13(1-2)): 27–47.

Morris, M., D. (1991). Morris factorial sampling plans for preliminary computational experiments, *Technometrics* (33): 161–174.

Nowak, W. (2009). Best unbiased ensemble linearization and the quasi-linear Kalman ensemble generator, *Water Resour. Res.* 45(W04431).

Oladyhskin, S., Class, H., Helmig, R., Nowak, W., de Barros, F. P. J. & Ashraf, M. (2011). Data-driven polynomial response surfaces as efficient tool for applied tasks under uncertainty, SAMSI Geosciences Applications Opening Workshop 2011.

Oladyshkin, S., Class, H., Helmig, R. & Nowak, W. (2011a). A concept for data-driven uncertainty quantification and its application to carbon dioxide storage in geological formations, *Advances in Water Resources* 34: 1508–1518.

Oladyshkin, S., Class, H., Helmig, R. & Nowak, W. (2011b). An integrative approach to robust design and probabilistic risk assessment for co_2 storage in geological formations, *Computational Geosciences* 15(3): 565–577.

Oladyshkin, S., de Barros, F. P. J. & Nowak, W. (2011). Global sensitivity analysis: a flexible and efficient framework with an example from stochastic hydrogeology, *Advances in Water Resources* DOI: 10.1016/j.advwatres.2011.11.001.

Oladyshkin, S. & Nowak, W. (2011). Data-driven uncertainty quantification using the arbitrary polynomial chaos expansion, *Reliability Engineering and System Safety* .

Plischke, E. (2010). An effective algorithm for computing global sensitivity indices (easi), *Reliability Engineering and System Safety* 95(4): 354–360.

Rubin, Y. (2003). *Applied Stochastic Hydrogeology*, Oxford University Press, Oxford.

Saltelli, A., Ratto, M. & Andres, T. (2008). *Global Sensitivity Analysis: The Primer*, John Wiley & Sons.

Siebert, W. M. (1986). Circuits, signals, and systems, *MIT Press: Cambridge, MA* pp. 410–411.

Siirila, E., R., Navarre-Sitchler, A., K., Maxwell, R., M. & McCray, J., E. (2011). A quantitative methodology to assess the risks to human health from co2 leakage into groundwater., *Advances in Water Resources* .

Sobol, I. M. (1990). On sensitivity estimation for nonlinear mathematical models, *Mathem. Mod.* 2(1): 112–118.

Soize, C. & Ghanem, R. (2004). Physical systems with random uncertainties: chaos representations with arbitrary probability measure, *SIAM J. Sci. Comput.* 26(2): 395–Ű410.

Stieltjes, T., J. (1884). Quelques recherches sur la théorie des quadratures dites méchaniques., *Oeuvres I* pp. 377–396.

Sudret, B. (2008). Global sensitivity analysis using polynomial chaos expansions, *Reliability Engineering and System Safety* 93(7): 964–979.

USEPA (December 1989). Risk assessment guidance for superfund volume i: Part a, human health manual., *Tech. Rep. Rep.EPA/540/1-89/002* .

USEPA (December 2001). Risk assessment guidance for superfund: Volume iii – part a, process for conducting probabilistic risk assessment, *Tech. Rep. Rep.EPA 540/R-02/002*, .

Villadsen, J. & Michelsen, M. L. (1978). *Solution of differential equation models by polynomial approximation*, Prentice-Hall.

Wackernagel, H. (1998). *Multivariate Geostatistics, An Introduction With Applications*, Second, Springer, Berlin.

Wan, X. & Karniadakis, G., E. (2006). Multi-element generalized polynomial chaos for arbitrary probability measures, *SIAM Journal of Scientific Computing* 28(3): 901–928.

Wiener, N. (1938). The homogeneous chaos, *Am. J. Math* 60: 897–936.

Winter, C. L., Guadagnini, A., Nychka, D. & Tartakovsky, D. M. (2006). Multivariate sensitivity analysis of saturated flow through simulated highly heterogeneous groundwater aquifers, *Journal of Computational Physics* 217(1): 166–175.

Witteveen, J. A. S. & Bijl, H. (2006). Modeling arbitrary uncertainties using gram-schmidt polynomial chaos, *44th AIAA Aerospace Sciences Meeting and Exhibit* Reno, Nevada: AIAA–2006–896.

Witteveen, J. A. S., Sarkar, S. & Bijl, H. (2007). Modeling physical uncertainties in dynamic stall induced fluidŰstructure interaction of turbine blades using arbitrary polynomial chaos, *Computers and Structures* 85: 866Ű–878.

Xiu, D. & Karniadakis, G. E. (2003). Modeling uncertainty in flow simulations via generalized polynomial chaos, *Journal of Computational Physics* 187: 137–167.

Zhang, D. & Lu, Z. (2004). An efficient, high-order perturbation approach for flow in random media via karhunen-loeve and polynomial expansions, *Journal of Computational Physics* 194: 773–794.